D0875486

INORGANIC SYNTHESES

Volume 24

Editor-in-Chief

JEAN'NE M. SHREEVE

Department of Chemistry
University of Idaho

INORGANIC SYNTHESES

Volume 24

A Wiley-Interscience Publication

JOHN WILEY & SONS

New York Chichester Brisbane Toronto Singapore

Published by John Wiley & Sons, Inc.

Library of Congress Catalog Number: 39-23015

ISBN 0-471-83441-6

Printed in the United States of America

10 9 8 7 6 5 4 3 2 1

This volume is dedicated to
Professor Malcolm M. Renfrew
on the occasion of his seventy-fifth birthday

PREFACE

Preparative chemistry is the heart of chemistry. The goal of *Inorganic Syntheses* is to provide both the neophyte and the seasoned veteran with safe and reliable methods to prepare useful compounds. The raison d'être of *Inorganic Syntheses* is to serve as a valuable treasure trove of precursors to yet undiscovered compounds and to demonstrate practical techniques to the undergraduate student. It is *not* to be a repository for the exotic, easily forgotten compounds that have no practical use. The inclusion of useful compounds only is an admirable goal and one that the Editor and the Editorial Board have striven valiantly to attain, for the utility of a particular synthesis is often in the eyes of the beholder. It is my sincere hope that prospective authors who may read this will keep in mind that utility of the product and safety of the procedure must be considered to be equally important prerequisites for publication in *Inorganic Syntheses*.

The real heroes in compiling one of these volumes are the checkers. Careless authors supply sloppily written manuscripts in incorrect format with incomplete information with respect to reaction vessels, reaction conditions, and means of characterization. Although this is an entirely voluntary process, our loyal checkers come through in admirable style. My deepest gratitude is owed to them. Additionally, there are five other magnificent colleagues who have given at least as much of their energy to this volume of *Inorganic Syntheses* as the Editor has. Without the considerable help and continuing interest of Duward F. Shriver, John C. Bailar, Jr., Therald Moeller, Thomas E. Sloan, and William Powell, my assignment would have been completed with considerably greater difficulty. My sincerest appreciation must be expressed to the two very competent women, Barbara Crabtree and LeNelle McInturff, who handled many of the details and who retyped many of the manuscripts.

Preparative chemistry is fun. Safe and reliable preparative chemistry is even more satisfying. I hope that the inorganic chemistry community will find the preparations in this volume both useful and safe.

JEAN'NE M. SHREEVE

Moscow, Idaho
September 1985

Previous volumes of *Inorganic Syntheses* are available. Volumes I–XVI can be ordered from R. E. Krieger Publishing Co., Inc., P.O. Box 9542, Melbourne, Florida 32901; Volume XVII is available from McGraw-Hill, Inc.; subsequent volumes can be obtained from John Wiley & Sons, Inc.

NOTICE TO CONTRIBUTORS AND CHECKERS

The *Inorganic Syntheses* series is published to provide all users of inorganic substances with detailed and foolproof procedures for the preparation of important and timely compounds. Thus the series is the concern of the entire scientific community. The Editorial Board hopes that all chemists will share in the responsibility of producing *Inorganic Syntheses* by offering their advice and assistance in both the formulation of and the laboratory evaluation of outstanding syntheses. Help of this kind will be invaluable in achieving excellence and pertinence to current scientific interests.

There is no rigid definition of what constitutes a suitable synthesis. The major criterion by which syntheses are judged is the potential value to the scientific community. For example, starting materials or intermediates that are useful for synthetic chemistry are appropriate. The synthesis also should represent the best available procedure, and new or improved syntheses are particularly appropriate. Syntheses of compounds that are available commercially at reasonable prices are not acceptable. We do not encourage the submission of compounds that are unreasonably hazardous, and in this connection, less dangerous anions generally should be employed in place of perchlorate.

The Editorial Board lists the following criteria of content for submitted manuscripts. Style should conform with that of previous volumes of *Inorganic Syntheses*. The introductory section should include a concise and critical summary of the available procedures for synthesis of the product in question. It should also include an estimate of the time required for the synthesis, an indication of the importance and utility of the product, and an admonition if any potential hazards are associated with the procedure. The Procedure should present detailed and unambiguous laboratory directions and be written so that it anticipates possible mistakes and misunderstandings on the part of the person who attempts to duplicate the procedure. Any unusual equipment or procedure should be clearly described. Line drawings should be included when they can be helpful. All safety measures should be stated clearly. Sources of unusual starting materials must be given, and, if possible, minimal standards of purity of reagents and solvents should be stated. The scale should be reasonable for normal laboratory operation, and any problems involved in scaling the procedure either up or down should be discussed. The criteria for judging the purity of the final product should be delineated clearly. The section on Properties should supply and discuss those physical and chemical characteristics that are relevant to judging the purity of the product and to permitting its handling and use in an intelligent manner. Under References, all pertinent literature citations should be listed in order. A style sheet is available from the Secretary of the Editorial Board.

The Editorial Board determines whether submitted syntheses meet the general specifications outlined above, and the Editor-in-Chief sends the manuscript to an independent laboratory where the procedure must be satisfactorily reproduced.

Each manuscript should be submitted in duplicate to the Secretary of the Editorial Board, Professor Jay H. Worrell, Department of Chemistry, University of South Florida, Tampa, FL 33620. The manuscript should be typewritten in English. Nomenclature should be consistent and should follow the recommendations presented in *Nomenclature of Inorganic Chemistry*, 2nd ed., Butterworths & Co., London, 1970, and in *Pure and Applied Chemistry*, Volume 28, No. 1 (1971). Abbreviations should conform to those used in publications of the American Chemical Society, particularly *Inorganic Chemistry*.

Chemists willing to check syntheses should contact the editor of a future volume or make this information known to Professor Worrell.

TOXIC SUBSTANCES AND LABORATORY HAZARDS

Chemicals and chemistry are by their very nature hazardous. Chemical reactivity implies that reagents have the ability to combine. This process can be sufficiently vigorous as to cause flame, an explosion, or, often less immediately obvious, a toxic reaction.

The obvious hazards in the syntheses reported in this volume are delineated, where appropriate, in the experimental procedure. It is impossible, however, to foresee every eventuality, such as a new biological effect of a common laboratory reagent. As a consequence, *all* chemicals used and *all* reactions described in this volume should be viewed as potentially hazardous. Care should be taken to avoid inhalation or other physical contact with all reagents and solvents used in procedures described in this volume. In addition, particular attention should be paid to avoiding sparks, open flames, or other potential sources that could set fire to combustible vapors or gases.

A list of 400 toxic substances may be found in the *Federal Register*, Vol. 40, No. 23072, May 28, 1975. An abbreviated list may be obtained from *Inorganic Syntheses*, Volume 18, p. xv, 1978. A current assessment of the hazards associated with a particular chemical is available in the most recent edition of *Threshold Limit Values for Chemical Substances and Physical Agents in the Workroom Environment* published by the American Conference of Governmental Industrial Hygienists.

The drying of impure ethers can produce a violent explosion. Further information about this hazard may be found in *Inorganic Syntheses*, Volume 12, p. 317. A hazard associated with the synthesis of tetramethyldiphosphine disulfide [*Inorg. Synth.*, **15**, 186 (1974)] is cited in *Inorganic Syntheses*, Volume 23, p. 199.

CONTENTS

Chapter One FLUORINE-CONTAINING COMPOUNDS

Chapter Three TRANSITION METAL ORGANOMETALLIC COMPOUNDS

Chapter Four TRANSITION METAL COMPOUNDS AND COMPLEXES

INORGANIC SYNTHESES

Volume 24

Chapter One

FLUORINE-CONTAINING COMPOUNDS

1. CHLORINE FLUORIDE

Chlorine fluoride is a versatile and very useful reagent.[1,2] Preparation from the elements is well established,[3] but since the reaction involves the use of elemental fluorine and either Monel or nickel reactors, it is a procedure not readily accessible to many laboratories. In addition, the reaction is exothermic. Two procedures that are excellent lower temperature routes to chlorine fluoride are described below.

■ **Caution.** *Chlorine fluorides are vigorous oxidizers and are toxic. In addition, hydrolysis can produce shock-sensitive chlorine oxides. Suitable shielding for high-pressure work is required. Protective clothing and face masks must be worn at all times.*

A. CHLORINE TRIFLUORIDE WITH CHLORINE

$$ClF_3 + Cl_2 \xrightarrow{CsF} 3ClF$$

Submitted by CARL J. SCHACK* and R. D. WILSON*
Checked by DARRYL D. DESMARTEAU†

*Rocketdyne Division, Rockwell International, Canoga Park, CA 91304.
†Department of Chemistry, Clemson University, Clemson, SC 29631.

Procedure

A clean, passivated (with ClF_3) 150-mL high-pressure stainless steel Hoke cylinder, equipped with a Hoke valve, is loaded with CsF (9 mmol) in the dry box. After connection to a passivated (with ClF_3) stainless steel, Teflon FEP vacuum line,[4] the evacuated cylinder is cooled to −196°, and ClF_3 [Ozark-Mahoning] (40 mmol) and Cl_2 (38 mmol) are condensed into it. The cylinder is first warmed to ambient temperature, then placed in an oven (150–155°) for 15 hr and finally recooled to ambient temperature. The products are separated by fractional condensation through a series of U-traps cooled to −142 and −196°. A very small amount of noncondensable gas is pumped away, and the ClF is collected in the trap at −196°. Unreacted starting materials are retained in the trap at −142° (methylcyclopentane/liquid N_2). The yield of ClF (113 mmol, 97.4%) may be increased to 99+%‡ by pyrolyzing the solids (CsF, $CsClF_2$, $CsClF_4$) left in the reactor with a heat gun and fractionating the evolved gases as above. The reactor and solids (with or without pyrolysis) may be used over again, and the ClF may be stored in the reactor until needed.

■ **Caution.** *The solid residue is highly oxidizing. Contact with organic materials should be avoided. Slow hydrolysis by exposure to the atmosphere in a well-ventilated hood destroys the material satisfactorily. Water can then be added cautiously.*

B. CHLORINE TRIFLUORIDE WITH MERCURY(II) CHLORIDE

$$2ClF_3 + HgCl_2 \xrightarrow{25^\circ} 4ClF + HgF_2$$

Submitted by A. WATERFELD* and R. MEWS*
Checked by DARRYL D. DESMARTEAU†

Procedure

■ **Caution.** *Mercury salts are poisonous. Chlorine fluorides are powerful oxidizers and fluorinators. Protective clothing and face masks must be worn at all times, and extreme care must be taken to avoid contact between the fluorides and oxidizable materials.*

‡The checker used one-half the amounts of starting materials in a Monel vessel and obtained essentially the same yield.

*Institut fur Anorganische Chemie, Tammanstrasse 4, D-3400 Göttingen, West Germany.
†Department of Chemistry, Clemson University, Clemson, SC 29631.

A 300-mL Monel Hoke cylinder that has been passivated with ClF_3[4] and equipped with a manometer [Dresser] (fluorine-resistant, 1000 psi range) is loaded with 40.2 g (148 mmol) of dry $HgCl_2$. Using a metal vacuum line, 27.4 g (296 mmol) of ClF_3 [Ozark-Mahoning] is condensed into the evacuated cylinder at −196°. After 14 days at room temperature, the internal pressure reaches 645 psi. The chlorine fluoride is distilled out of the cylinder, which is held at −125° (30–60° petroleum ether/liquid N_2) via a stainless steel vacuum line into a passivated Monel Hoke cylinder at −196°. The yield is 29.5 g (542 mmole, 91.5%).‡ The mercury salt remaining in the cylinder is 87% HgF_2 based on a loss of 4.3 g. The HgF_2 may be used for further syntheses or destroyed as prescribed.[5] The chlorine fluoride is sufficiently pure for most preparative purposes (contains traces of ClO_2F and Cl_2). For final purification, literature methods are recommended.[6]

Properties

Chlorine fluoride is colorless as a solid (mp −155°) and pale yellow as a liquid (bp −100.1°). The IR spectrum of the gas in a stainless steel cell with AgCl windows shows a band centered at 772 cm^{-1} with P and R branches but no Q branch.

References

1. L. Stein, in *Halogen Chemistry,* Vol. 1, V. Gutmann (ed.), Academic Press, New York, 1967, p. 134.
2. C. J. Schack and K. O. Christe, *Israel J. Chem.*, **17,** 20 (1978).
3. O. Ruff, E. Ascher, J. Fischer, and F. Laass, *Z. Anorg. Allgem. Chem.*, **176,** 258 (1928).
4. K. O. Christe, R. D. Wilson, and C. J. Schack, *Inorg. Synth.*, **24,** 3 (1986).
5. D. F. Shriver, *The Manipulation of Air-Sensitive Compounds,* McGraw-Hill, New York, 1969.
6. C. J. Schack and R. D. Wilson, *Synth. Inorg. Metal-Org. Chem.*, **3,** 393 (1973).

2. CHLORYL FLUORIDE

$$6NaClO_3 + 4ClF_3 \rightarrow 6NaF + 2Cl_2 + 3O_2 + 6ClO_2F$$

Submitted by KARL O. CHRISTE,* RICHARD D. WILSON,* and CARL J. SCHACK*
Checked by D. D. DESMARTEAU†

‡The checker used one-half the amounts of starting materials in a 150 mL Monel cylinder and obtained the same yield.

*Rocketdyne, A Division of Rockwell International Corp., Canoga Park, CA 91304.
†Department of Chemistry, Clemson University, Clemson, SC 29631.

Chloryl fluoride is the most common chlorine oxyfluoride. It is always encountered in reactions of chlorine mono-, tri-, or pentafluorides with oxides, hydroxides, or poorly passivated surfaces. It was first obtained[1] in 1942 by Schmitz and Schumacher by the reaction of ClO_2 with F_2. Other methods involve the reaction of $KClO_3$ with either BrF_3[2] or ClF_3.[3,4] The simplest method[5] involves the reaction of $NaClO_3$ with ClF_3, resulting in the highest yields and products that can readily be separated.

Procedure

■ **Caution.** *The hydrolysis of ClO_2F can produce shock-sensitive ClO_2.[6] Therefore, the use of a slight excess of ClF_3 is recommended for the synthesis to suppress any ClO_2 formation. Chlorine trifluoride is a powerful oxidizer and ignites most organic substances on contact. The use of protective face shields and gloves is recommended when working with these materials.*

In the dry box, dry sodium chlorate (30 mmol, 3.193 g) is loaded into a 30-mL high-pressure stainless steel Hoke cylinder equipped with a stainless steel Hoke valve. The cylinder is connected to a stainless steel–Teflon FEP vacuum manifold (Fig. 1) that has been well passivated with ClF_3 [Ozark-Mahoning] until the ClF_3, when condensed at $-196°$, shows no color. The cylinder is then evacuated and ClF_3 (21.5 mmol) is condensed into the cylinder at $-196°$. The cylinder is allowed to warm to room temperature and is kept at this temperature for 1 day. The cylinder is then cooled back to $-196°$, and during subsequent warm-up of the cylinder the volatile products are separated by fractional condensation under dynamic vacuum through a series of U-traps kept by liquid N_2 slush baths at $-95°$ (toluene), $-112°$ (CS_2), and $-126°$ (methylcyclohexane). The trap at $-95°$ contains only a trace of chlorine oxides, the trap at $-112°$ contains most of the ClO_2F (29 mmol), and the trap at $-126°$ (7 mmol) contains mainly Cl_2 and some ClO_2F. The yield of ClO_2F is almost quantitative based on the limiting reagent $NaClO_3$ (29.4 mmol, 98%).‡ The purity of the material is checked by infrared spectroscopy in a well-passivated Teflon or metal cell equipped with AgCl windows. The product should not show any impurities. A small amount of chlorine oxides, which can be readily detected by their intense color if present or if formed during handling of ClO_2F, can readily be removed by conversion to ClO_2F with elemental F_2 or will decompose to Cl_2 and O_2 during storage at ambient temperature. Chloryl fluoride can be stored in a metal vessel at room temperature for long time periods without significant decomposition.

‡The checker used one-third of the stated scale and obtained ClO_2F in a yield of 95%.

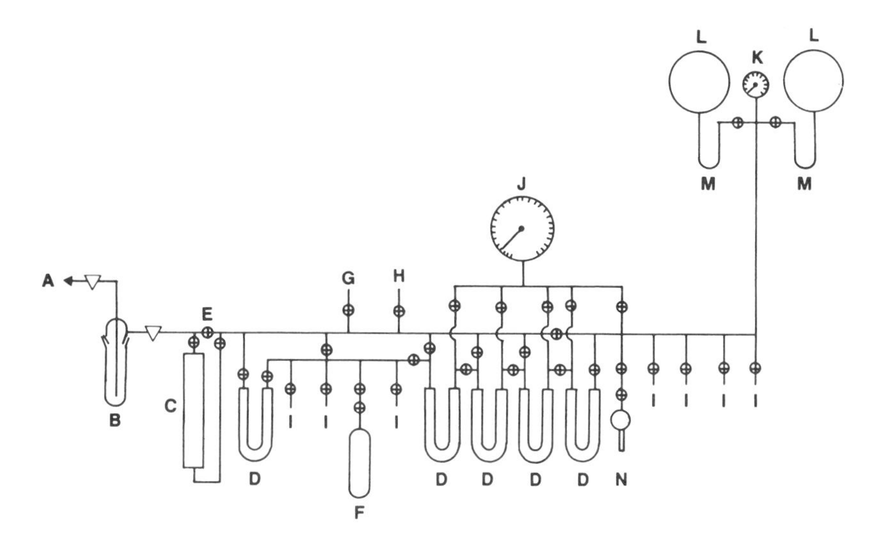

Fig. 1. Typical metal-Teflon vacuum system used for handling strongly oxidizing or corrosive fluorine compounds. As the vacuum source A, a good mechanical pump (10^{-4} torr or better) is normally sufficient. The use of a fluorocarbon oil, such as Halocarbon [Halocarbon Products] as a pump oil is strongly recommended for safety reasons. *B,* glass waste trap with glass or Teflon stopcocks and a detachable bottom; only fluorocarbon grease [Halocarbon Products] should be used for the stopcocks and joint; the trap is kept cold by a Dewar flask with liquid nitrogen; great care must be taken, and a face shield and heavy leather gloves must be worn when pulling off the cold lower half of the waste trap for disposal of the trapped material by evaporation in a fume hood. The glass waste trap can be connected to the metal line by a glass-metal joint, a graded glass-metal seal, or most conveniently by a quick-coupling compression fitting with Viton O-ring seals. *C,* scrubber for removal of fluorine; the scrubber consists of a glass tower packed with alternating layers of NaCl and soda lime that are held in place by plugs of glass wool at either end. The valves *E* are arranged in such a manner that the scrubber can be by-passed during routine operation. *D,* Teflon-FEP (fluoro-ethylene-propylene copolymer [Zeus]) or PFA (polyperfluoroether) U-traps made from ½- or ¾-in. o.d. commercially available heavy wall tubing; all metal lines are made from either 316 or 321 ⅜-in. o.d. stainless steel or Monel tubing, except for the lines from the U-traps to the Heise gage [Dresser] *J,* for which ¼-in. o.d. tubing is preferred; stainless steel bellows valves *E,* such as Hoke Model 4200 series, are used throughout the entire line; metal-metal or metal-Teflon connections are all made with either flare or compression (Swagelok [Crawford Fitting] or Gyrolok) fittings. *F,* lecture bottle of ClF_3 [Ozark-Mahoning] used for passivation of the vacuum line. *G,* He gas inlet. *H,* F_2 gas inlet. *I,* connectors for attaching reaction vessels, reagent containers, etc. *J,* Heise Bourdon tube pressure gage (0–1000 mm ± 0.1%). *K,* crude pressure gage (0–5 atm). *L,* 2-L steel bulbs used for either measuring or storing larger amounts of gases. *M,* ⅜-in. o.d. metal U-tubes to permit condensation of gases into the storage bulbs *L. N,* infrared cell for gases, Teflon body with condensing tip, 5 cm path length, AgCl windows. The four U-traps *D,* connected in series, constitute the fractionation train used routinely for the separation of volatile materials by fractional condensation employing slush baths of different temperatures. The volumes of each section of the vacuum line are carefully calibrated by PVT measurements using a known standard volume.

Properties[6]

Chloryl fluoride is a colorless liquid boiling at $-6°$. The IR spectrum of the gas[4] shows the following major bands: 1271 (vs), 1106 (ms), 630 (s), and 547 (ms) cm^{-1}. The ^{19}F NMR spectrum[7] of the liquid at $-80°$ consists of a singlet at 315 ppm downfield from external $CFCl_3$.

References

1. H. Schmitz and H. J. Schumacher, *Z. Anorg. Allgem. Chem.*, **249,** 238 (1942).
2. A. A. Woolf, *J. Chem. Soc.*, **1954,** 4113.
3. A. Engelbrecht and H. Atzwanger, *J. Inorg. Nucl. Chem.*, **2,** 348 (1956).
4. D. F. Smith, G. M. Begun, and W. H. Fletcher, *Spectrochim. Acta,* **20,** 1763 (1964).
5. K. O. Christe, R. D. Wilson, and C. J. Schack, *Inorg. Nucl. Chem. Lett.*, **11,** 161 (1975).
6. K. O. Christe and C. J. Schack, *Adv. Inorg. Chem. Radiochem.*, **18,** 319 (1976).
7. K. O. Christe, J. F. Hon, and D. Pilipovich, *Inorg. Chem.*, **12,** 84 (1973).

3. CHLORINE FLUOROSULFATE

$$SO_3 + ClF \rightarrow ClOSO_2F$$

Submitted by CARL J. SCHACK* and RICHARD D. WILSON*
Checked by W. TOTSCH†

Chlorine fluorosulfate is a very reactive compound that is useful in numerous applications as a chlorinating agent, a fluorosulfating agent, and an oxidizer.[1] Originally $ClOSO_2F$ was prepared by Gilbreath and Cady[2] by reactions of Cl_2 and FO_2SOOSO_2F at 125° for 5 days. Since the peroxide is not readily available and its preparation[3] requires the use of elemental fluorine, this synthesis is not accessible to many laboratories. A useful alternative to this method involves the direct addition of ClF to SO_3. This procedure provides $ClOSO_2F$ in very high yield. This synthesis may be conducted in stainless steel[3] or glass,[4,5] is rapid, and can be used on any scale with good results.

*Rocketdyne, A Division of Rockwell International Corp., Canoga Park, CA 91304.
†Materials and Molecular Research Division, Lawrence Berkeley Laboratory and Chemistry Department, University of California, Berkeley, CA 94720.

Procedure

■ **Caution.** *The starting materials and product are strong oxidizers and are toxic. They must not be allowed to touch the skin or organic materials, such as grease, other than halocarbon type. Gloves, face shields, and other protective devices must be utilized where contact with these compounds is possible.*
Sulfur trioxide [Sargent-Welch] is available in sealed borosilicate glass ampules in 2-lb quantities. The ampule should be opened and handled only in a hood in accordance with the manufacturer's data sheet. (Place the ampule in a pan with sand or vermiculite, scratch the neck with a file, and, wearing rubber gloves and a face shield, snap the top of the neck off. Transfer the unused contents to a bottle with a fluoropolymer cap.)

A clean, dry, 150-mL stainless steel Hoke cylinder and a valve for it are tared. Using a funnel, the cylinder is loaded with SO_3 (0.388 mol) through the cylinder opening. The valve is inserted into the opening and tightened, and the outside of the cylinder is rinsed with water to remove possible external SO_3/H_2SO_4 contamination. After wiping dry, the weight of the cylinder and SO_3 is determined. The cylinder is connected to a stainless steel vacuum manifold, equipped with a pressure gauge, through a ~-shaped 0.25-in. diameter, stainless steel connector about 1 ft long. This type of connector provides flexibility and thereby permits shaking of the cylinder during the reaction with ClF. Also attached to the vacuum manifold is a supply of ClF [Ozark-Mahoning].[6] The cylinder is cooled to $-78°$ and evacuated. While the cylinder is warmed from $-78°$, the addition of ClF is begun by bleeding it in from the manifold and maintaining a pressure of about 1 atm.

At first the uptake of ClF may be slow due to the fact that the SO_3 is still solid, but when it melts the rate of ClF addition increases, especially if the reaction cylinder is shaken. Most of the ClF is added at about $-30°$ by intermittently immersing the reactor in a cold bath. After 2 hr, the uptake of ClF (0.40 mol) becomes slow. A slight further excess of ClF (0.02 mol) is added by cooling the cylinder to below $-78°$. It is left overnight at ambient temperature.

Purification of the product is accomplished by fractional condensation through a series of Teflon FEP U-traps cooled to -78 and $-196°$ while the reaction cylinder is kept at about $-30°$. After removal of the noncondensable gases, only occasional pumping is required. This minimizes the passage of $ClOSO_2F$ through the trap at $-78°$. The trap cooled to $-196°$ contains Cl_2, $FClO_2$, SO_2F_2, unreacted ClF, and a little $ClOSO_2F$. This material is disposed of by transferring it to a removable cold trap and venting this trap behind a shield in a hood. The cylinder at $-30°$ contains any HSO_3F by-product formed. Pure $ClOSO_2F$ (0.372 mole, 96% yield) is obtained in the trap cooled to $-78°$.

Alternatively, when the reaction is completed, the cylinder is cooled to $-78°$, and the volatile material is removed. The product can be kept in the cylinder at ambient temperature and purified by fractional condensation as needed. Using a 1-L cylinder, this reaction has been run on a 6-mole scale with similar results. For small amounts of $ClOSO_2F$, the ClF may simply be condensed onto the SO_3 at $-196°$ and the closed cylinder warmed slowly to ambient temperature in a prechilled Dewar flask.[4]

Properties[2,4]

Chlorine fluorosulfate is a pale yellow liquid that solidifies at $-84.3°$ and boils at 43.4°. The IR spectrum of the gas[7] has bands at 1481 (vs), 1248 (vs), 855 (vs), 831 (ms, sh), 703 (ms), 572 (ms), and 529 (m) cm^{-1} (stainless steel cell with AgCl windows). Chlorine fluorosulfate is the only known practical intermediate for the synthesis of chlorine perchlorate, $ClOClO_3$,[8] and it is easily photolyzed in Pyrex at room temperature to provide FO_2SOOSO_2F in high yield.[9] Chlorine fluorosulfate reacts violently with water.

References

1. C. J. Schack and K. O. Christe, *Israel J. Chem.*, **17,** 20 (1978).
2. W. P. Gilbreath and G. H. Cady, *Inorg. Chem.*, **2,** 496 (1963).
3. J. M. Shreeve and G. H. Cady, *Inorg. Synth.*, **7,** 124 (1963).
4. C. J. Schack and R. D. Wilson, *Inorg. Chem.*, **9,** 311 (1970).
5. C. V. Hardin, C. T. Ratcliffe, L. R. Anderson, and W. B. Fox, *Inorg. Chem.*, **9,** 1938 (1970).
6. C. J. Schack and R. D. Wilson, *Inorg. Synth.*, **24,** 3 (1986).
7. K. O. Christe, C. J. Schack, and E. C. Curtis, *Spectrochim. Acta*, **26A,** 2367 (1970).
8. C. J. Schack and D. Pilipovich, *Inorg. Chem.*, **9,** 1387 (1970).
9. C. J. Schack and K. O. Christe, *Inorg. Nucl. Chem. Lett.*, **14,** 293 (1978).

4. SULFUR CHLORIDE PENTAFLUORIDE

$$SF_4 + ClF \xrightarrow{CsF} SClF_5$$

Submitted by CARL J. SCHACK,* RICHARD D. WILSON,* and MICHAEL G. WARNER†
Checked by A. WATERFELD‡

*Rocketdyne, A Division of Rockwell International Corp., Canoga Park, CA 91304.
†Jacobs Engineering Group, Inc., Pasadena, CA 91101.
‡Department of Chemistry, University of California, Berkeley, CA 94720.

Sulfur chloride pentafluoride has utility in the preparation of numerous inorganic and organic derivatives.[1] A procedure for its synthesis has appeared previously in this series.[2] This method involved the in situ generation of ClF by a hot flow tube reaction followed by a second hot flow tube reaction of the ClF and SF_4 to furnish $SClF_5$. These reactions are time consuming to set up and sometimes difficult to master. A far simpler synthesis is the reaction of SF_4 and ClF at ambient temperature in the presence of CsF.[3] The reaction proceeds rapidly and in high yield with a minimum of by-product formation, thus allowing easy separation of the products.

Procedure

■ **Caution.** *The reagents and products are toxic and oxidizing. They must not be allowed to contact the skin or organic materials such as grease, other than halocarbon type. Gloves, face shields, and other protective devices must be utilized where contact with these compounds is possible.*

To a clean, passivated (treated with ClF_3 or ClF vapor at subatmospheric pressure for 1 hr or more), high-pressure, 30-mL stainless steel Hoke cylinder equipped with a valve is added powdered CsF (10 mmol) in a dry box. The CsF should have been dried previously by fusion in a platinum crucible followed by both cooling and grinding in a dry box. The cylinder is connected to a stainless steel-Teflon FEP (fluorinated-ethylene-propylene copolymer) vacuum system,[4] evacuated, and cooled to $-196°$. Based on PVT measurements, SF_4 [Matheson] (7.23 mmol) and ClF[5] (7.59 mmol) [Ozark-Mahoning] are added successively. The closed cylinder is allowed to warm to ambient temperature and to stand for 2 hr. Products are then separated by fractional condensation through a series of U-traps cooled at $-126°$ (liquid nitrogen/methylcyclohexane slush) and $-196°$. Passing into the trap at $-196°$ are small amounts of SOF_2, SO_2F_2, $SClF_5$, and ClF. Remaining in the trap at $-126°$ is pure $SClF_5$ (7.01 mmol, 97% yield).* Much larger amounts of $SClF_5$ may be prepared using this quantity of catalyst and only slightly longer reaction times. The catalyst can be reused.

Properties[2,3,6]

Sulfur chloride pentafluoride is colorless as a solid, liquid, or gas. It melts at $-64°$ and boils at $-19.1°$. The IR spectrum of the gas[7,8] has bands at (int.); 908 (vs), 854 (vs), 802 (m), 713 (m), 706 (s), 602 (vs), and 578 (m) cm^{-1}. It

*The checker, using SF_4/ClF/CsF in 42.3:53.3:13.4 mmol ratio, found $SClF_5$ (36.4 mmol, 85.5%). This lower yield was attributed to some SOF_2 impurity in the SF_4 used. Other products were SOF_4, Cl_2, and SF_6. If SF_5OCl is formed from SOF_4 and ClF, it will remain as a contaminant in the $SClF_5$.

is stable to acids but is rapidly hydrolyzed by alkalis. The compound is storable at ambient temperature in stainless steel and is thermally stable at 250°.

References

1. J. M. Shreeve, in *Sulfur in Organic and Inorganic Chemistry*, Vol. 4, A. Senning (ed.), Marcel Dekker, New York, 1982, p. 131.
2. F. Nyman, H. L. Roberts, and I. Seaton, *Inorg. Synth.*, **8,** 160 (1966).
3. C. J. Schack, R. D. Wilson, and M. G. Warner, *Chem. Commun.*, **1969,** 1110.
4. K. O. Christe, R. D. Wilson, and C. J. Schack, *Inorg. Synth.*, **24,** 3 (1986).
5. C. J. Schack and R. D. Wilson, *Inorg. Synth.*, **24,** 1 (1986).
6. R. D. W. Kemmitt and D. W. A. Sharp, *Adv. Fluorine Chem.*, **4,** 230 (1965).
7. L. H. Cross, H. L. Roberts, P. Goggin, and L. A. Woodward, *Trans. Faraday Soc.*, **56,** 945 (1960).
8. J. E. Griffiths, *Spectrochim. Acta*, **23A,** 2145 (1967).

5. (FLUOROCARBONYL)IMIDOSULFUROUS DIFLUORIDE

$$4SF_4 + Si(NCO)_4 \rightarrow 4FC(O)NSF_2 + SiF_4$$

Submitted by JOSEPH S. THRASHER*
Checked by DIETER LENTZ†

Since its initial preparation from the reaction of sulfur tetrafluoride with inorganic isocyanates of silicon, phosphorus, or sulfur, (fluorocarbonyl)imidosulfurous difluoride, $FC(O)NSF_2$,[1] has become one of the primary starting materials in the field of sulfur-nitrogen-fluorine chemistry. Nitrogen fluoride sulfide, NSF, and thiazyl trifluoride, NSF_3, the two key substances in this field of chemistry,[2] are both readily prepared from $FC(O)NSF_2$. Nitrogen fluoride sulfide is best prepared from the pyrolysis of the mercurial formed between HgF_2 and $FC(O)NSF_2$,[3] while oxidative fluorination of $FC(O)NSF_2$ with AgF_2 gives NSF_3.[4]

The method described below is convenient for preparing $FC(O)NSF_2$ in 100-g quantities starting from tetraisocyanatosilane and sulfur tetrafluoride. This procedure requires a stainless steel pressure vessel and a glass vacuum line.

*Department of Chemistry, The University of Alabama, University, AL 35486.

†Institut fur Anorganische und Analytische Chemie, Freie Universitat Berlin, Fabeckstr. 34–36, D-1000 Berlin 33.

Procedure

■ **Caution.** *Sulfur tetrafluoride and SiF_4 are both toxic gases. The toxicity of $FC(O)NSF_2$ is unknown, but it also should be regarded as hazardous. All operations should be carried out in a well-constructed vacuum line or in a well-ventilated hood.*

Tetraisocyanatosilane[5] (30.1 mL; 220 mmol) is loaded into a 500-mL stainless steel Hoke cylinder in an inert atmosphere. The cylinder is then attached to a glass vacuum line, and the $Si(NCO)_4$ is degassed by several freeze-thaw cycles. With the reaction vessel held at −196°, an excess of sulfur tetrafluoride [Matheson] (900 mmol) is added by vacuum transfer. The small excess of SF_4 is used because the reaction product reacts further with $Si(NCO)_4$ to give $SF_2{=}NC(O)NCO$.[6] The vessel that contains the reaction mixture is then placed in an ice-water bath, which is allowed to warm slowly to room temperature. After 24 hr, the reaction vessel is cooled to −85° (1-propanol/liquid nitrogen) and reattached to the vacuum line, where both the by-product SiF_4 and unreacted SF_4 are transferred under dynamic vacuum into a trap at −196°. When little or no more SiF_4 is observed to condense in the trap at −196°, the remaining volatile contents of the reaction vessel are transferred to the vacuum line for trap-to-trap distillation. The trap at −78° stops primarily $FC(O)N{=}SF_2$, with SF_4, SOF_2, and SiF_4 as contaminants. This fraction is further purified by redistillation through a series of traps at −25, −78, and −196°. The fraction retained in the −78° trap is generally found to be spectroscopically pure $FC(O)NSF_2$ (108.6 g, 94% yield), but further fractionations through the same series of traps can be carried out if necessary. The light-sensitive product is stored in Pyrex glass vessels in the dark below 0°. The by-product SiF_4 is vented to an efficient fume hood or stored in an appropriate pressure vessel.

Properties

(Fluorocarbonyl)imidosulfurous difluoride is a pungent, colorless liquid, bp 48.8° (extr.), mp −94.7°.[1] When heated at 190°, it decomposes to COF_2 and NSF.[7] This decomposition also takes place at temperatures as low as 0° in the presence of fluoride ion, for example, CsF.[8] IR (gas)[1]: 1850 (vs) ($\nu_{C=O}$), 1350 (vs) ($\nu_{S=N}$), 1160 (vs) (ν_{CF}), 1132 (s) (ν_{CF}), 865 (s), 764 (vs) (ν_{SF}), 727 (vs) (ν_{SF}) cm^{-1}. ^{19}F NMR[9] (ext. CCl_3F; 30°): δ_{SF} 42.5 (s), δ_{CF} 21.0 (s); (−80°) δ_{SF} 35.7 (d), δ_{CF} 19.5 (t) (J_{F-F} = 4 Hz).

References

1. A. F. Clifford and C. S. Kobayashi, *Inorg. Chem.*, **4**, 571 (1965).
2. O. Glemser and R. Mews, *Angew. Chem. Int. Ed. Engl.*, **19**, 883 (1980).

3. O. Glemser, R. Mews, and H. W. Roesky, *Chem. Ber.*, **102,** 1523 (1969); R. Mews, K. Keller, and O. Glemser, *Inorg. Synth.*, **24,** 14 (1986).
4. A. F. Clifford and J. W. Thompson, *Inorg. Chem.*, **5,** 1424 (1966); R. Mews, K. Keller, and O. Glemser, *Inorg. Synth.*, **24,** 12 (1986).
5. J. S. Thrasher, *Inorg. Synth.*, **24,** 99 (1986).
6. A. F. Clifford, J. S. Harmon, and C. A. McAuliffe, *Inorg. Nucl. Chem. Lett.*, **8,** 567 (1972).
7. A. F. Clifford, C. S. Kobayashi, and J. H. Stanton, Abstract of paper presented before the Symposium on Covalent Inorganic Fluorine Compounds, 148th National Meeting of the American Chemical Society, Chicago, IL., Sept. 1964.
8. J. K. Ruff, *Inorg. Chem.*, **5,** 1787 (1966).
9. U. Biermann and O. Glemser, *Chem. Ber.*, **100,** 3795 (1967).

6. ACYCLIC SULFUR NITROGEN FLUORINE COMPOUNDS

Submitted by RUDIGER MEWS,* KLAUS KELLER,* and OSKAR GLEMSER*
Checked by K. SEPPELT† and J. THRASHER‡

Key substances in sulfur-nitrogen-fluorine chemistry are thiazyl fluoride (sulfur nitride fluoride), NSF, and thiazyl trifluoride (sulfur nitride trifluoride), NSF_3. Almost all known derivatives in this field of chemistry are obtainable through these two molecules.[1] The first preparation of NSF_3 by fluorination of S_4N_4 with AgF_2 is still the simplest and most efficient,[2] but it is less attractive for a general preparation because of the unstable nature of S_4N_4. For NSF, several procedures are mentioned in the literature.[3–6] The easiest way to obtain very pure thiazyl fluoride is the thermal decomposition of $Hg(NSF_2)_2$.[6] The mercury compound also serves as the starting material for the preparation of haloimidosulfurous difluorides.

A. THIAZYL TRIFLUORIDE (NITROGEN TRIFLUORIDE SULFIDE)

$$FC(O)NSF_2 + 2AgF_2 \rightarrow COF_2 + NSF_3 + 2AgF$$

*Institut für Anorganische Chemie der Universität, Tammanstrasse 4, D-3400 Göttingen, W. Germany.
†Institut für Anorganische und Analytische Chemie, Freie Universität Berlin, Fabeckstrasse 34–36, D-1000 Berlin 33.
‡Department of Chemistry, Clemson University, Clemson, SC 29631.

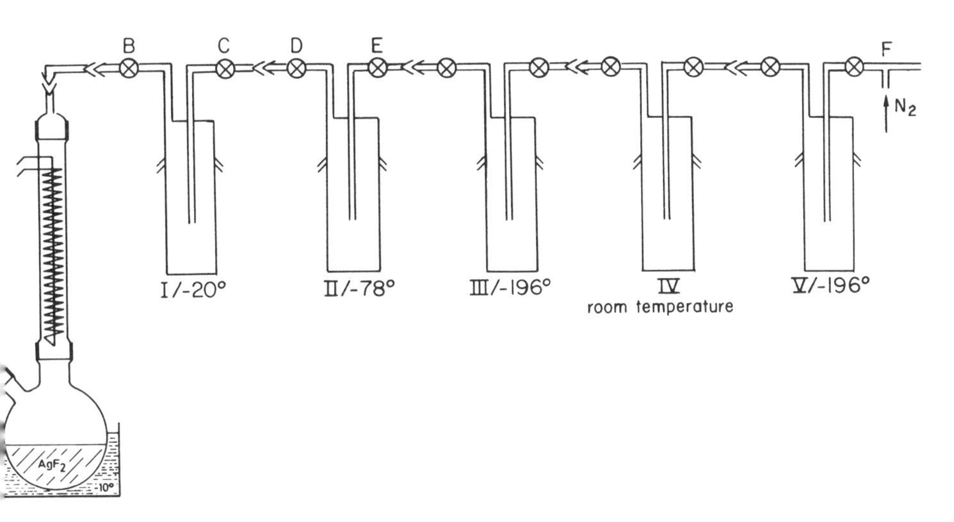

Fig. 1. Apparatus for the synthesis of NSF_3 and for separation of products by trap-to-trap fractionation.

Procedure

■ **Caution.** *All compounds used in this procedure are extremely toxic and moisture sensitive. Contact with skin or mucous membranes should be rigorously avoided.*

A 1-L two-necked flask with a reflux condenser is connected by a polyethylene tube to a system of five traps maintained at −20° (I), −78° (II), −196° (III), room temperature (IV), and −196° (V), as shown in Fig. 1. The temperatures are reached by cooling methanol with Dry Ice or liquid nitrogen.

■ **Caution.** *For the preparation of cooling baths from flammable organic solvents, liquid nitrogen must be used. A slow-flowing stream of dry nitrogen must be present at all times to prevent condensation of liquid air in traps that are at −196° and open to the atmosphere.*

The reaction vessel is loaded with 800 g of AgF_2* (5.48 moles) and cooled by an ice/salt bath to −10°, and $FC(O)NSF_2$[7,8] (320 g, 2.44 moles) is added in 10-mL portions (about 15 g) with a syringe through the free second neck A. The stopper is lifted as little as possible to prevent contamination with moisture.

*AgF_2 is readily prepared from AgCl or AgF and elemental fluorine in a flow system (prefluorinated Cu tube) at 100–250°. The residue of this reaction might be recycled under these conditions. Silver(II) fluoride is available commercially from Ozark-Mahoning Inc.

Silver(II) fluoride and $FC(O)NSF_2$ are mixed by shaking the vessel or by mechanical stirring. The temperature of the flask should not exceed room temperature. At higher temperatures, larger amounts of SF_6 are formed.

After addition of the last portion of $FC(O)NSF_2$, the reaction vessel is held at 25° for another hour and then flushed with a slow stream of dry nitrogen introduced via neck *A* to transfer the reaction products quantitatively into the cooled traps. Then stopcock *B* is closed, and the reaction vessel is disconnected from the trap system. Trap I contains unreacted $FC(O)NSF_2$ and some NSF_3. By allowing trap I to warm to room temperature, the product is transferred to trap II which is at −78°. The $FC(O)NSF_2$ that remains in trap I may be used in further reactions. Traps I and II are closed (at *C, D,* and *E*). Trap IV is cooled with liquid N_2, and trap III is warmed slowly to −30°. Carbonyl fluoride evaporates from trap III and condenses in trap IV, while some NSF_3 remains in trap III. The contents of traps II and III are combined. For further purification the trap containing the crude product is connected to two wash flasks (containing 200 mL of water and 200 mL of 5% aqueous $KMnO_4$, respectively), followed by a drying tube filled with granulated P_4O_{10} and three traps cooled to −20, −78, and −196°, respectively. The pure product is collected in the trap at −78°. Sometimes it is necessary to change the contents of the flasks after about half of the product is purified. Dry nitrogen is required to prevent condensation of air. Yields up to 70% (175 g) NSF_3 are obtained. This procedure can be scaled down easily to 1/10 or even lower.

Anal.[2] Calcd. for NSF_3: N, 21.53; S, 49.26. Found: N, 21.12; S, 49.37.

Properties

Thiazyl trifluoride (nitrogen trifluoride sulfide) is a colorless gas with a pungent odor, bp −27°, mp −72.6°. The compound is thermally stable up to >200°, is resistant to water, and is attacked only by strong nucleophiles or electrophiles.[1] IR (gas) 1523 (m), ν_{SN}; 815 (vs), $\nu_{as_{SF}}$; 773 (s), $\nu_{s_{SF}}$; 525 (w), $\delta_{s_{SF_3}}$; 432 (w), δ_{SF_3}; 346 (w), δ_{NSF}; ^{19}F NMR[9] ($CFCl_3$ ext. ref.) ϕ(SF) 70.0 ppm (tr).[2]

B. BIS(IMIDOSULFUROUS DIFLUORIDATO-*N*)MERCURY(II)

$$2FC(O)NSF_2 + HgF_2 \xrightarrow{10^\circ} Hg(NSF_2)_2 + 2COF_2$$

Procedure

■ **Caution.** *All compounds used in this procedure are extremely toxic. Decomposition products (remaining as residues in the condenser, the reaction*

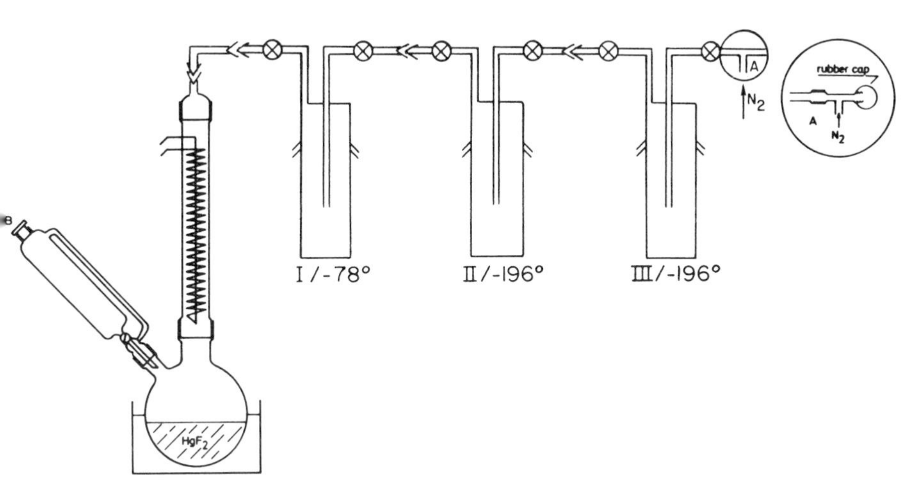

Fig. 2. Apparatus for the synthesis of $Hg(NSF_2)_2$.

flask, or the traps) often react explosively with water. Initial rinsing with CCl_4 *is recommended for cleaning. Surplus* $FC(O)NSF_2$, *recovered from the reaction, should be fractionated carefully and stored at low temperature until further use. On standing at room temperature, explosive decomposition products form rapidly.*

A two-necked 250-mL flask equipped with a dropping funnel and a condenser is connected to three traps cooled to −78° (Dry Ice), −196°, and −196° as shown in Fig. 2. The diameter of the second trap should be at least 5 cm to prevent plugging from COF_2. Because moisture must be carefully excluded, all glassware must be dried in an oven (150°). First the traps and the reflux condenser are put together while hot (fluorine- and heat-resistant grease should be used; for example, homogeneous 2:1 mixtures of Fluorolube Gr-90 [Fisher] and Kel-F wax 200 [Halocarbon]), and then they are cooled to room temperature in a stream of dry nitrogen through *A*. The two-necked flask, filled with HgF_2, is connected to the condenser, and finally the hot dropping funnel is added. After the funnel has reached room temperature, it is filled with $FC(O)NSF_2$ by means of a syringe. When the funnel is closed at *B*, the T-joint is opened at *A* to prevent excess pressure. The traps are cooled to the temperature indicated.

The nitrogen flow is continued to protect trap III. The reaction vessel is kept at 10° with running water. Into 110 g of HgF_2* (0.461 mole), 100 g of $FC(O)NSF_2$[8] (0.76 mole) is dropped slowly. After 24 hr, the reaction flask and the condenser are disconnected from the trap system, joined to two other traps (cooled with liquid N_2), and evacuated (oil pump) at room temperature for 30 min to remove

*It is possible to use commercially available HgF_2 [Ozark-Mahoning]; however, better results are obtained from freshly prepared HgF_2 (from $HgCl_2$ and F_2 at 100–250°).

unreacted $FC(O)NSF_2$. The remaining residue contains about 80% of the initial $FC(O)NSF_2$ as $Hg(NSF_2)_2$. For most purposes (preparation of NSF, $XNSF_2$), this product is sufficiently pure. Rather pure $Hg(NSF_2)_2$ can be prepared by using a twofold stoichiometric excess of $FC(O)NSF_2$ in this procedure. The analytically pure mercury compound recrystallizes from hot CH_2Cl_2 under dry nitrogen in colorless monoclinic needles. However, during this procedure, extensive decomposition is observed.

Anal. Calcd. for $Hg(NSF_2)_2$: F, 20.6; Hg, 54.4; N, 7.59; S, 17.4. Found: F, 20.1; Hg, 53.7; N, 7.7; S, 17.4.

Properties

Bis(imidosulfurous difluoridato-*N*)mercury(II) is a colorless, moisture-sensitive solid that decomposes slowly at room temperature. Storage below $-10°$ is recommended. IR (solid): 1313 (vs), 680 (vs), 574 (s), 550 (m) cm^{-1}.

C. THIAZYL FLUORIDE (NITROGEN FLUORIDE SULFIDE)

$$Hg(NSF_2)_2 \xrightarrow[\text{vac.}]{100\text{–}140°} 2NSF + HgF_2$$

Procedure

A 250-mL flask is loaded with 100–150 g (0.18–0.27 mole) of $Hg(NSF_2)_2$ (content 66%) and connected via a bent joint with a stopcock (*A*) to three traps and an oil pump (Fig. 3). Because moisture must be rigorously excluded, the

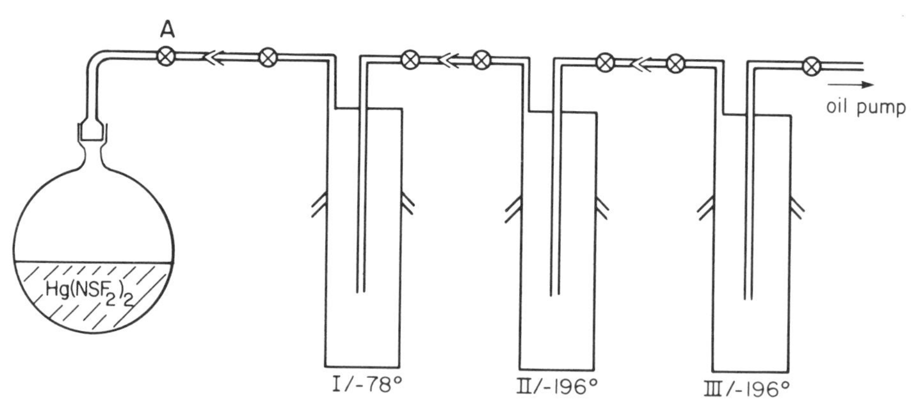

Fig. 3. Apparatus for thermal decomposition of $Hg(NSF_2)_2$ to give NSF.

thoroughly dried traps are put together directly from an oven (150°) as in procedure B, and evacuated while *A* is closed. Initially only trap III is cooled. After pumping for 1 hr, traps I and II are cooled to −78 and −196°, respectively. Then *A* is opened slowly, and the reaction flask is evacuated at room temperature and heated slowly in an oil bath to 110°. After 3 hr, the temperature is raised to 140°. Within 8–10 hr the decomposition is complete. If only a certain amount of NSF is needed, the decomposition can be interrupted at any time. Then the flask is cooled to room temperature, the stopcock at joint *A* is closed, and the flask is stored below −30° for further use.

The procedure given above may be repeated until all of the $Hg(NSF_2)_2$ is decomposed. Almost pure NSF is collected in trap II. Further purification is possible by fractional condensation [−78°, −130° (petroleum ether/liquid N_2), −196°] under vacuum (0.1 torr). The product is collected in the trap at −130° in almost quantitative yield.

Anal. Calcd.[2] for NSF: F, 29.20; S, 49.26. Found: F, 28.54; S, 48.44.

Properties

Thiazyl fluoride (nitrogen fluoride sulfide) is a colorless gas that condenses to a yellow liquid (bp 0.4°, mp −89°). The compound decomposes readily at room temperature and should be stored at −78°, where trimerization occurs slowly. It reacts violently with water. Thiazyl fluoride is very reactive toward electrophiles and nucleophiles. IR (gas)[9]: 1372 (s), 641 (vs), 366 (s) cm^{-1}. ^{19}F NMR[5,10]: ϕ(SF) ($CFCl_3$ int. ref.) 234 ppm.

References

1. O. Glemser and R. Mews, *Angew. Chem.*, **92,** 904 (1980); *Angew. Chem. Int. Ed. (Engl.)*, **19,** 883 (1980).
2. O. Glemser and H. Schröder, *Z. Anorg. Allgem. Chem.*, **284,** 97 (1956); O. Glemser, H. Meyer, and A. Haas, *Chem. Ber.*, **97,** 1704 (1964).
3. O. Glemser, H. Schröder, and H. Haeseler, *Z. Anorg. Allgem. Chem.*, **279,** 28 (1955).
4. A. F. Clifford and C. S. Kobayashi, *Inorg. Chem.*, **4,** 571 (1965). J. K. Ruff, *ibid.*, **5,** 1787 (1966).
5. R. Appel and E. Lassmann, *Chem. Ber.*, **104,** 2246 (1971).
6. O. Glemser, R. Mews, and H. W. Roesky, *Chem. Ber.*, **102,** 1523 (1969).
7. A. F. Clifford and J. W. Thompson, *Inorg. Chem.*, **5,** 1424 (1966).
8. J. S. Thrasher, *Inorg. Synth.*, **24,** 10 (1986).
9. H. Richert and O. Glemser, *Z. Anorg. Allgem. Chem.*, **307,** 328 (1961); F. Königer, A. Müller, and O. Glemser, *J. Mol. Structure*, **46,** 29 (1978).
10. O. Glemser and H. Richert, *Z. Anorg. Allgem. Chem.*, **307,** 313 (1961).

7. HALOIMIDOSULFUROUS DIFLUORIDES, $XNSF_2$

Submitted by R. MEWS,* K. KELLER,* and O. GLEMSER*
Checked by K. SEPPELT† and J. THRASHER‡

Haloimidosulfurous difluorides $XNSF_2$ [X = F,[1] Cl,[2,3] Br[2,3] (I)[3]] are prepared by the formal addition of X—F across the NS triple bond of thiazyl fluoride (sulfur nitride fluoride), NSF. An alternative method is the Si—N bond cleavage of $XN(SiMe_3)_2$ (X = Cl, Br) by SF_4.[4] This addition is easily realized by halogenating salts containing the NSF_2^- anion. These salts are prepared by adding NSF to metal fluorides (e.g., CsF[2]) or, more conveniently, by cleaving the CN bond in (fluorocarbonyl)imidosulfurous difluoride [$FC(O)NSF_2$]. For the preparation of $XNSF_2$, either $CsNSF_2$[2] or $Hg(NSF_2)_2$[3,4] can be used as an intermediate, but the latter is more readily accessible in high purity.

Chloroimidosulfurous difluoride and bromoimidosulfurous difluoride, $ClNSF_2$ and $BrNSF_2$, which may be used extensively for further syntheses, are formed in high yield from $Hg(NSF_2)_2$ and the elemental halogen. The iodo compound is very unstable. The fluoro derivative has not been synthesized in large quantities, and very little is known about its chemistry.

A. CHLOROIMIDOSULFUROUS DIFLUORIDE

$$Hg(NSF_2)_2 + 2Cl_2 \rightarrow 2ClNSF_2 + HgCl_2$$

Procedure

■ **Caution.** *Mercury compounds are highly toxic, and chlorine is poisonous. These materials should be handled in fume hoods or vacuum lines. Contact with skin or mucous membranes should be rigorously avoided. Because all products are hydrolyzed violently, all operations should be performed under rigorously anhydrous conditions.*

A Pyrex U-tube (B) (2 cm i.d., 15-cm legs), equipped with ground joints as shown in Fig. 1, is filled with 80 g (0.14 mole) of crude $Hg(NSF_2)_2$[5] (which

*Institut für Anorganische Chemie der Universität, Tammannstrasse 4, D-3400 Göttingen, West Germany.

†Institut für Anorganische und Analytische Chemie, Freie Universität Berlin, Fabeckstrasse 34–36, 1000 Berlin 33.

‡Department of Chemistry, Clemson University, Clemson, SC 29631.

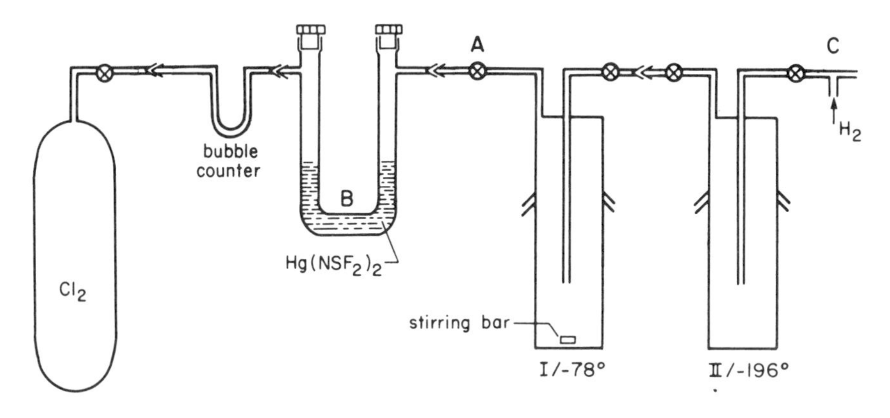

Fig. 1. Apparatus for synthesis of $ClNSF_2$.

contains 66% NSF_2^- ion). The tube is connected to trap I (−78°) and trap II (−196°). A slow stream of nitrogen is maintained at C to prevent condensation of air in trap II. Trap I contains a Teflon-coated stirring bar. Dry chlorine (40–50 g, 0.6–0.7 mole) is passed over the $Hg(NSF_2)_2$ in the U-tube over a period of 6 hr. The flow rate is controlled by a bubble counter filled with Kel-F oil No. 3 [Applied Science Laboratories].

When all of the chlorine has been added, trap I contains about 50–60 g of a yellow liquid (Cl_2 and $ClNSF_2$). Stopcock *A* is closed, and the U-tube is removed. While it is being stirred, the mixture in trap I is warmed slowly to +10°. Most of the chlorine distills into trap II. For further purification, the product in trap I is condensed onto 10 g of Hg at −196° and stirred at 10° until the liquid becomes colorless. This procedure may need to be repeated. Final purification of the product is accomplished by fractional distillation [−78° (methanol/Dry Ice), −120° (pentane/liquid N_2), and −196°] under high vacuum (10^{-2}–10^{-3} torr). The trap at −120° contains pure $ClNSF_2$ (27 g, 0.23 mole, 80%).

Anal.[2] Calcd. for $ClNSF_2$: N, 11.72; Cl, 29.7; S, 26.8; F, 31.8. Found: N, 11.25; Cl, 29.2; S, 27.1; F, 30.1.

Properties

Chloroimidosulfurous difluoride is a colorless moisture-sensitive liquid (bp 24°). It is thermally stable to 100°. The N—Cl bond is readily cleaved by photolysis. IR spectrum (gas): 1291 (w), 1200 (s), 752 (vs), 692 (vs), 644.5 (m), 548.5 (m), and 405 (m), cm^{-1}. ^{19}F NMR spectrum ($CFCl_3$) ϕ(SF) 46.3 ppm.

B. BROMOIMIDOSULFUROUS DIFLUORIDE

$$\text{(a)}\ Hg(NSF_2)_2 + 2Br_2 \rightarrow HgBr_2 + 2BrNSF_2$$
$$\text{(b)}\ HgF_2 + 2FC(O)NSF_2 + 2Br_2 \rightarrow HgBr_2 + 2COF_2 + 2BrNSF_2$$

Procedure

■ **Caution.** *Mercury compounds are highly toxic, and bromine is poisonous. These materials should be handled in fume hoods or vacuum lines. Contact with skin or mucous membranes should be rigorously avoided.*

(a) A 500-mL Hoke stainless steel cylinder containing 60 g of crude $Hg(NSF_2)_2$ (66%, 0.215 mole NSF_2^- anion) is attached to a vacuum line, cooled to −196°, and evacuated. Bromine (34.7 g, 0.217 mole) is distilled into the cylinder. The mixture is warmed to room temperature and agitated for 24 hr. As shown in Fig. 2, the cylinder is attached to two thoroughly dried traps [taken from a hot oven (150°) and cooled to room temperature with a stream of dry nitrogen flowing through them]. Into trap I, a stirring bar and 10 g of Hg are placed. To remove the last traces of moisture, trap II is cooled to −196° and the system is evacuated (oil pump) for 0.5 hr. Then trap I is also cooled to −196°. The valve (*A*) of the cylinder is opened very slowly, and under dynamic vacuum the crude product is transferred into the first trap.

After all volatile materials are condensed into trap I, stopcocks *B* and *C* are closed, and the cylinder and the pump are removed. By introducing dry nitrogen through stopcock C via a T-adapter, the traps are brought to atmospheric pressure.

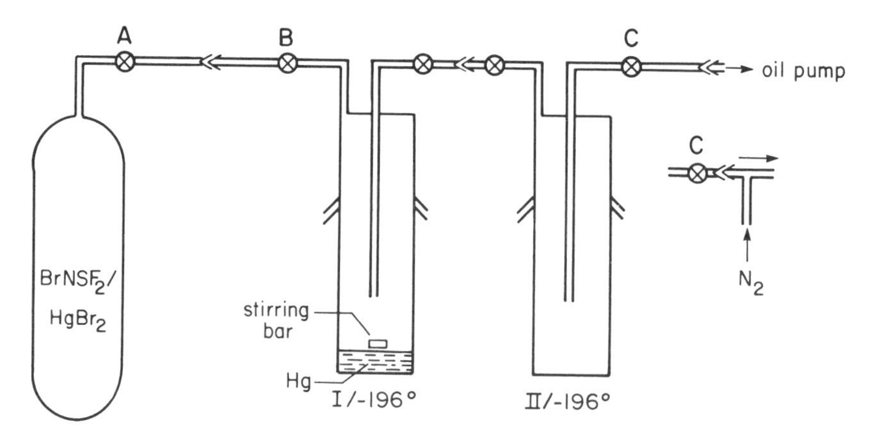

Fig. 2. Apparatus for synthesis of $BrNSF_2$.

Trap I is warmed to -20 to $-30°$ while trap II remains cooled to $-196°$ to protect the product in trap I from hydrolysis. A slow stream of dry nitrogen is maintained at the T to prevent condensation of liquid air in trap II.

The mixture in trap I is stirred for 3–4 hr at -20 to $-30°$, causing the excess bromine to react with the mercury. This process is continued until a pale yellow liquid remains. A highly pure product is obtained by fractional condensation by using traps at $-30°$ (methanol/liquid N_2), $-78°$, and $-196°$. The pure $BrNSF_2$ (24.7 g, 0.15 mole; 70%) is found in the trap at $-78°$.

(b) Similar results are obtained for the preparation of $BrNSF_2$, when $FC(O)NSF_2$[6] is allowed to react with a slight excess of Br_2 in the presence of HgF_2. Pure mercury(II) fluoride is prepared readily by direct fluorination of other mercury halides in a flow system (prefluorinated Cu tube, 5 cm i.d.) at 100–200°. Fluorination is started slowly at 100°, and the temperature is raised gradually. Mercury(II) fluoride is available commercially [Ozark-Mahoning].

(Fluorocarbonyl)imidosulfurous difluoride,[6] $FC(O)NSF_2$ (40 g, 0.305 mole) and 49.5 g (0.309 mole) of Br_2 are condensed at $-196°$ onto 50 g (0.21 mole) of HgF_2 in a 500-mL steel Hoke cylinder, warmed to room temperature, and agitated for 40 hr. The COF_2 that is formed can be distilled out of the cylinder at $-78°$ into a storage tube and retained for future use or destroyed by bubbling into an aqueous sodium hydroxide solution in the fume hood. The residue in the cylinder is handled as in part (a). The yield of pure $BrNSF_2$ is 49.5 g; 92%.

Anal. Calcd. for $BrNSF_2$: Br, 48.7; F, 23.2; N, 8.53; S, 19.55. Found: Br, 48.5; F, 22.8; N, 8.5; S, 19.2.

Properties

Bromoimidosulfurous difluoride, $BrNSF_2$, is a slightly yellow liquid that boils at 56.5°. The compound is light-sensitive and not very stable thermally. It can be stored in Pyrex glass vessels at $-78°$. It reacts explosively with water. IR spectrum (gas): 1215 (s), 1180 (m), 745 (vs), 689 (vs), and 468 (m), cm^{-1}. ^{19}F NMR spectrum ($CFCl_3$): ϕ(SF) 57.6 ppm.

References

1. O. Glemser, R. Mews, and H. W. Roesky, *Chem. Commun.*, **1969,** 914.
2. J. K. Ruff, *Inorg. Chem.*, **5,** 1787 (1966).
3. O. Glemser, R. Mews, and H. W. Roesky, *Chem. Ber.*, **102,** 1523 (1969).
4. K. Seppelt and W. Sundermeyer, *Angew. Chem.*, **81,** 785 (1969); *Angew. Chem. Int. Ed. (Engl.)*, **8,** 771 (1969).
5. R. Mews, K. Keller, and O. Glemser, *Inorg. Synth.*, **24,** 14 (1986).
6. J. S. Thrasher, *Inorg. Synth.*, **24,** 10 (1986).

8. CESIUM FLUOROXYSULFATE (CESIUM FLUORINE SULFATE)

$$Cs^+ + SO_4^{2-} + F_2 \rightarrow F^- + CsSO_4F \downarrow$$

Submitted by EVAN H. APPELMAN*
Checked by KARL O. CHRISTE,† RICHARD D. WILSON,† and CARL J. SCHACK†

The fluoroxysulfate ion is a true hypofluorite, O_3SOF^-. It is formed by the fluorination of sulfate ion with molecular fluorine in aqueous solution.[1] To obtain a high product yield, it is best to choose a cation that forms a highly soluble sulfate, since reaction efficiency is enhanced by high sulfate concentration. On the other hand, if the product is to be removed easily from the reaction mixture, the same cation must form a relatively insoluble fluoroxysulfate. These conditions are met by the Cs^+ ion, and $CsSO_4F$ is the easiest fluoroxysulfate salt to prepare and isolate in good yield.

The fluoroxysulfate ion is analogous to and isoelectronic with the long-known compound fluorine perchlorate, ClO_4F. But whereas ClO_4F is a dangerously unstable gas, cesium fluoroxysulfate is a rather stable salt that can be stored for long periods of time and used conveniently on the benchtop (but see cautionary note that follows). It possesses formidable oxidizing and fluorinating capabilities, which give it a variety of potential applications as a synthetic and analytical reagent for inorganic and organic chemistry.[2–4]

A 20-g preparation is described in this article. Scaling down poses no problem, but substantial scaling up may lead to difficulties in introduction of fluorine at a sufficiently rapid rate. If the rate of fluorine introduction is not scaled up with the size of the preparation, prolongation of the preparation time may lead to loss by decomposition.

Most of the time and labor involved in the preparation are taken up in assembling the necessary apparatus. The actual reaction takes about an hour, while another 90 min or so is required for filtration, preliminary drying, and transfer of the sample. The preparation is concluded by overnight drying of the product under vacuum.

■ **Caution.** *A certain amount of hazard accompanies any procedure that makes use of molecular fluorine. The experimenter should familiarize him- or*

*Chemistry Division, Argonne National Laboratory, Argonne, IL 60439. Work performed under the auspices of the Office of Basic Energy Sciences, Division of Chemical Sciences, U.S. Department of Energy, under Contract W-31-109-Eng-38.

†Rocketdyne, a Division of Rockwell International Corporation, Canoga Park, CA 91304.

herself with the precautions necessary for safe handling of fluorine before undertaking the synthesis.[5] *In addition, cesium fluoroxysulfate is thermodynamically unstable. Mild detonations have occasionally occurred during the handling of 100-mg portions of the salt. The causes of these detonations were not always clear, although they usually took place in the course of either deliberate or inadvertent crushing or scraping of the sample. It is very likely that the detonations were brought about by impurities on the surfaces of containers or spatulas. We have now worked routinely with 10–20-g quantities of* $CsSO_4F$ *over the past several years and have never observed any kind of detonation involving a large amount of the material. Nevertheless,* $CsSO_4F$ *must be regarded as a substance whose kinetic stability is not well understood. Use of gloves and a plastic face shield are advisable when manipulating the salt. Cesium fluoroxysulfate is also a powerful oxidant, and it reacts violently with a number of organic solvents, including dimethyl sulfoxide, dimethylformamide, pyridine, and ethylene diamine (1,2-diaminoethane).*

Special Materials and Equipment

Fluorine is used as a 20% mixture with nitrogen. Such mixtures may be obtained from any of several specialty gas suppliers. Prices vary substantially and should be compared. The cylinder should be equipped with a pressure regulator that is designed expressly for fluorine service and that incorporates a nitrogen purge assembly.*

Commercial fluorine usually contains too much HF for satisfactory use of a glass flowmeter. The simplest solution is the use of a flowmeter that consists of a metal float moving in a channel machined in a Kel-F block [Brooks]. A convenient flowmeter range is 0–1 L/min of air.

Plumbing between the regulator and flowmeter and between the flowmeter and the delivery tube is most conveniently effected with ¼-in. o.d. copper tubing. Connections can be of the swage or compression type, or the 45° flare connections commonly used for refrigeration lines may be used. Brass, stainless steel, aluminum, or Monel fittings are satisfactory. Connections to pipe fittings may be sealed with Teflon tape *only*—never with pipe dope or lacquer.

Since HF is formed as a by-product of the reaction, glass should not come into contact with the reaction mixture, and the reaction vessel, fluorine delivery tube, thermocouple sheath, and filtration apparatus should all be made of Teflon. The delivery tube is made from ¼-in. o.d. × 3/16-in. i.d. FEP tubing [Zeus]. One end of the tube should be heated and drawn to a narrow tip (ca. 1 mm i.d.).

*A typical regulator of this sort is the Matheson Co. Model B15F679. The standard model is usually provided with a 0–300 psig inlet pressure gauge; a higher pressure range (500 or 1000 psig) should be requested when ordering.

A right-angle bend should be made in the tube by heating it approximately 10 in. from the tip, and the far end should be cut off about 4 in. from the bend.

The reaction vessel is a Teflon FEP tube ⅞ in. i.d. × 0.045 in. wall × 9 in. long, made by heat-sealing one end of a length of the appropriate tubing [Zeus]. This may be accomplished by heating the tubing with a high-temperature hot air gun or over a cool flame, drawing it down, and pinching the end shut with a hot pair of pliers. A sheath for the thermocouple used to measure reaction temperature is made in similar fashion from a length of FEP tubing 5⁄32 in. o.d. × 3⁄32 in. i.d. The sheath should be some 3-in. longer than the reaction tube so that the sheathed thermocouple can be used as a stirrer.

The filtration apparatus consists of a commercially available 47-mm Teflon filter funnel containing a Teflon filter disk of fine-porosity (5–10 μm) [Savillex].* A 1½-in. NPT male thread should be cut into the top ½ in. of the chimney of this funnel. When it is necessary to pressurize the funnel with nitrogen, a Teflon transfer cap* is threaded onto the free end of the chimney, and a plastic line to a nitrogen supply can readily be attached to the cap.

A minor inconvenience of this funnel is that its lower portion is square rather than tapered. As a result, filtrate tends to be trapped beneath the filter disk, and the funnel must be shaken to remove it.

Stirring rods used during the filtration should be either solid Teflon or Teflon-coated. The dried product may be handled with porcelain, glass, or Teflon-coated metal spatulas. The product should be stored in a Teflon vessel [Savillex].

For drawing dry nitrogen through the filter during the preliminary drying step, an air pump of moderate free-air capacity is needed.†

Procedure

■ **Caution.** *The reaction must be carried out in a well-ventilated fume hood.*

It is first necessary to calibrate the flow of the 20% F_2/80% N_2 mixture. This can be done by passing it into a 1 *M* KI solution for a specified period of time and titrating the I_3^- formed with standard thiosulfate. A flow of 3.5–4 mmol F_2/min should be achieved and the corresponding flowmeter setting noted.

Now 50 mL of 2 *M* aqueous Cs_2SO_4 is put into the Teflon reaction tube, and the tube is immersed in an ice/salt eutectic in a cylindrical Dewar flask. An unsilvered Dewar is most convenient for observing the reaction. A Teflon-sheathed

*Filter funnel is Catalog No. 4750-47-6; fine filter disks are Catalog No. 1131; transfer cap is Catalog No. 501-25 (¼-in. tubing connection) or 501-37 (⅜-in. tubing connection). Savillex makes screw-capped Teflon PFA vials in a wide variety of sizes. (Many of the Savillex products and some sizes of Teflon FEP tubing are available through general laboratory suppliers.)

†A suitable pump is the Gast Model 0211, which has a free-air capacity of 1.3 cu ft/min.

thermocouple and the fluorine delivery tube are inserted, and the fluorine flow is commenced and adjusted to provide 3.5–4 mmol F_2/min. The temperature of the reaction mixture is monitored with an appropriate thermocouple readout or potentiometer. The heat generated by the reaction will normally maintain the temperature between −5 and −10°. If the temperature falls below −10°, the mixture may freeze. To prevent this, the ice/salt mixture should be diluted. Conversely, if the temperature rises to near 0°, the mixture should be "stiffened" with ice and salt. After the reaction has proceeded for 20–30 min, it will be necessary to stir the mixture regularly to prevent it from foaming out of the tube. The sheathed thermocouple can be used as a manual stirrer. Introduction of fluorine is continued for 1 hr. Then the cylinder valve is closed, and the pressure in the regulator is allowed to bleed down to atmospheric. A flow of nitrogen through the line is immediately commenced via the purge assembly. The reaction mixture is removed promptly from the ice/salt bath to prevent it from freezing, and it is filtered under suction through the Teflon filter assembly, which should have been prechilled to 0°. The filter cake is stirred thoroughly with two successive 20-mL portions of ice-cold water, as much as possible of the supernatant liquid being sucked off each time. Excessive washing must be avoided, since the product is significantly soluble in water. Filtrate and washings should be disposed of promptly and carefully, since they contain both HF and a powerful oxidant.

Now the transfer cap is screwed onto the funnel chimney, and a supply of clean, dry nitrogen is attached. The outlet of the funnel is connected to the suction side of the air pump, and nitrogen is drawn through the filter cake while maintaining ~2 psig of positive pressure on the inlet of the funnel. After 25 min, the funnel is opened, and the filter cake is broken up with a spatula, care being taken not to damage the filter disk.

■ **Caution.** *Face shield and gloves are recommended.*

Passage of nitrogen through the filter is resumed for an additional 25 min. The product is then transferred to a 60-mL wide-mouthed Teflon bottle [Savillex], any lumps being broken up with a spatula. The uncovered bottle is placed in a vacuum desiccator over a desiccant such as anhydrous $CaSO_4$, and the product is dried overnight under dynamic vacuum, using a liquid nitrogen or Dry Ice trap between the pump and desiccator. The final product is approximately 18 g of white or slightly off-white material, which assays as ~98% $CsSO_4F$ by iodometric titration. The yield is ~70% based on the Cs_2SO_4 taken, or ~30% based on the fluorine used.

The dried product may be stored at room temperature in a screw-cap Teflon bottle, but slow decomposition (1–2% per month) is sometimes observed. To assure the greatest possible integrity of the sample, the following storage procedure is recommended. The *loosely* capped Teflon bottle containing the $CsSO_4F$ is placed in a vacuum desiccator, which is evacuated and backfilled with dry nitrogen. The desiccator is opened, and the sample bottle is closed tightly. The sample bottle is then placed in an outer glass screw-cap bottle containing desiccant, and the entire package is stored in a freezer. The package is warmed to room temperature before opening.

Fluoroxysulfate reactions may be carried out routinely in glass vessels, but the consequences of the possible production of small quantities of HF should be considered.

Analytical[1]

Routine analysis of $CsSO_4F$ is carried out by adding an excess of 0.1–0.2 *M* aqueous KI to a weighed portion of the solid, acidifying the solution slightly, and titrating the I_3^- formed with standard thiosulfate, using starch as an endpoint indicator:

$$SO_4F^- + 3I^- \rightarrow SO_4^{2-} + F^- + I_3^-$$
$$I_3^- + 2S_2O_3^{2-} \rightarrow 3I^- + S_4O_6^{2-}$$

The solutions will darken slowly at the endpoint, due to gradual reduction of peroxydisulfate, $S_2O_8^{2-}$, which is present as an impurity in the fluoroxysulfate. For the most accurate results, the drift should be measured and extrapolated back to the time of mixing. The amount of $S_2O_8^{2-}$ present may be determined by making the titrated solution 1 *M* in KI, flushing it with N_2, and allowing it to stand covered and in the dark for 15 min. Titration of the additional I_3^- formed gives a measure of the amount of peroxydisulfate in the $CsSO_4F$:

$$S_2O_8^{2-} + 3I^- \rightarrow I_3^- + 2SO_4^{2-}$$

Typical fluoroxysulfate titer extrapolated to time zero: 7.9 meq/g (calcd. for $CsSO_4F$: 8.07 meq/g).

Properties

Except for the very gradual decomposition that has been noted, cesium fluoroxysulfate is stable at room temperature. It is not, however, thermodynamically stable, and at ~100° it undergoes rapid, though not especially violent, decomposition to $CsSO_3F$ and O_2.[1]

Cesium fluoroxysulfate is soluble in water to the extent of about 0.5 *M* at 25°C. The aqueous solutions are not stable, decomposing with a ~15-min half-time to yield a mixture of HF, HSO_4^-, SO_3F^-, HSO_5^-, H_2O_2, and O_2.[2] In alkaline solution, decomposition is almost instantaneous, and some OF_2 is produced. A wide variety of inorganic substrates are oxidized by aqueous SO_4F^-, although at varying rates.[2,3] Particularly noteworthy is the very rapid oxidation of Ag(I) to Ag(II). Because of this, certain substrates, such as Cr(III), that do not react directly with SO_4F^- can be oxidized smoothly in the presence of catalytic amounts of Ag^+.[3]

Cesium fluoroxysulfate is soluble in acetonitrile to the extent of about 0.07–0.08 *M* at room temperature.[4] The solutions show little or no decomposition over the course of many hours, and acetonitrile has been found to be a convenient solvent in which to examine the reactions of SO_4F^- with organic substrates, although heterogeneous systems have also been used.[4,6–10] The ^{19}F NMR spectrum of $CsSO_4F$ in acetonitrile shows a single line with a shift of + 132 ppm downfield from $CFCl_3$.[1]

References

1. E. H. Appelman, L. J. Basile, and R. C. Thompson, *J. Am. Chem. Soc.*, **101,** 3384 (1979).
2. R. C. Thompson and E. H. Appelman, *Inorg. Chem.*, **19,** 3248 (1980).
3. R. C. Thompson and E. H. Appelman, *Inorg. Chem.*, **20,** 2114 (1981).
4. E. P. Ip, C. D. Arthur, R. E. Winans, and E. H. Appelman, *J. Am. Chem. Soc.*, **103,** 1964 (1981).
5. *Matheson Gas Data Book,* 6th ed., Matheson Co., Inc., East Rutherford, NJ, 1980.
6. S. Stavber and M. Zupan, *J. Chem. Soc., Chem. Commun.*, **1981,** 148.
7. S. Stavber and M. Zupan, *J. Fluorine Chem.*, **17,** 597 (1981).
8. S. Stavber and M. Zupan, *J. Chem. Soc., Chem. Commun.*, **1981,** 795.
9. S. Stavber and M. Zupan, *J. Chem. Soc., Chem. Commun.*, **1983,** 563.
10. E. H. Appelman, L. J. Basile, and R. Hayatsu, *Tetrahedron,* **40,** 189 (1984).

9. SELENIUM TETRAFLUORIDE, SELENIUM DIFLUORIDE OXIDE (SELENINYL FLUORIDE), AND XENON BIS[PENTAFLUOROOXOSELENATE(VI)]

Submitted by KONRAD SEPPELT,* DIETER LENTZ,* and GERHARD KLÖTER*
Checked by CARL J. SCHACK†

Preparations for seleninyl fluoride and selenium tetrafluoride that require only SeO_2 and SF_4 are very convenient and are identical to that used for the synthesis of TeF_4.[1] Other methods of preparation include partial fluorination of elemental

*Institut fur Anorganische und Analytische Chemie, Freie Universitat Berlin, Fabeckstrasse, 34–36, 1000 Berlin 33.
†Rocketdyne Division, Rockwell International, 6633 Canoga Ave., Canoga Park, CA 91304.

selenium with fluorine or chlorine trifluoride.[2] The present procedure does not require the use of fluorinating reagents that are difficult to handle and highly oxidizing.

Because of this straightforward synthesis of seleninyl fluoride, xenon bis[pentafluorooxoselenate(VI)] can now be prepared easily. The latter was prepared initially by the reaction of XeF_2 and $H[SeF_5O]$,[3] which results from a time-consuming multistep synthesis.[4]

A. SELENIUM TETRAFLUORIDE AND SELENIUM DIFLUORIDE OXIDE

$$SeO_2 + SF_4 \rightarrow SeOF_2 + SOF_2$$
$$SeO_2 + 2SF_4 \rightarrow SeF_4 + 2SOF_2$$

The system $SeO_2/SeOF_2/SeF_4/SF_4/SOF_2$ is interconnected, and the product that forms depends only on the ratio of starting materials. Thus, if an excess of SeO_2 is used, $SeOF_2$ is formed in high yield, whereas if SF_4 is in excess, SeF_4 is formed in excellent yield. Also, $SeOF_2$ is converted by SF_4 to SeF_4, and SeO_2 with SeF_4 forms pure $SeOF_2$.[5]

Procedure

■ **Caution.** *All compounds used in these syntheses are highly poisonous. Contact with skin or nasal passages should be avoided. Gloves should be worn to preclude HF burns. All selenium compounds should be regarded as toxic.*
Selenium dioxide [Alfa Products] (22.2 g, 0.2 mole) is placed in a 200-mL stainless steel Hoke pressure vessel. The vessel is connected to a glass vacuum line[6] and is evacuated and cooled to −196°. Sulfur tetrafluoride [Matheson] (19.4 g, 0.18 mole for preparation of $SeOF_2$; 64.8 g, 0.6 mole for preparation of SeF_4) is condensed into it. The closed vessel is heated for 12 hr at 120° with stirring or shaking of the reaction mixture.

After the reaction is allowed to cool to ambient temperature, it is further cooled to −196°. The reaction vessel is then connected to a trap-to-trap system with traps cooled to −78 and −196°. It is allowed to warm to room temperature while the contents are fractionated slowly under a vacuum $\leq 10^{-1}$ torr. In the trap at −196° are found SOF_2 and excess SF_4. The pure product, either $SeOF_2$ or SeF_4, is found in nearly quantitative yield in the trap at −78°.* Since either

*For $SeOF_2$, the checker used 72.5 mmol SeO_2 and 65.5 mmol SF_4 and heated the mixture at 120° for 36 hr without stirring (yield 93%). For SeF_4, 14.9 mmol SeO_2 and 45 mmol SF_4 under the same conditions gave a 85% yield of SeF_4. A stainless steel–Teflon FEP vacuum line was used.

product attacks glass slowly at room temperature, storage should be in Kel-F or FEP (fluoroethylene-propylene copolymer) [Zeus] containers kept in a dry box.

Properties

Selenium tetrafluoride is a colorless liquid (mp $-13.8°$; bp 106°). Since it is an excellent solvent, it is often yellow. It is hydrolyzed readily and attacks Pyrex glass at room temperature, especially if water is present. IR gas spectrum: 733 (s), 622 (s), 361 (m) cm^{-1}.

Anal. Calcd. for SeF_4: Se, 50.9; F, 49.0. Found: Se, 51.4; F, 48.7.

Seleninyl fluoride is a colorless liquid (mp 15°; bp 126°). It is less sensitive to moisture and attacks glass somewhat more slowly than SeF_4. The IR gas spectrum shows the characteristic Se=O stretching vibration at 1037 cm^{-1}. ^{19}F NMR: ϕ33.5 ppm, $J_{^{77}Se\text{-}^{19}F} = 837$ Hz.

Anal. Calcd. for $SeOF_2$: Se, 59.4; F, 28.6. Found: Se, 59.1; F, 29.1.

The relatively high boiling points of both materials indicate rather strong fluorine bridging and fluorine exchange. The molecular structures of SeF_4 and $SeOF_2$ are analogous to those of SF_4 and SOF_2.[7] The fluorine exchange of the two different pairs of fluorine atoms in SeF_4 can be frozen out at $-140°$ and in dilute solutions only as shown by the ^{19}F NMR spectrum: ϕ_A 37.7 ppm, ϕ_B 12.1 ppm, $J_{AB} = 26$ Hz.[8] Both compounds exhibit Lewis acid character in reactions with alkali metal fluorides to form salts of SeF_5^- and $SeOF_3^-$, respectively.[9,10] With Lewis acids, SeF_4 forms the cation SeF_3^+.[11]

B. XENON BIS[PENTAFLUOROOXOSELENATE(VI)]

$$3XeF_2 + 2SeOF_2 \rightarrow Xe(OSeF_5)_2 + 2Xe$$

Procedure

■ **Caution.** *All selenium compounds are toxic. Gloves must be worn.* All compounds are handled in a dry box. Xenon difluoride[12] (8.5 g, 0.05 mole) is placed in a 100-mL stainless steel pressure vessel [Hoke]. Seleninyl fluoride (4.4 g, 0.033 mole) is placed into a Kel-F or FEP (fluoroethylene propylene copolymer) test tube–shaped tube and inserted into the reaction vessel in an upright position. The vessel is closed and then cooled to $-78°$ (methanol/Dry Ice) and placed in a horizontal position on a mechanical shaker. After being shaken for 24 hr at room temperature, the vessel is attached to a vacuum line and cooled to $-20°$, and the volatile material (Xe) is transferred under dynamic vacuum into a trap cooled to $-196°$. The vessel is filled with dry argon (or nitrogen) and opened in the dry box. The impure pale yellow liquid is transferred into a glass sublimator. Sublimation at room temperature (10^{-2} torr) onto a cold finger (−20°) gives pure $Xe(OSeF_5)_2$ (yield 3.5 g, 41%).*

Anal. Calcd. for $F_{10}O_2Se_2Xe$: F, 37.4; Se, 29.7; Xe, 25.6. Found: F, 37.1; Se, 31.0; Xe, 25.6.

Properties

Xenon bis[pentafluorooxoselenate(VI)] is a pale yellow crystalline solid (mp 69°) that is easily hydrolyzed. It is readily characterized by its ^{19}F NMR spectrum in $CFCl_3$ solution (downfield shifts are positive); a typical AB_4 pattern: ϕ_A 80.5 ppm, ϕ_B 69.4 ppm, J_{AB} = 234 Hz. In addition, ^{77}Se satellites and ^{129}Xe satellites are observed $J_{^{77}Se\text{-}F_A}$ = 1323 Hz, $J_{^{77}Se\text{-}F_B}$ = 1318 Hz, and $J_{^{129}Xe\text{-}F_B}$ = 38 Hz.[3] The ^{129}Xe NMR spectrum shows the expected nine-line spectrum due to coupling with the eight equivalent equatorial fluorine atoms at +3131 ppm (downfield) from elemental xenon.[13] IR spectrum (solid): 787 (m), 725 (vs), 700 (s), 612 (s), 550 (m), 430 (s) cm^{-1}. Its molecular structure has been determined by X-ray crystallography.[14] Pyrolysis of the compound yields $F_5SeOSeF_5$,[15] whereas photolysis gives the corresponding peroxide $F_5SeOOSeF_5$.[16]

References

1. K. Seppelt, *Inorg. Synth.*, **20,** 33 (1980); A. L. Oppegard, W. C. Smith, E. C. Muetterties, and V. A. Engelhardt, *J. Am. Chem. Soc.*, **82,** 3835 (1960).
2. E. E. Aynsley, R. D. Peacock, and P. L. Robinson, *J. Chem. Soc.*, **1952,** 1231; G. A. Clark, M. Nojina, and J. Kerekes, *J. Am. Chem. Soc.*, **96,** 925 (1974).
3. K. Seppelt, *Angew. Chem.*, **84,** 715 (1972); *Angew. Chem., Int. Ed. Engl.*, **11,** 723 (1972).
4. K. Seppelt, *Inorg. Synth.*, **20,** 38 (1980).
5. R. D. Peacock, *J. Chem. Soc.*, **1953,** 3617.
6. F. Haspel-Hentrich and J. M. Shreeve, *Inorg. Synth.*, **24,** 58 (1986).
7. I. C. Bowater, R. D. Brown, and F. R. Burden, *J. Mol. Spectry.*, **23,** 272 (1967); **28,** 446 (1968).
8. K. Seppelt, *Z. Anorg. Allgem. Chem.*, **416,** 12 (1975).
9. K. O. Christe, E. C. Curtis, C. J. Schack, and D. Pilipovich, *Inorg. Chem.*, **11,** 1679 (1972).
10. R. Paetzold and K. Aurich, *Z. Anorg. Allgem. Chem.*, **348,** 94 (1966).
11. N. Bartlett and P. L. Robinson, *J. Chem. Soc.*, **1961,** 3417.
12. J. L. Wachs and M. S. Matheson, *Inorg. Synth.*, **8,** 260 (1966).

*The checker takes a Teflon FEP U-tube equipped with Gyrolock fittings [Hoke] and a magnetic stirring bar and, in the dry box, adds XeF_2 (2.466 g; 14.56 mmol). After evacuation on the vacuum line and cooling the U-tube to −196°, $SeOF_2$ (~18 mmol) is distilled into the tube. The cold bath is removed, and as the vessel warms the $SeOF_2$ melts and runs down onto the solid XeF_2. Bubbling begins, and soon only a liquid is present. The rate of gas evolution (Xe) is moderated by intermittent cooling. After 2 hr, no more gas is evolved. The reaction mixture is stirred overnight. All the volatile materials are then removed under vacuum at 20°. After 3 hr, a white solid [$Xe(OSeF_5)_2$] remains. Based on ^{19}F NMR, it is pure (1.58 g, 63%).

13. K. Seppelt and H. H. Rupp, *Z. Anorg. Allgem. Chem.*, **409,** 338 (1974).
14. L. K. Templeton, D. H. Templeton, K. Seppelt, and N. Bartlett, *Inorg. Chem.*, **15,** 2718 (1976).
15. K. Seppelt, *Chem. Ber.*, **106,** 157 (1973).
16. K. Seppelt and D. Nöthe, *Inorg. Chem.*, **12,** 2727 (1973).

10. TELLURIUM CHLORIDE PENTAFLUORIDE

$$TeF_4 + ClF \rightarrow TeClF_5$$
$$TeCl_4 + 5ClF \rightarrow TeClF_5 + 4Cl_2$$

Submitted by M. MURCHIE* and J. PASSMORE*
Checked by K. SEPPELT†

The syntheses described provide straightforward, relatively high-yield routes to reasonably pure $TeClF_5$.[1] The chemistry of the analogous $SClF_5$ has been studied extensively,[2] and the chemistry of $TeClF_5$ is likely to be fruitful.

The advantages of the TeF_4/ClF reaction are that one-fifth as much chlorine monofluoride is necessary as when $TeCl_4$ is used and there is less chlorine to separate from $TeClF_5$. The advantage of the $TeCl_4$/ClF reaction is that $TeCl_4$ is readily available commercially, whereas TeF_4 is not. The purity of $TeClF_5$ in both cases depends largely upon the purity of ClF and the method of purifying the $TeClF_5$.

Procedure

■ **Caution.** *Tellurium compounds are highly toxic.*[3] *They should be handled with care in a well-ventilated fume hood and treated in the same way as toxic inorganic fluorides (e.g., HF, ClF, AsF_5).*[4]
Tellurium tetrafluoride[5] is moisture-sensitive and should be handled in a moisture-free environment, as has been fully described.[6] The tellurium tetrachloride [Alfa Inorganics] may be used without further purification. Chlorine monofluoride [Ozark-Mahoning] is best manipulated in a preconditioned metal vacuum line.[6] The reaction vessel is pretreated with F_2 or SF_4, followed by treatment overnight with ClF. The reaction vessel is constructed by either sealing one end of a length

*Department of Chemistry, University of New Brunswick, Fredericton, New Brunswick, Canada E3B 6E2.

†Freie Universität Berlin, Institut für Anorganische und Analytische Chemie, Fabeckstrasse 34–36, 1000 Berlin 33.

of Teflon or Kel-F tubing [Warehoused Plastics] (12.7 mm o.d., 8.0 mm i.d.) or boring out a length of Teflon rod (19.1 mm o.d., 14.3 mm i.d.). These tubes are connected to the vacuum line via a Whitey valve (1KS4) [Crawford Fitting] and a brass or Teflon (machined from Teflon rod) reducing Swagelok junction. Although chlorine monofluoride can generally be used without further purification, it has been found that distilling at $-130°$ and collecting the more volatile fraction produces improved yields of $TeClF_5$ of higher purity.

Method 1

Tellurium tetrafluoride (0.72 g, 3.5 mmol) is placed in the 25-mL Teflon vessel inside the dry box. The vessel and contents are evacuated for 1 hr. Chlorine monofluoride (0.42 g, 7.9 mmol) is condensed in small aliquots (aliquots of 0.23 g of ClF have been used; this amount corresponds to approximately 100 mL of ClF at a pressure of 760 torr) onto the tellurium tetrafluoride at $-196°$. The reactants are warmed to room temperature and left for at least 20 min. This procedure is repeated until all of the ClF is consumed. The reactants are allowed to stand overnight at 25°. The solid TeF_4 is gradually consumed, and a yellow liquid is formed. The product is separated by holding the Teflon vessel at $-78°$ and allowing any volatile materials to distill into another vessel held at $-196°$ for a period of about 1 min. The more volatile fraction consists of unreacted ClF and traces of SiF_4, $TeClF_5$, and fluorocarbon impurities. Upon warming the Teflon vessel that contains the less volatile fraction to room temperature, clear liquid, $TeClF_5$, is observed.

Calcd. based on the weight of TeF_4: 0.91 g. Found, 0.71 g (78% yield).

Method 2

Chlorine monofluoride (1.7 g, 31.2 mmol) and tellurium tetrachloride (1.5 g, 5.4 mmol) are allowed to react in the same manner as described for the reaction between tellurium tetrafluoride and chlorine monofluoride. A relatively pure sample of $TeClF_5$ can be obtained by holding the Teflon vessel at room temperature and condensing about two-thirds of the yellow liquid (this will include the more volatile species) into another vessel, held at $-196°$. The last traces of chlorine in the remaining liquid may be removed by condensing the liquid into a vessel containing a large excess of mercury. The vessel and contents are then agitated for 3 hr, resulting in a colorless liquid, $TeClF_5$, which can be distilled out under vacuum.

Weight of $TeClF_5$: Calcd. based on the weight of $TeCl_4$: 1.39 g. Found: 1.19 g (86% yield).

Properties

Tellurium chloride pentafluoride is a colorless liquid (mp −28°, bp 13.5°).[1] It may be stored in dry metal, glass, or Teflon and is unreactive toward mercury at room temperature. It is best characterized by its infrared[7,8] and Raman spectra.[8] The IR spectrum shows strong peaks at 727 (s), 410 (s), 317 (vs), and 259 (s) cm^{-1}. The Raman spectrum has strong lines at 708(31), 659(100), 413(77), and 407(23) cm^{-1}. ^{19}F NMR spectrum in CCl_3F: ϕ_A −43.99 ppm (quintet), ϕ_B −4.63 ppm (doublet), J_{AB} = 170.9 Hz.[7] The sample was a 4.3 *M* solution of $TeClF_5$ in CCl_3F in a 5-mm tube.

References

1. C. Lau and J. Passmore, *Inorg. Chem.*, **13,** 2278 (1974).
2. (a) T. Kitazume and J. M. Shreeve, *J. Chem. Soc. Chem. Commun.*, **1976,** 982; (b) T. Kitazume and J. M. Shreeve, *J. Am. Chem. Soc.*, **99,** 3690 (1977); (c) G. Kleemann and K. Seppelt, *Angew. Chem. Int. Ed.*, **17,** 516 (1978).
3. M. E. Green and A. Turk, *Safety in Working with Chemicals,* Macmillan, New York, 1978, p. 98.
4. (a) R. Y. Eagers, *Toxic Properties of Inorganic Fluorine Compounds,* Elsevier, Amsterdam, 1969; (b) C. F. Reinhardt, W. G. Hume, A. L. Linch, and J. M. Wetherhold, *Am. Ind. Hygiene Assoc. J.* **27,** 166 (1966); (c) F. A. Hohorst and J. M. Shreeve, *Inorg. Synth.*, **11,** 143 (1968).
5. K. Seppelt, *Inorg. Synth.*, **20,** 33 (1980).
6. M. Murchie and J. Passmore, *Inorg. Synth.*, **24,** 76 (1986).
7. G. W. Fraser, R. D. Peacock, and P. M. Watkins, *J. Chem. Soc. Chem. Commun.*, **1968,** 1257.
8. W. V. F. Brooks, M. Eshaque, C. Lau, and J. Passmore, *Can. J. Chem.*, **54,** 817 (1976).

11. PENTAFLUOROOXOTELLURATES(VI)

Submitted by F. SLADKY*
Checked by CARL J. SCHACK†

In recent years, a large number of compounds containing the F_5TeO group attached to a variety of elements across the periodic table have been synthesized. The F_5TeO ligand is characterized by a group electronegativity close to that of fluorine, strictly nonbridging behavior, and the ability to stabilize elements in their high oxidation states. The compounds $HOTeF_5$, $B(OTeF_5)_3$, and $Xe(OTeF_5)_2$

*Institut für Anorganische und Analytische Chemie der Universität, Innrain 52a, A-6020 Innsbruck, Austria.
†Rocketdyne Division, Rockwell International, Canoga Park, CA 91304.

are the main precursors for the preparation of pentafluorooxotellurates(VI). Depending on the chemical nature of the reactants, the following procedures can be employed: (a) acid displacement by the strong acid $HOTeF_5$; (b) metathetical reaction between fluorides and the extraordinarily strong Lewis acid $B(OTeF_5)_3$; and (c) oxidative addition of F_5TeO radicals created thermally or photolytically from $Xe(OTeF_5)_2$:[1]

$$\text{(a)}\ E—X + HOTeF_5 \rightarrow HX + E—OTeF_5$$

$$\text{(b)}\ 3E—F + B(OTeF_5)_3 \rightarrow BF_3 + 3E—OTeF_5$$

$$\text{(c)}\ E—X + Xe(OTeF_5)_2 \rightarrow Xe + EX(OTeF_5)_2$$

■ **Caution.** *All tellurium compounds are toxic. When ingested they give rise at least to a long-lasting garlic odor of the breath and sweat. Fluorosulfuric acid is very reactive toward all organic materials and like all pentafluorooxotellurates(VI) must not be allowed to contact the skin. All manipulations should be carried out in a vacuum system, a dry box, or a fume hood.*

A. HYDROGEN PENTAFLUOROOXOTELLURATE(VI)

$$(HO)_6Te + 5HSO_3F \rightarrow 5H_2SO_4 + HOTeF_5$$

Procedure

Orthotelluric acid (H_6TeO_6, 230 g, 1 mole)‡ is added to a 12-fold molar excess of HSO_3F (1200 g, 690 mL) in a 2-L glass vessel. The HSO_3F is distilled (60°/15 torr) before use. The vessel is equipped with a short water-cooled reflux condenser (preferably of fused quartz or Teflon). The upper end of the condenser, topped with a Teflon-coated thermometer, leads into a declining glass tube (about 2 cm in diameter and 40 cm long) that is connected to a two-necked 500-mL flask equipped with a P_4O_{10} drying tube. All joints are best protected by Kel-F grease or Teflon sleeves. The gradually forming solution is stirred magnetically and heated at reflux for 1 hr with an electric heating mantle with a stir-through opening at the bottom. Some etching and evolution of SiF_4 are observed. The condenser is emptied of water, and crude $HOTeF_5$ is distilled up to about 140° into the receiving flask, which is cooled to −23° by a CCl_4 slush. Some $HOTeF_5$ tends to solidify within the glass tube. It is driven over by gentle heating. Final purification is accomplished by adding about a threefold excess by weight of

‡The checker carried out the procedure at 20% of the scale described with satisfactory results.

98% sulfuric acid to the $HOTeF_5$ in the receiving flask.* This mixture, sometimes containing two liquid phases, is heated at reflux for several hours or until the lower layer has disappeared. All F_5TeO by-products, mainly F_5TeOSO_2F, which greatly depresses the melting point of $HOTeF_5$ are thereby hydrolyzed to $HOTeF_5$. Pure hydrogen pentafluorooxotellurate(VI) is sublimed under vacuum from the sulfuric acid solution. Yield: 215 g (90%).

Anal. Calcd. for $HOTeF_5$: Te, 53.2; F, 39.7. Found: Te, 53.6; F, 38.5.

Properties

Hydrogen pentafluorooxotellurate(VI) is a colorless, glassy-appearing solid that melts at 39° and boils at 60°.[2] The compound is moisture-sensitive and is best handled by using a conventional vacuum line (vapor pressure 180 torr/25°).

The gas-phase IR spectrum has strong absorption bands at 3625 (ν_{O-H}), 1024, 1015 (δ_{O-H}), 741, 734 (ν_{Te-O}, ν_{Te-F}), and 328 (δ_{F-Te-O}) cm^{-1}. The ^{19}F NMR spectrum shows a highly solvent-dependent A_4B pattern. CH_3CN solution: ϕ_A −47.3 ppm, ϕ_B −40.4 ppm, J_{AB} = 190 Hz. CCl_4 solution: ϕ_A −47.0 ppm, ϕ_B −44.5 ppm, J_{AB} = 182 Hz (rel. $CFCl_3$).

B. BORON TRIS[PENTAFLUOROOXOTELLURATE(VI)]

$$BCl_3 + 3HOTeF_5 \rightarrow 3HCl + B(OTeF_5)_3$$

Procedure

With the aid of a vacuum line, $HOTeF_5$ (31 g, 129 mmol)† is sublimed onto BCl_3 (5 g, 42.7 mmol) in a 150-mL Pyrex glass vessel. The reaction proceeds at about −60°. The evolved HCl, contaminated with small amounts of BCl_3, is expanded into a 2-L glass bulb. The gas is condensed back into the reaction vessel, and the HCl is pumped off at −112° (CS_2 slush). This process is repeated several times until no further gas evolution is observed. To remove included gas, the product is melted and pumped upon for a few seconds. Pure boron tris[pentafluorooxotellurate(VI)] (31 g, 42.6 mmol) is obtained in about quantitative yield.

Anal. Calcd. for $B(OTeF_5)_3$: Te, 52.7; F, 39.2. Found: Te, 52.1; F, 39.3.

*Instead of heating the distillate at reflux with H_2SO_4 to destroy impurities, the checker purified the products trapped at −23° by fractional condensation through a series of traps cooled at −30, −78, and −196°. The fraction at −78° was pure TeF_5OH (83% yield).

†The checker carried out the procedure at 20% of the scale described with satisfactory results.

Properties

Boron tris[pentafluorooxotellurate(VI)] is a moisture-sensitive, colorless solid that tends to crystallize in large hexagonal prisms that melt at 37°. The compound is thermally stable to about 130° and can be handled in a dry box or by vacuum sublimation.[3]

The infrared spectrum of solid $B(OTeF_5)_3$ shows absorption bands at 1330, 615, 430 (ν_{B-O}) and 740, 725, 705 (ν_{Te-F}, ν_{Te-O}) cm^{-1}. Its ^{19}F NMR spectrum with $CFCl_3$ as solvent exhibits an A_4B pattern, with ϕ_A −44.4 ppm, ϕ_B −48.2 ppm, J_{AB} = 181 Hz.

C. XENON(II) BIS[PENTAFLUOROOXOTELLURATE(VI)]

$$XeF_2 + 2HOTeF_5 \rightarrow 2HF + Xe(OTeF_5)_2$$

Procedure

A weighable 10-mL Kel-F or FEP [Zeus] (fluoroethylene propylene copolymer) reaction vessel is best suited for this reaction. The compound $HOTeF_5$ is condensed onto XeF_2[4] (3 g, 17.7 mmol)* in about 3-g (12.5-mmol) portions, and the reaction mixture is warmed to a uniform melt. Subsequently, the HF that has formed and the unreacted $HOTeF_5$ are pumped off at 0° and discarded. Alternatively, the $HOTeF_5$ can be freed of HF by trapping at −78° and reused. The above procedure must be repeated about six times, corresponding to a molar excess of $HOTeF_5$ of approximately 100%, to shift the equilibrium completely to formation of the product. This displacement is best monitored by weighing until constancy is reached. Yield: 10.6 g (96%).

Anal. Calcd. for $Xe(OTeF_5)_2$: Te, 41.9; F, 31.2. Found: Te, 41.0; F, 31.5.

Properties

Xenon(II) bis[pentafluorooxotellurate(VI)] is a slightly yellow crystalline solid that melts at 38–40°. The compound is moisture-sensitive but soluble without reaction in CCl_4 or CH_3CN. It is thermally stable to about 140° and can be vacuum-sublimed and handled in glass apparatus.[5]

■ **Caution.** *Explosive reactions occur with ethanol, acetone, benzene, and other related organic compounds.*

*The checker carried out the procedure at 60% of the scale described. The TeF_5OH is added to the XeF_2 (4:1, 100% excess) in two additions. The HF is pumped away at −78° six or seven times for several days, leaving all the TeF_5OH behind.

The IR spectrum of a solution of $Xe(OTeF_5)_2$ in CCl_4 shows absorption bands at 780, 705, 628 (ν_{Te-O}, ν_{Te-F}), and 475 (ν_{Xe-O}) cm^{-1}. Its ^{19}F NMR spectrum in the same solvent shows an A_4B pattern ϕ_A −43.4 ppm, ϕ_B −48.2 ppm, J_{AB} = 182 Hz.

References

1. A. Engelbrecht and F. Sladky, *Int. Rev. Sci., Ser. 2,* **3,** 167 (1975); K. Seppelt, *Acc. Chem. Res.,* **12,** 211 (1979); A. Engelbrecht and F. Sladky, *Adv. Inorg. Chem. Radiochem.,* **24,** 211 (1981).
2. A. Engelbrecht and F. Sladky, *Monatsh. Chem.,* **96,** 159 (1965); A. Engelbrecht, W. Loreck, and W. Nehoda, *Z. Anorg. Allgem. Chem.,* **366,** 88 (1968).
3. F. Sladky, H. Kropshofer, and O. Leitzke, *J. Chem. Soc., Chem. Commun.,* **1973,** 134; H. Kropshofer, O. Leitzke, P. Peringer, and F. Sladky, *Chem. Ber.,* **114,** 2644 (1981); J. F. Sawyer and G. J. Schrobilgen, *Acta Cryst.,* **B38,** 1561 (1982).
4. S. M. Williamson, F. Sladky, and N. Bartlett, *Inorg. Synth.,* **11,** 147 (1968).
5. F. Sladky, *Monatsh. Chem.,* **101,** 1559 (1970).

12. TUNGSTEN TETRAFLUORIDE OXIDE

$$2WF_6 + SiO_2 \xrightarrow{HF} 2WOF_4 + SiF_4$$

Submitted by WILLIAM W. WILSON* and KARL O. CHRISTE*
Checked by ROLAND BOUGON†

Tungsten tetrafluoride oxide can be prepared by numerous methods, such as the fluorination of WO_3 at 300°,[1] slow hydrolysis of WF_6,[2] the direct fluorination of W in the presence of O_2 at 300°,[3] the reaction of WF_6 with WO_3 at 400°,[4] the reaction of $WOCl_4$ with HF,[5,6] or oxygen-fluorine exchange between WF_6 and B_2O_3.[6] The method given below is a modification of the method of Paine and McDowell, who used stoichiometric amounts of SiO_2 and WF_6 in anhydrous HF for the controlled hydrolysis of WF_6.[2] In our experience,[7] the use of stoichiometric amounts of SiO_2 and WF_6 leads to the formation of some $[H_3O]^+[WOF_5]^-$ and $[H_3O]^+[W_2O_2F_9]^-$ as by-products that are difficult to separate from WOF_4. This problem can, however, be minimized by the use of an excess of WF_6. Tungsten tetrafluoride oxide is a starting material for the syntheses of numerous WOF_5^- salts.

*Rocketdyne, A Division of Rockwell International Corp., Canoga Park, CA 91304.
†Centre d'Études Nucléaires de Saclay, 91191 Gif sur Yvette, France.

Procedure

■ **Caution.** *Anhydrous HF causes severe burns. Protective clothing and safety glasses should be worn when working with liquid HF.*

Quartz wool [Preiser Scientific] (1.048 g, 17.44 mmol) is placed into a ¾-in. o.d. Teflon FEP (fluoroethylene propylene copolymer) ampule [Zeus] equipped with a Teflon-coated magnetic stirring bar and a stainless steel valve. The ampule is connected to a metal-Teflon vacuum system[8] and evacuated, and dry[9] HF [Matheson] (19 g) and WF_6 [Alfa] (22.102 g, 74.21 mmol) are condensed into the ampule at −196°. The contents of the ampule are allowed to warm to room temperature and are kept at this temperature for 15 hr with stirring. All material volatile at room temperature is pumped off (10^{-4} torr) for 12 hr, leaving behind 9.723 g of a white solid (weight calcd. for 34.89 mmol WOF_4 is 9.624 g). This crude product usually still contains $[H_3O]^+[W_2O_2F_9]^-$ (IR spectrum of the solid pressed as a AgCl disk: 3340, 3100, 1625, 1040, 1030, 908 cm^{-1}) and can be purified by vacuum (10^{-4} torr) sublimation in an ice water–cooled Pyrex sublimator at 55°, resulting in 4.245 g of sublimate. The purity of the sublimate is verified by vibrational spectroscopy of the solid (IR spectrum as a AgCl disk: 1054 (vs), 733 (s), 666 (vs), and 550 (vs) cm^{-1}. Raman: 1058 (10), 740 (1.9), 727 (6.3), 704 (0+), 668 (0+), 661 (0.9), 559 (0+), 518 (0.7), 325 (sh), 315 (sh), 311 (5), 260 (0+), 238 (0.7), 212 (0.5), 185 (0+) cm^{-1}.)[10]

Anal. Calcd. for WOF_4: W, 66.65; F, 27.55. Found: W, 66.5; F, 27.7.

Properties

Tungsten tetrafluoride oxide is a white hygroscopic solid (mp 104.7 at 25 torr, bp 185.9°) which can be sublimed readily. It is soluble in HF and in propylene carbonate. The ^{19}F NMR spectrum in propylene carbonate solution consists of a singlet at 65.2 ppm downfield from external $CFCl_3$, with two satellites with J_{WF} = 69 Hz.[11]

References

1. G. H. Cady and G. B. Hargreaves, *J. Chem. Soc.*, **1961,** 1563.
2. R. T. Paine and R. S. McDowell, *Inorg. Chem.*, **13,** 2367 (1974).
3. H. Meinert, L. Friedrich, and W. Kohl, *Z. Chem.*, **15,** 492 (1975).
4. F. N. Tebbe and E. L. Muetterties, *Inorg. Chem.*, **7,** 172 (1968).
5. O. Ruff, F. Eisner, and W. Heller, *Z. Anorg. Allgem. Chem.*, **52,** 256 (1907).
6. R. C. Burns, T. A. O'Donnell, and A. B. Waugh, *J. Fluorine Chem.*, **12,** 505 (1978).
7. W. W. Wilson and K. O. Christe, *Inorg. Chem.*, **20,** 4139 (1981).
8. K. O. Christe, R. D. Wilson, and C. J. Schack, *Inorg. Synth.*, **24,** 3 (1986).
9. K. O. Christe, W. W. Wilson, and C. J. Schack, *J. Fluorine Chem.*, **11,** 71 (1978).
10. I. R. Beattie, K. M. S. Livingston, D. J. Reynolds, and G. A. Ozin, *J. Chem. Soc. A*, **1970,** 1210.
11. R. Bougon, T. Bui Huy, and P. Charpin, *Inorg. Chem.*, **14.** 1822 (1975).

13. TETRAFLUOROAMMONIUM SALTS

Submitted by KARL O. CHRISTE,* WILLIAM W. WILSON,* CARL J. SCHACK,* and RICHARD D. WILSON*
Checked by R. BOUGON†

Since $[NF_4]^+$ is a coordinatively saturated complex fluoro cation, the syntheses of its salts are generally difficult.[1] A limited number of salts can be prepared directly from NF_3, and these salts can then be converted by indirect methods into other $[NF_4]^+$ salts that are important for solid propellant NF_3-F_2 gas generators or reagents for the electrophilic fluorination of aromatic compounds.

The two direct methods for the syntheses of $[NF_4]^+$ salts are based on the reaction of NF_3 with either $[KrF]^+$ salts[2]

$$NF_3 + [KrF][AsF_6] \rightarrow [NF_4][AsF_6] + Kr$$

or F_2 and a strong Lewis acid in the presence of an activation energy source E.[3]

$$NF_3 + F_2 + XF_n \xrightarrow{E} [NF_4][XF_{n+1}]$$

For the chemist interested in synthesis, the second method[3] is clearly superior, due to its high yields, relative simplicity, and scalability.

Four different activation energy sources have been used for the direct synthesis of $[NF_4]^+$ salts:

1. Heat[4–7]: $[NF_4][BiF_6]$, $[NF_4][SbF_6]$, $[NF_4][AsF_6]$, $[NF_4]_2[TiF_6 \cdot nTiF_4]$
2. Glow discharge[8,9]: $[NF_4][AsF_6]$, $[NF_4][BF_4]$
3. UV photolysis[10,11]: $[NF_4][SbF_6]$, $[NF_4][AsF_6]$, $[NF_4][PF_6]$, $[NF_4][GeF_5]$, $[NF_4][BF_4]$
4. Bremsstrahlung[12]: $[NF_4][BF_4]$

Of these, the thermal synthesis of $[NF_4][SbF_6]$[4–7] is most convenient (Synthesis A) and provides the starting material required for the synthesis of other $[NF_4]^+$ salts by indirect methods. For the synthesis of pure $[NF_4]^+$ salts on a small scale, low-temperature UV photolysis is preferred (Synthesis B).[11]

The following indirect methods for the interconversion of $[NF_4]^+$ salts are known:

*Rocketdyne, A Division of Rockwell International Corp., Canoga Park, CA 91304.
†Centre d'Études Nucléaires de Saclay, 91191 Gif sur Yvette, France.

1. Metathesis reaction:

$$n[NF_4][XF_6] + [Cs]_n[MF_{m+n}] \xrightarrow{\text{solvent}} [NF_4]_n[MF_{m+n}] + n[Cs][XF_6]\downarrow$$

soluble soluble soluble insoluble

where typically X = Sb and the solvent is anhydrous HF or BrF_5. This method is limited to anions that are stable in the given solvent and results in an impure product. Typical compounds prepared in this manner include $[NF_4][BF_4]$,[6,13,14] $[NF_4][HF_2]$ (Synthesis C),[15] $[NF_4][SO_3F]$,[16] $[NF_4][ClO_4]$,[15] and $[NF_4]_2[MF_6]$ (M = Sn,[17] Ti,[18] Ni,[19] Mn[20]) (Synthesis D).

2. Reaction of solid $[NF_4][HF_2]\cdot x$HF with a weak Lewis acid: When the MF_{m+n}^{n-} anion is unstable in a solvent, such as HF, and the Lewis acid MF_m is volatile, the equilibrium

$$n[NF_4][HF_2]\cdot x\text{HF} + MF_m \rightleftarrows [NF_4]_n[MF_{m+n}] + n(x+1)\text{HF}$$

can be shifted to the right by the use of an excess of MF_m and continuous removal of HF with the excess of MF_m. Typical salts prepared in this manner include $[NF_4]_2[SiF_6]$[21] (Synthesis E) and $[NF_4][MF_7]$ (M = U, W,[22] Xe[23]).

3. Reaction of $[NF_4][HF_2]$ with a nonvolatile polymeric Lewis acid: When, in the metathesis (1), all the materials except $[NF_4][XF_6]$ are insoluble, product separation becomes impossible. This problem is avoided by digesting the Lewis acid in a large excess of $[NF_4][HF_2]$ in HF solution, followed by thermal decomposition of the excess $[NF_4][HF_2]$ at room temperature.

$$[NF_4][HF_2] + MF_n \xrightarrow{\text{HF}} [NF_4][MF_{n+1}] + \text{HF}$$

Salts prepared in this manner include $[NF_4][MOF_5]$ (M = U,[24] W[25]) (Synthesis F), $[NF_4][AlF_4]$,[26] and $[NF_4][Be_2F_5]$.[26]

4. Displacement reaction: Displacement of a weaker Lewis acid by a stronger Lewis acid can be carried out easily, as demonstrated for $[NF_4][PF_6]$.[11]

$$[NF_4][BF_4] + PF_5 \rightarrow [NF_4][PF_6] + BF_3$$

5. Rearrangement reaction: When $[NF_4][GeF_5]$ is treated with anhydrous HF, the following equilibrium is observed:

$$2[NF_4][GeF_5] \underset{+GeF_4}{\overset{+HF}{\rightleftarrows}} [NF_4]_2[GeF_6] + GeF_4$$

This equilibrium can be shifted to the right by repeated treatments of $[NF_4][GeF_5]$ with HF and GeF_4 removal, and to the left by treatment of $[NF_4]_2[GeF_6]$ with GeF_4.[11]

A. TETRAFLUOROAMMONIUM HEXAFLUOROANTIMONATE(V)

$$NF_3 + 2F_2 + SbF_3 \xrightarrow[70\ atm]{250°} [NF_4][SbF_6]$$

Procedure

■ **Caution.** *High-pressure fluorine reactions should be carried out only behind barricades or in a high-pressure bay using appropriately pressure-temperature–rated nickel or Monel reactors that have been well passivated with several atmospheres of F_2 at the described reaction temperature. Stainless steel reactors should be avoided owing to the potential of metal fires. All $[NF_4]^+$ salts are moisture-sensitive and must be handled in a dry atmosphere. They are strong oxidizers—contact with organic materials and fuels must be avoided.*

A prepassivated (with ClF_3), single-ended 95-mL Monel cylinder [Hoke, rated for 5000 psi working pressure], equipped with a Monel valve [Hoke 3232 M4M or equivalent], is loaded in the dry nitrogen atmosphere of a glove box with SbF_3 [Ozark-Mahoning] (31 mmol). The cylinder is connected to a metal vacuum system,[27] evacuated, vacuum leak tested, and charged with NF_3 [Air Products and Chemicals] (65 mmol) and F_2 [Air Products and Chemicals] (98 mmol) by condensation at −196°. The barricaded cylinder is heated for 5 days to 250°. The cylinder is allowed to cool by itself to ambient temperature and is then cooled to −196°. The unreacted F_2 and NF_3 are pumped off at −196° (the pump must be protected by a fluorine scrubber[28]); during the subsequent warm-up of the cylinder to ambient temperature, $[NF_4][SbF_6]$ (10.1 g, 31 mmol, 100% yield based on SbF_3) is left behind as a solid residue. The product is either scraped out of the cylinder in the dry box or, more conveniently, dissolved in anhydrous HF that has been dried over BiF_5.[13] Small amounts of $Ni[SbF_6]_2$ and $Cu[SbF_6]_2$, formed as impurities in the attack of the Monel reactor by F_2 and SbF_5, are only sparingly soluble in HF and are removed from the $[NF_4][SbF_6]$ solution by filtration using a porous Teflon filter [Pallflex]. If desired, the SbF_3 starting material can be replaced by SbF_5 [Ozark-Mahoning] without changing the remaining procedure.

Anal.[29] Calcd. for $[NF_4][SbF_6]$: NF_3, 21.80; Sb, 37.38. Found: NF_3, 21.73; Sb, 37.41.

Properties

Tetrafluoroammonium hexafluoroantimonate(V) is a hygroscopic, white crystalline solid that is stable to about 270°.[4,5,30] It is highly soluble in anhydrous HF (259 mg per g of HF at −78°)[13] and moderately soluble in BrF_5. Its ^{19}F NMR spectrum[4] in anhydrous HF solution consists of a triplet of equal intensity at 214.7 ppm downfield from $CFCl_3$ (J_{NF} = 231 Hz) for $[NF_4]^+$. The vibrational spectra[5] of the solid exhibit the following major bands: IR (pressed AgCl disk): 1227 (mw), 1162 (vs), 675 (vs), 665 (vs), 609 (m) cm^{-1}. Raman 1160(0.6), 1150(0.2), 843(7.0), 665(1), 648(10), 604(3.9), 569(0.9), 437(1.5), and 275(3.8) cm^{-1}

B. TETRAFLUOROAMMONIUM TETRAFLUOROBORATE(III)

$$NF_3 + F_2 + BF_3 \xrightarrow[-196^\circ]{h\nu} [NF_4][BF_4]$$

Procedure

■ **Caution.** *Ultraviolet goggles should be worn for eye protection when working with higher power UV lamps, and the work should be carried out in a fume hood. $[NF_4][BF_4]$ is a strong oxidizer; contact with organic materials, fuels, and moisture must be avoided.*

The low-temperature UV photolysis reaction is carried out in a quartz reactor with a pan-shaped bottom and a flat top consisting of a 7.5-cm diameter optical grade quartz window (Fig. 1). The vessel has a side arm connected by a Teflon O-ring joint to a Fischer-Porter Teflon valve to facilitate removal of solid reaction products. The depth of the reactor is about 4 cm, and its volume is about 140 mL. The UV source consists of a 900-W, air-cooled, high-pressure mercury arc (General Electric Model B-H6) positioned 4 cm above the flat reactor surface. The bottom of the reactor is kept cold by immersion in liquid N_2. Dry, gaseous N_2 is used as a purge gas to prevent condensation of atmospheric moisture on the flat top of the reactor. As a heat shield, a 6-mm-thick quartz plate is positioned between the UV source and the top of the reactor.

Premixed NF_3 [Air Products] and BF_3 [Matheson] (27 mmol of each) are condensed into the cold bottom of the quartz reactor. Fluorine [Air Products] (9 mmol) is added, and the mixture is photolyzed at −196° for 1 hr. After termination of the photolysis, volatile material is pumped out of the reactor (through a scrubber[28]) during its warm-up to room temperature. The nonvolatile white solid residue (1.0 g) is pure $[NF_4][BF_4]$. Instead of the pan-shaped reactor, a simple round quartz bulb can be used with a $[NF_4][BF_4]$ yield of about 0.3 g/hr.

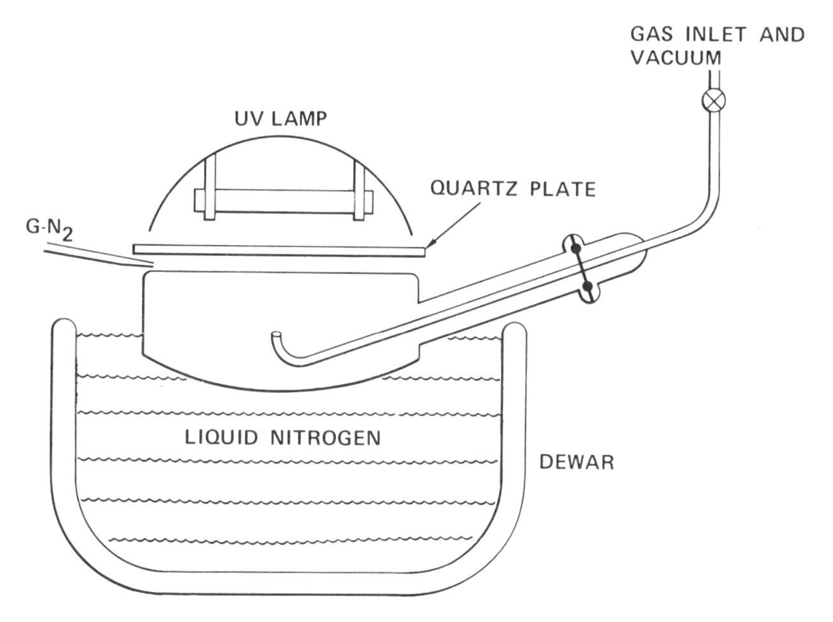

Fig. 1. Apparatus for synthesis of [NF_4][BF_4].

Anal.[29] Calcd. for [NF_4][BF_4]: NF_3, 40.16; B, 6.11. Found: NF_3, 40.28; B, 6.1.

Properties

Tetrafluoroammonium tetrafluoroborate(III) is a hygroscopic white crystalline solid that is stable up to about 150°.[9,11,12,30] It is highly soluble in anhydrous HF and moderately soluble in BrF_5. Its ^{19}F NMR spectrum in anhydrous HF solution consists of a sharp triplet of equal intensity at ϕ 220 ppm downfield from $CFCl_3$ (J_{NF} = 230 Hz) for $[NF_4]^+$ and an exchange broadened singlet at ϕ -158 ppm upfield from $CFCl_3$ for $[BF_4]^-$. The vibrational spectra of the solid exhibit the following major bands: IR (pressed AgCl disk): 1298 (ms), 1222 (mw), 1162 (vs), 1057 (vs), 609 (s), and 522 (s) cm^{-1}. Raman: 1179(0.6), 1148(0.6), 1130(0+), 1055(0.2), 884(0+), 844(10), 772(3.2), 609(6.3), 524(0.4), 443(2.6), and 350(0.9) cm^{-1}

C. TETRAFLUOROAMMONIUM (HYDROGEN DIFLUORIDE)

$$[NF_4][SbF_6] + CsF \xrightarrow[-78^\circ]{HF} Cs[SbF_6]\downarrow + [NF_4][HF_2]$$

Procedure

■ **Caution.** *Anhydrous HF causes severe burns. Protective clothing should be worn when working with this material. The HF solutions of* $[NF_4]^+$ *salts are strongly oxidizing; contact with fuels must be avoided.*

A mixture of dry CsF [Kawecki Berylco] (2.361 g, 15.54 mmol) and $[NF_4][SbF_6]$ (5.096 g, 15.64 mmol) is placed inside the dry box into trap I of the leak-checked and passivated (with ClF_3 and dry HF[13]) Teflon FEP Monel metathesis apparatus shown in Fig. 2. The CsF is dried by fusion in a platinum crucible, immediately transferred to the dry box, cooled, and finely ground. The apparatus is attached to a metal Teflon vacuum system[27] by two flexible corrugated Teflon tubes [Penntube Plastics], and the connections are vacuum leak-checked and passivated. The system is repeatedly exposed to anhydrous HF [Matheson], until the HF is colorless when frozen out at −196° in a Teflon U-trap of the vacuum system to avoid contamination of the product with any chlorine fluorides that may be adsorbed onto the walls of the metal vacuum system. Anhydrous HF[13] (16.2 g, 810 mmol) is added to trap I, and the mixture is stirred magnetically for 1 hr at room temperature. The metathesis apparatus is cooled with powdered Dry Ice to −78° for 1 hr and then inverted. The HF solution that contains the

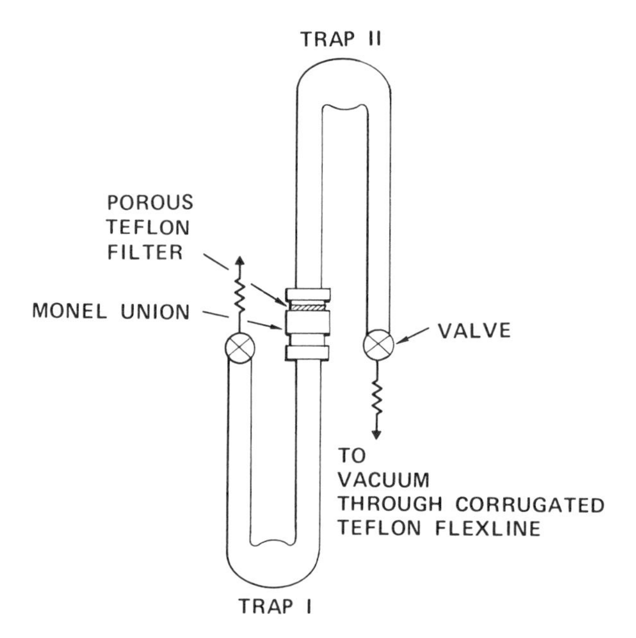

Fig. 2. Apparatus for syntheses of $[NF_4][HF_2]$, $[NF_4]_2[MnF_6]_2$, $[NF_4]_2[SiF_6]$, *and* $[NF_4][WOF_5]$.

$[NF_4][HF_2]$ is separated from the $Cs[SbF_6]$ precipitate by filtration. To facilitate the filtration step, trap I is pressurized with 2 atm of dry N_2 after inversion. A pressure drop in trap I indicates the completion of the filtration step. If desired, repressurization of trap I may be repeated to minimize the amount of mother liquor held up in the filter cake. The desired HF solution of $[NH_4][HF_2]$ is collected in trap II. It contains about 94% of the original $[NF_4]^+$ values, with the remainder being adsorbed on the $Cs[SbF_6]$ filter cake. The $[NF_4][HF_2]$ solution has a purity of about 97 mole % and contains small amounts of $Cs[SbF_6]$ (solubility of $Cs[SbF_6]$ in HF at $-78°$ is 1.8 mg/g HF)[13] and $[NF_4][SbF_6]$ (if a slight excess of $[NF_4][SbF_6]$ has been used in the reaction to suppress, by the common ion effect, the amount of dissolved $Cs[SbF_6]$).

An unstable solid having the composition $[NF_4][HF_2]\cdot n HF$ ($n = 2$–10) can be prepared by pumping off as much HF as possible below 0°.

Properties

Tetrafluoroammonium (hydrogen difluoride) is stable in HF solution at room temperature but decomposes to NF_3, F_2, and HF on complete removal of the solvent.[15] The ^{19}F NMR spectrum of the solution shows a triplet of equal intensity of ϕ 216.2 ppm downfield from $CFCl_3$ with $J_{NF} = 230$ Hz. The Raman spectrum of the HF solution has bands at 1170 (w), 854 (vs), 612 (m), and 448 (mw) cm^{-1}.

D. BIS(TETRAFLUOROAMMONIUM) HEXAFLUOROMANGANATE(IV)

$$2[NF_4][SbF_6] + Cs_2[MnF_6] \xrightarrow[-78°]{HF} 2Cs[SbF_6] + [NF_4]_2[MnF_6]$$

Procedure

■ **Caution.** *Anhydrous HF can cause severe burns; protective clothing should be worn when working with this solvent. $[NF_4]_2[MnF_6]$ is a strong oxidizer; contact with water and fuels must be avoided.*

The same apparatus is used as for synthesis C. In the dry N_2 atmosphere of a glove box, a mixture of $[NF_4][SbF_6]$ (37.29 mmol) and $Cs_2[MnF_6]$[20] (18.53 mmol) is placed in the bottom of a prepassivated (with ClF_3) Teflon FEP (fluoroethylene propylene copolymer) double U-tube metathesis apparatus. Dry HF[13] (20 mL of liquid) is added at $-78°$ on the vacuum line,[27] and the mixture is warmed to 25° for 30 min with stirring. The mixture is cooled to $-78°$ and pressure-filtered at this temperature. The HF solvent is pumped off at 30° for 12 hr, resulting in 14 g of a white filter cake (mainly $Cs[SbF_6]$) and 6.1 g of a yellow filtrate residue having the approximate composition (weight %): $[NF_4]_2[MnF_6]$, 92; $[NF_4][SbF_6]$, 4; $Cs[SbF_6]$, 4. Yield of $[NF_4]_2[MnF_6]$ is 87% based on $Cs_2[MnF_6]$.

Properties

Bis(tetrafluoroammonium) hexafluoromanganate(IV) is a yellow crystalline solid that is stable at 65° but slowly decomposes at 100° to NF_3, F_2, and MnF_3.[20] It is highly soluble in anhydrous HF and reacts violently with water. Its ^{19}F NMR spectrum in anhydrous HF solution shows a broad resonance at ϕ 218 ppm below $CFCl_3$ due to $[NF_4]^+$. The vibrational spectra of the solid show the following major bands: IR (pressed AgCl disk): 1221 (mw), 1160 (vs), 620 (vs), and 338 (s) cm^{-1}. Raman: 855 (m), 593 (vs), 505 (m), 450 (w), and 304 (s) cm^{-1}

E. BIS(TETRAFLUOROAMMONIUM) HEXAFLUOROSILICATE(IV)

$$2[NF_4][HF_2]\cdot nHF + SiF_4 \rightarrow [NF_4]_2[SiF_6] + 2(n+1)HF$$

Procedure

■ **Caution.** *Anhydrous HF can cause severe burns, and protective clothing should be worn when working with liquid HF. All $[NF_4]^+$ salts are strong oxidizers, and contact with fuels and water must be avoided.*

A solution of $[NF_4][HF_2]$ (27 mmol) in anhydrous HF[13] is prepared at −78° by synthesis C. Most of the HF solvent is pumped off during warm-up toward 0° until the first signs of decomposition of $[NF_4][HF_2]$ are noted from the onset of gas evolution. The resulting residue is cooled to −196°, and SiF_4 [Matheson] (33 mmol) is added. The mixture is allowed to warm to ambient temperature while providing a volume of about 1 L in the vacuum line for expansion. During warm-up of the apparatus, the SiF_4 evaporates first and, upon melting of the $[NF_4][HF_2]\cdot nHF$ phase, a significant reduction in the SiF_4 pressure is noted, resulting in a final pressure of about 400 torr. A clear, colorless solution is obtained without any sign of solid formation. The material volatile at 0° is pumped off and separated by fractional condensation through traps at −126 and −196°. The SiF_4 portion (about 22 mmol), trapped at −196°, is condensed back into the reactor, which contains a white fluffy solid. After this mixture has been kept at 25° for 24 hr, all volatile material is pumped off at 25° and the SiF_4 is separated again from the HF. The solid residue is treated again with the unreacted SiF_4 at

25° for 14 hr. The materials volatile at 25° are pumped off again. They contain less than 1 mmol of HF at this point. The solid residue is heated in a dynamic vacuum to 50° for 28 hr or until no further HF evolution is noticeable. The white solid residue (about 3.8 g, 80% yield) has the approximate composition (weight %): $[NF_4]_2[SiF_6]$ 95.0, $Cs[SbF_6]$ 2.2, $[NF_4][SbF_6]$ 2.3.

Properties

Bis(tetrafluoroammonium) hexafluorosilicate(IV) is a white crystalline solid that is stable at 25° but slowly decomposes at 90° to NF_3, F_2, and SiF_4.[21] The vibrational spectra of the solid show the following major bands: IR (pressed AgCl disk): 1223 (mw), 1165 (vs), 735 (vs,br), 614 (m), 609 (mw), 478 (s), and 448 (w) cm^{-1}. Raman: 1164 (1.5), 895 (0+), 885 (0+), 859 (10), 649 (3.2), 611 (5.8), 447,441 (3.8), and 398 (1) cm^{-1}.

F. TETRAFLUOROAMMONIUM PENTAFLUOROOXOTUNGSTATE(VI)

$$[NF_4][HF_2] + WOF_4 \xrightarrow[25^\circ]{HF} [NF_4][WOF_5] + HF$$

Procedure

■ **Caution.** *Anhydrous HF can cause severe burns. Protective clothing should be worn when working with liquid HF. All $[NF_4]^+$ salts are strong oxidizers, and contact with fuels and water should be avoided.*

A solution of 20 mmol of $[NF_4][HF_2]$ in 16 mL of dry HF[13] is prepared at −78° by synthesis C and pressure-filtered into the second half of the metathesis double U-tube containing 14.6 mmol of WOF_4.[31] The mixture is stirred with a magnetic stirring bar for 30 min at 25°. The volatile material is pumped off at 25° for 12 hr. The solid residue (about 5 g, 86% yield based on WOF_4) has the approximate composition (weight %): $[NF_4][WOF_5]$ 96, $Cs[SbF_6]$ 2, $[NF_4][SbF_6]$ 2.

Properties

Tetrafluoroammonium pentafluorooxotungstate(VI) is a white crystalline solid that is stable at 55° but slowly decomposes at 85° to yield NF_3, OF_2, WF_6, and $[NF_4][W_2O_2F_9]$.[25] The vibrational spectra of the solid show the following major bands: IR (pressed AgCl disk): 1221 (mw), 1160 (vs), 991 (vs), 688 (vs), 620 (vs,br), and 515 (vs) cm^{-1}. Raman: 1165 (0.7), 996 (10), 852 (8.4), 690 (5.4), 613 (4.9), 446 (1.6), 329 (6.8), and 285 (0.5) cm^{-1}.

References

1. K. O. Christe, W. W. Wilson, and E. C. Curtis, *Inorg. Chem.*, **22,** 3056 (1983).
2. A. A. Artyukhov and S. S. Koroshev, *Koord. Khim.*, **3,** 1478 (1977); K. O. Christe, W. W. Wilson, and R. D. Wilson, *Inorg. Chem.*, **23,** 2058 (1984).
3. K. O. Christe, J. P. Guertin, and A. E. Pavlath, U.S. Pat. 3503719 (1970).
4. W. E. Tolberg, R. T. Rewick, R. S. Stringham, and M. E. Hill, *Inorg. Chem.*, **6,** 1156 (1967).
5. K. O. Christe, R. D. Wilson, and C. J. Schack, *Inorg. Chem.*, **16,** 937 (1977).
6. K. O. Christe, C. J. Schack, and R. D. Wilson, *J. Fluorine Chem.*, **8,** 541 (1976).
7. W. W. Wilson and K. O. Christe, *J. Fluorine Chem.*, **15,** 83 (1980).
8. K. O. Christe, J. P. Guertin, and A. E. Pavlath, *Inorg. Nucl. Chem. Lett.*, **2,** 83 (1966); *Inorg. Chem.*, **5,** 1921 (1966).
9. S. M. Sinel'nikov and V. Ya. Rosolovskii, *Dokl. Akad. Nauk SSSR,* **194,** 1341 (1970).
10. K. O. Christe, R. D. Wilson, and A. E. Axworthy, *Inorg. Chem.*, **12,** 2478 (1973).
11. K. O. Christe, C. J. Schack, and R. D. Wilson, *Inorg. Chem.*, **15,** 1275 (1976).
12. C. T. Goetschel, V. A. Campanile, R. M. Curtis, K. R. Loos, D. C. Wagner, and J. N. Wilson, *Inorg. Chem.*, **11,** 1696 (1972).
13. K. O. Christe, W. W. Wilson, and C. J. Schack, *J. Fluorine Chem.*, **11,** 71 (1978).
14. W. E. Tolberg, private communication.
15. K. O. Christe, W. W. Wilson, and R. D. Wilson, *Inorg. Chem.*, **19,** 1494 (1980).
16. K. O. Christe, R. D. Wilson, and C. J. Schack, *Inorg. Chem.*, **19,** 3046 (1980).
17. K. O. Christe, C. J. Schack, and R. D. Wilson, *Inorg. Chem.*, **16,** 849 (1977).
18. K. O. Christe and C. J. Schack, *Inorg. Chem.*, **16,** 353 (1977).
19. K. O. Christe, *Inorg. Chem.*, **16,** 2238 (1977).
20. W. W. Wilson and K. O. Christe, *Inorg. Synth.*, **24,** 48 (1986).
21. W. W. Wilson and K. O. Christe, *J. Fluorine Chem.*, **19,** 253 (1982).
22. W. W. Wilson and K. O. Christe, *Inorg. Chem.*, **21,** 2091 (1982).
23. K. O. Christe and W. W. Wilson, *Inorg. Chem.*, **21,** 4113 (1982).
24. W. W. Wilson, R. D. Wilson, and K. O. Christe, *J. Inorg. Nucl. Chem.*, **43,** 1551 (1981).
25. W. W. Wilson and K. O. Christe, *Inorg. Chem.*, **20,** 4139 (1981).
26. K. O. Christe, W. W. Wilson, and C. J. Schack, *J. Fluorine Chem.*, **20,** 751 (1982).
27. K. O. Christe, R. D. Wilson, and C. J. Schack, *Inorg. Synth.*, **24,** 3 (1986).
28. G. H. Cady, *Inorg. Synth.*, **8,** 165 (1966).
29. R. Rushworth, C. J. Schack, W. W. Wilson, and K. O. Christe, *Anal. Chem.*, **53,** 845 (1981).
30. K. O. Christe, R. D. Wilson, and I. B. Goldberg, *Inorg. Chem.*, **18,** 2578 (1979).
31. W. W. Wilson and K. O. Christe, *Inorg. Synth.*, **24,** 37 (1986).

14. CESIUM HEXAFLUOROMANGANATE(IV)

$$2CsF + MnCl_2 + 7F_2 \rightarrow Cs_2[MnF_6] + 2ClF_5$$

Submitted by WILLIAM W. WILSON* and KARL O. CHRISTE*
Checked by ROLAND BOUGON†

*Rocketdyne, A Division of Rockwell International Corp., Canoga Park, CA 91304.
†Centre d'Études Nucléaires de Saclay, 91191 Gif sur Yvette, France.

Several methods have been described in the literature for the syntheses of alkali metal hexafluoromanganates(IV). The reactions of $K_2[MnO_4]$,[1] MnO_2, and KF mixtures, or $KMnO_4$ and 30% H_2O_2[2] with aqueous HF produce $K_2[MnF_6]$. However, the yields and product purities are low. Pure alkali metal hexafluoromanganates(IV) are obtained in high yield by fluorination with F_2 in a flow system of either $MnCl_2$ + 2MCl at 375 to 400°,[3,4] MnF_3 + 2KF in a rotating Al_2O_3 tube at 600°,[5] or $MnCl_2$ + 2KCl at 280°,[6] or by the fluorination of a $KMnO_4$/KCl mixture with BrF_3.[7] The method described below is based on the fluorination of a stoichiometric mixture of CsF and $MnCl_2$ in a static system at 400°.[8] Hexafluoromanganate(IV) salts have interesting spectroscopic properties,[9,10] and $Cs_2[MnF_6]$ is a starting material for the metathetical synthesis of $(NF_4)_2[MnF_6]$.[11]

Procedure

■ **Caution.** *Safety barricades must be used for carrying out high-pressure fluorination reactions. The ClF_5-ClF_3 by-products are strong oxidizers. Contact with fuel, water, or reducing agents must be avoided.*

Commercially available $MnCl_2 \cdot 4H_2O$ [Alfa] is dehydrated by heating in a Pyrex flask to 255° under vacuum (10^{-4} torr) for 24 hr. The completeness of the dehydration step is verified by recording the infrared spectrum, which should not show any water bands. Commercially available CsF [Kawecki Berylco] is dried by fusion in a platinum crucible and is immediately transferred to the dry box.

A mixture of finely ground dry CsF (7.717 g, 50.80 mmol) and $MnCl_2$ (3.150 g, 25.40 mmol) is placed inside the dry box into a prepassivated (with ClF_3) 95-mL high-pressure Monel cylinder (Hoke Model 4HSM, rated for 5000 psi working pressure) equipped with a Monel valve (Hoke, Model 3212M4M). The cylinder is attached to a metal-Teflon vacuum system,[12] evacuated, and cooled to −196° with liquid N_2. Fluorine* (262 mmol) is condensed into the cylinder. The cylinder is disconnected from the vacuum line, heated in an oven to 400° for 36 hr, and then cooled again to −196° on the vacuum line. Unreacted F_2 is pumped off at −196° through a fluorine scrubber,[12] and the ClF_5-ClF_3 byproducts are pumped off during the warm-up of the cylinder toward room temperature. The yellow solid residue (11.045 g, 100% yield) is pure $Cs_2[MnF_6]$.

Anal. Calcd. for $Cs_2[MnF_6]$: Cs, 61.14; Mn, 12.63. Found: Cs, 61.2; Mn, 12.5.

*[Air Products and Chemicals] Prior to use, the fluorine should be passed through a NaF scrubber to remove any HF present, which would promote attack of the Monel reactor as evidenced by the formation of some Cs_2NiF_6.

Properties

Cesium hexafluoromanganate(IV) is a stable yellow solid that decomposes only slowly in moist air. The IR spectrum of the solid as a dry powder pressed between AgCl plates shows the following major absorptions: 620 (vs) (antisymmetric stretch) and 338 (s) (antisymmetric deformation) cm^{-1}. The Raman spectrum of the solid shows bands at 590 (vs) (symmetric in-phase stretch), 502 (m) (symmetric out-of-phase stretch), and 304 (s) (symmetric deformation) cm^{-1}.[8] The compound $Cs_2[MnF_6]$ crystallizes at room temperature in the cubic $K_2[PtCl_6]$ system with $a = 8.92$ Å.[4]

References

1. R. F. Weinland and O. Lauenstein, *Z. Anorg. Chem.*, **20,** 40 (1899).
2. H. Bode, H. Jenssen, and F. Bandte, *Angew. Chem.*, **65,** 304 (1953).
3. E. Huss and W. Klemm, *Z. Anorg. Allgem. Chem.*, **262,** 25 (1950).
4. H. Bode and W. Wendt, *Z. Anorg. Allgem. Chem.*, **269,** 165 (1952).
5. C. B. Root and R. A. Sutula, Proc. 22nd Ann. Power Sources Conf., U.S. Army Electronics Command, 1968, p. 100.
6. T. L. Court, Ph.D. thesis, University of Nottingham, England, 1971.
7. A. G. Sharpe and A. A. Woolf, *J. Chem. Soc.*, **1951,** 798.
8. K. O. Christe, W. W. Wilson, and R. D. Wilson, *Inorg. Chem.*, **19,** 3254 (1980).
9. C. D. Flint, *J. Mol. Spectry.*, **37,** 414 (1971).
10. S. L. Chodos, A. M. Black, and C. D. Flint, *J. Chem. Phys.*, **65,** 4816 (1976).
11. K. O. Christe, W. W. Wilson, C. J. Schack, and R. D. Wilson, *Inorg. Synth.*, **24,** 45 (1986).
12. K. O. Christe, R. D. Wilson, and C. J. Schack, *Inorg. Synth.*, **24,** 3 (1986).

15. AMMONIUM PENTAFLUOROMANGANATE(III) AND POTASSIUM PENTAFLUOROMANGANATE(III) HYDRATE

$$KMnO_4 + 4(CH_3CO)_2CH_2 + 4AHF_2 \rightarrow A_2[MnF_5] + 2AF + KF + 4H_2O + 2[(CH_3CO)_2CH]_2 \quad (A = NH_4, K)$$

Submitted by MANABENDRA N. BHATTACHARJEE* and MIHIR K. CHANDHURI*
Checked by HELEN M. MARSDEN†

Although ammonium pentafluoromanganate(III), $(NH_4)_2[MnF_5]$, and potassium pentafluoromanganate(III) monohydrate, $K_2[MnF_5] \cdot H_2O$, have been known for

*Department of Chemistry, North-Eastern Hill University, Shillong 793 003, India.
†Department of Chemistry, University of Idaho, Moscow, ID 83843.

quite some time, there has been no easy and direct general method for their syntheses. The usual general method involves the reaction between MnF_3 in hydrofluoric acid and an alkali metal fluoride.[1] However, MnF_3 is difficult to prepare and is unstable. A general method has been developed for the synthesis of the title compounds[2] directly from $KMnO_4$ that does not require MnF_3 or the use of HF. The synthesis involves the reduction of a concentrated solution of $KMnO_4$ with 2,4-pentanedione in the presence of AHF_2 (A = NH_4 or K), which acts as a fluorinating agent. The method is fast and can be scaled up to larger quantities if desired.

■ **Caution.** *Ammonium (hydrogen difluoride), $NH_4[HF_2]$, and potassium (hydrogen difluoride), KHF_2, are highly hygroscopic and hydrolyze to give HF. Care should be taken in handling them so that they do not come in contact with the skin.*

A. AMMONIUM PENTAFLUOROMANGANATE(III), $(NH_4)_2[MnF_5]$

Procedure

A 2.0g (12.67-mmol) sample of potassium permanganate and 6.5 g (113.75 mmol) of ammonium (hydrogen difluoride)[3] are mixed by powdering together in an agate mortar. The finely mixed powder is dissolved in a *minimum* volume of water, and the mixture is filtered. The filtrate is collected in a 150-mL polythene beaker, and 15.4 mL (154 mmol) of 2,4-pentanedione is added to the solution with constant stirring. An exothermic reaction occurs to give rose-pink microcrystalline $(NH_4)_2[MnF_5]$.[4,5] The mother liquor becomes virtually colorless or slightly yellow. The compound is separated by centrifuging in a polythene centrifuge tube and is washed 2 or 3 times with heptane and then twice with alcohol. It is dried under vacuum. The yield of $(NH_4)_2[MnF_5]_2$ is 2.2 g (93.6%).*

Anal. Calcd. for $H_8F_5N_2Mn$: N, 15.06; Mn, 29.53; F, 51.06. Found: N, 15.16; Mn, 29.66; F, 51.12. The IR spectrum shows bands at 614 (ν_3) and 564 (s) (ν_4) Mn—F; 3040 (s) (ν_1), 3157 (m) (ν_3) and 1400 (s) (ν_4) N—H. The μ_{eff} at 302 K is 3.19 BM.

B. POTASSIUM PENTAFLUOROMANGANATE(III) HYDRATE, $K_2[MnF_5]\cdot H_2O$

Procedure

Potassium pentafluoromanganate(III) monohydrate,[4,5] $K_2[MnF_5]\cdot H_2O$ is synthesized in a manner analogous to that described for the ammonium salt. In a typical

*Checker obtained 1.15 g (48%).

reaction of 2.0 g (12.6 mmol) of $KMnO_4$, 6.0 g (76.8 mmol) of $K[HF_2]$,[3] and 15.4 mL (154 mmol) of 2,4-pentanedione, carried out in a 150-mL polythene beaker, a yield of 3.08 g (99%)† of potassium pentafluoromanganate(III) hydrate, $K_2[MnF_5]\cdot H_2O$, is obtained.

Anal. Calcd. for $H_2F_5K_2MnO$: K, 31.77; Mn, 22.32; F, 38.60. Found: K, 31.81; Mn, 22.41; F, 38.70. The IR spectrum shows bands at 616 (m) (ν_3) and 565 (s) (ν_4) Mn—F; 3460 (s) (ν_{OH}) and 1635 (m) (δ_{HOH}) H_2O. The μ_{eff} at 302 K is 3.30 BM.

Properties

Both $(NH_4)_2[MnF_5]$ and $K_2[MnF_5]\cdot H_2O$ are rose-pink compounds that are unstable in water and attack glass slowly in the presence of moist air. However, the compounds can be stored for prolonged periods in sealed polythene bags and checked periodically by estimation of manganese. They are insoluble in common organic solvents.

References

1. E. Muller and P. Koppe, *Z. Anorg. Chem.*, **68,** 160 (1910).
2. M. N. Battacharjee, M. K. Chaudhuri, H. S. Dasgupta, and D. T. Khathing, *J. Chem. Soc. Dalton,* **1981,** 2587.
3. M. K. Chaudhuri and P. K. Choudhury, *Chem. Ind. (London)*, **1979,** 88.
4. R. D. Peacock and D. W. A. Sharp, *J. Chem. Soc.*, **1959,** 2762.
5. A. J. Edwards, *J. Chem. Soc. A,* **1971,** 2653.

16. BIS(TRIFLUOROMETHYL)MERCURY

$$Hg(CF_3COO)_2 \xrightarrow[120\text{–}180°]{K_2CO_3} Hg(CF_3)_2 + 2CO_2$$

Submitted by REINT EUJEN*
Checked by GERARD GOMES‡ and JOHN A. MORRISON‡

†Checker obtained 2.28 g (73%).

*FB 9, Anorganische Chemie, Universität Wuppertal, Postfach 100127, 5600 Wuppertal 1, FRG.

‡Department of Chemistry, University of Illinois at Chicago, P.O. Box 4348, Chicago, IL 60680.

The syntheses of trifluoromethylated organometallics require special methods since Grignard-type reagents CF_3MgX or CF_3Li are not readily available because of fluoride elimination.

Bis(trifluoromethyl)mercury is useful in the preparation of numerous CF_3 derivatives, especially of group IV elements.[1] It was first prepared in 1949 by irradiation of CF_3I and Hg in the presence of Cd.[2] Alternative routes include radiofrequency discharge methods, for example, reaction of CF_3 radicals with HgX_2 or elemental mercury.[3] Preparative scale quantities are best obtained by decarboxylation of mercury trifluoroacetate in the presence of carbonate.[4]

Procedure

■ **Caution.** *Mercury compounds are highly poisonous. Contact of either the solids or the solution with the skin should be avoided. The entire procedure should be carried out in a well-ventilated hood. Protective gloves and a face shield should be worn. Trifluoroacetic acid is a corrosive strong acid. Contact with skin or mucous membranes should be prevented. Disposal of residues must meet the safety requirements for toxic heavy metal derivatives.*

In a 2-L round-bottomed flask equipped with a magnetic stirrer, 433 g (2 moles) of red HgO [Alfa] is dissolved in 320 mL of trifluoroacetic acid [PCR]. Enough water (~50 mL) is added to prevent crystallization of the white $Hg(CF_3COO)_2$. Excess acid and water are removed by means of a rotary evaporator. The solid residue is finely powdered in a dry atmosphere (glove bag), dried under vacuum (10^{-3} torr) at 120°, carefully mixed with 500 g of K_2CO_3 (dried at 200° under vacuum), and placed in a 120 × 7-cm i.d. glass tube equipped with a 100-mm flange at the open end (Fig. 1). The reaction tube is attached to an oil pump via a by-passed silicon-oil bubbler and a CO_2/acetone slush bath. The pressure in the system may be adjusted by means of a leak valve that is open to the air. The reaction mixture is heated to 100° at 10^{-3} torr (open by-pass) for another

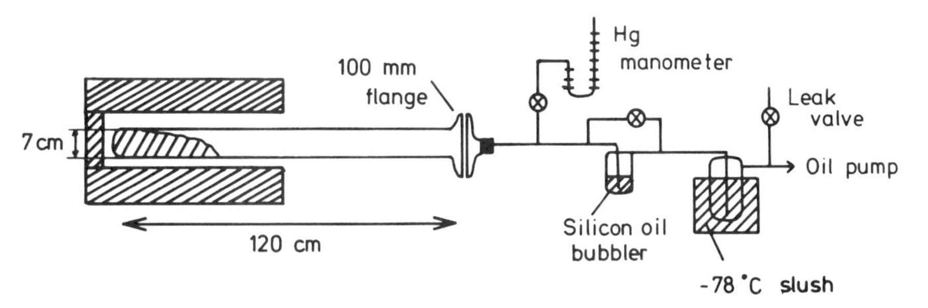

Fig. 1. Setup for the synthesis of $Hg(CF_3)_2$.

24 hr. After closing the by-pass valve the pressure is increased to ~25 torr, and the temperature is raised to 120°, at which point evolution of CO_2 begins. The furnace temperature is increased to 180° over a period of 3 days, the temperature being adjusted to maintain a constant evolution of CO_2. The $Hg(CF_3)_2$ that is formed condenses as white crystals on the cooler part of the reaction tube. The end of the reaction is indicated by reduction of the rate of liberation of CO_2 and increasing formation of elemental mercury. After cooling, the tube is opened to the air, and the product is loosened from the wall with a heat gun, transferred to a sublimation apparatus, and sublimed under vacuum at 30°. Separation from larger amounts of elemental mercury may be achieved by dissolving in diethyl ether prior to sublimation. A typical yield is 370 g of $Hg(CF_3)_2$ (55%); for small-scale preparations, yields up to 90% may be obtained.*

Properties

Bis(trifluoromethyl)mercury forms colorless, volatile crystals of pungent odor that melt at 163°. The crystals contain linear $F_3C—Hg—CF_3$ units with inter-molecular F—Hg contacts.[5] They are soluble in organic solvents, such as alcohols, ethers, halocarbons, and hydrocarbons, and also in water with slow decomposition. The ^{19}F NMR spectrum of a solution in CH_3CN consists of a singlet at −37.8 ppm (upfield from $CFCl_3$) with ^{199}Hg satellites ($^2J_{HgF}$ 1300 Hz); major IR bands are at 1145 (vs), 1070 (vs), 713 (m), and 272 (s) cm^{-1}.[5]

References

1. R. J. Lagow, R. Eujen, L. L. Gerchman, and J. A. Morrison, *J. Am. Chem. Soc.*, **100,** 1722 (1978); R. Eujen and R. J. Lagow, *J. Chem. Soc. Dalton Trans.*, **1978,** 541.
2. H. J. Emeléus and R. N. Haszeldine, *J. Chem. Soc.*, **1949,** 2953.
3. R. J. Lagow, L. L. Gerchman, R. A. Jacob, and J. A. Morrison, *J. Am. Chem. Soc.*, **97,** 518 (1975); R. Eujen and R. J. Lagow, *Inorg. Chem.*, **14,** 3128 (1975).
4. I. L. Knunyants, Ya. F. Komissarov, B. L. Dyatkin, and L. T. Lantseva, *Izv. Akad. Nauk SSSR, Ser. Khim.*, **4,** 943 (1973).
5. D. J. Brauer, H. Bürger, and R. Eujen, *J. Organometal. Chem.*, **135,** 281 (1977).

*The checkers repeated the reaction on a 0.25-size batch using 0.5 mole of HgO. They use a 500-mL round-bottomed flask positioned vertically, and the temperature is controlled by a sand bath. The flask is connected to the remainder of the apparatus shown in Fig. 1 by a 24/40 standard taper joint situated where the 100-mm flange is depicted. The initiation of the reaction requires temperatures slightly higher than the 120° described when the tube reactor is used. The yields of $Hg(CF_3)_2$ vary from 55 to 75%.

17. BIS(TRIFLUOROMETHYL)CADMIUM·1,2-DIMETHOXYETHANE

$$(CH_3)_2Cd + \text{excess } (CF_3)_2Hg \xrightarrow{\text{glyme}} (CF_3)_2Cd\cdot\text{glyme}$$

Submitted by C. D. ONTIVEROS* and J. A. MORRISON*
Checked by R. HANI† and R. A. GEANANGEL†

The incorporation of fluoroalkyl groups into organic or inorganic substrates frequently results in materials that have mechanical, chemical, or biological properties that are much more desirable for a given application than those of the parent compound.[1] For example, the (5-trifluoromethyl) pyrimidine nucleosides are known to inhibit tumor growth significantly.[2] These species are readily synthesized by the reaction of the appropriate halouracil derivative with the organometallic compound $(CF_3)Cu$.[3]

As has been shown recently, the trifluoromethyl derivatives of the group 12 (IIB) elements seem to be particularly suitable reagents for the syntheses of organometallic compounds containing perfluoroalkyl ligands.[4–9] The mercurial $(CF_3)_2Hg$, for example, reacts with the halides of several main group elements to form compounds such as $(CF_3)SnBr_3$ and $(CF_3)_3GeBr$.[4–6] Ligand exchange reactions using the mercurial, however, typically require elevated temperatures (>100°) and long reaction times (several days). Under these relatively harsh conditions, products that are more fully substituted with CF_3 groups are often thermally unstable, and decomposition is frequently observed.[4–6]

The cadmium analog $(CF_3)_2Cd\cdot$glyme, in which $(CF_3)_2Cd$ is lightly stabilized as the Lewis base adduct, has been found to be a much more powerful ligand exchange reagent than the mercurial.[7–10] For example, the reaction of $SnBr_4$ with $(CF_3)_2Cd\cdot$glyme occurs at ambient temperature and affords the fully substituted tin derivative $(CF_3)_4Sn$ in 66% yield.[7] Similarly, GeI_4, SnI_2, $Br_2Pd(PEt_3)_2$, and $Co(CO)CpI_2$ react with $(CF_3)_2Cd\cdot$glyme to yield $(CF_3)_4Ge$, $(CF_3)_2Sn$, $(CF_3)_2Pd(PEt_3)_2$, and $(CF_3)_2Co(CO)Cp$, respectively.[7–10]

Although early ^{19}F NMR studies appeared to indicate that $(CF_3)_2Hg$ and $(CH_3)_2Cd$ undergo ligand exchange reactions in basic solvents like pyridine, no products from the reaction had been isolated.[11,12] The procedure that follows depends upon the observation that the methyl, trifluoromethyl, and mixed methyl-

*Department of Chemistry, University of Illinois at Chicago, Chicago, IL 60680.
†Department of Chemistry, University of Houston—University Park, Houston, TX 77004.

trifluoromethyl species are all present in an equilibrium mixture that is formed when $(CF_3)_2Hg$ and $Cd(CH_3)_2$ react in basic solvents. The compound $(CF_3)_2Cd{\cdot}$glyme is the only presently known Lewis base adduct that does not undergo dissociation under vacuum; thus, the isolation and purification of $(CF_3)_2Cd{\cdot}$glyme is readily accomplished.

Procedure

■ **Caution.** *The substance $(CF_3)_2Hg$ is a white crystalline and sublimable solid that should be handled in an efficient hood. Its toxicity is unknown, but upon prolonged exposure $(CF_3)_2Hg$ can cause eye irritation as well as irritation to the nasal membranes.*

Initially, $(CF_3)_2Hg$ is purified by sublimation at ambient temperature and 10^{-3} torr, allowing the sublimate to condense onto a surface cooled to $-10°$ (ice/salt). In order to ensure the successful synthesis of $(CF_3)_2Cd{\cdot}$glyme, $(CF_3)_2Hg$, as obtained from the thermal decarboxylation of $(CF_3COO)_2Hg$,[13] should be sublimed at least twice prior to use. Glyme (1,2-dimethoxyethane) is dried over sodium benzophenone ketyl and degassed by using at least two freeze-pump-thaw cycles.* Dimethylcadmium can be prepared from the reaction of the methyl Grignard reagent with cadmium(II) bromide or can be obtained commercially [Alfa].

■ **Caution.** *Dimethylcadmium is a noxious smelling, toxic liquid that should be handled with care in an efficient hood.*

Freshly sublimed $(CF_3)_2Hg$ (4.20 g) is placed in a clean, dry, 50-mL round-bottomed flask (24/40 joint), along with a magnetic stirring bar. The flask is fitted with a vacuum stopcock to one end of which a 24/40 joint is affixed; an 18/9 ball joint is sealed to the other end of the stopcock. The flask is attached to a standard vacuum line[14] by means of a ball-and-socket connection. Approximately 10 mL of purified glyme is vacuum-distilled into the reaction vessel. As the contents of the flask are allowed to warm from $-196°$, they should be stirred magnetically for several minutes to ensure complete dissolution of the $(CF_3)_2Hg$. After the $(CF_3)_2Hg$ has dissolved, 1.44 g of $(CH_3)_2Cd$ is vacuum-distilled into the reaction vessel, and the solution is stirred for 2.3 hr at ambient temperature. During this time, the solution turns cloudy slowly as the $(CF_3)_2Cd{\cdot}$glyme begins to form. Upon completion of the reaction, all volatile material is removed under vacuum (which requires about 12 hr), leaving behind a snow white free-flowing powder, part of which adheres to the walls of the flask.

The vessel is then removed from the vacuum line and placed in a glove box or polyethylene glove bag, and the product is scraped out by means of a long-

*The checkers report that sodium benzophenone ketyl does not always remove peroxide. They feel that solvents such as glyme and tetrahydrofuran should be pretreated with CuCl.

handled metal spatula. If the solid seems slightly damp or is gray at this point, the reaction vessel should be returned to the vacuum line and treated with an additional small portion of dried glyme (<5 mL). The contents of the vessel are then pumped on until the solid is thoroughly dry and the product is treated as described above. Yield 1.97 g (57%).

■ **Caution.** *$(CF_3)_2Cd$·glyme, if not throughly dry and pure, may spontaneously ignite in air. It is recommended that $(CF_3)_2Cd$·glyme be handled under inert conditions at all times.*

Other Lewis base adducts of $(CF_3)_2Cd$ can be synthesized easily by either of two methods. In the first method, a procedure analogous to that presented above but utilizing THF (tetrahydrofuran), py (pyridine), or diglyme [1,1′-oxybis(2-methoxyethane)] as solvent and reagent yields the Lewis base adducts $(CF_3)_2Cd$·2THF, $(CF_3)_2Cd$·2py, or $(CF_3)_2Cd$·diglyme in yields of approximately 50, 70, and 70%, respectively. The only significant variable is the time required for the reaction to proceed. Using THF as solvent, equilibrium is attained after ~19 hr, while the reactions involving the more basic pyridine and diglyme proceed in less than 15 min. The alternative and simpler method involves direct Lewis base exchange. Bis(trifluoromethyl)cadmium·glyme, for example, dissolved in excess pyridine gives $(CF_3)_2Cd$·2py in 96% yield.

Properties

Bis(trifluoromethyl)cadmium·1,2-dimethoxyethane is a white air- and moisture-sensitive powder, mp 81° (dec.). Fluorine NMR spectra of a pure sample dissolved in CH_2Cl_2 contain a single sharp resonance 42.2 ppm deshielded from external TFA, with ^{111}Cd and ^{113}Cd satellites: $^2J_{F-^{111}Cd} = 461$ Hz, $^2J_{F-^{113}Cd} = 493$ Hz. ^{19}F NMR data for other Lewis base adducts are given in Table I. The ^1H NMR spectra contain resonances due to complexed glyme at δ 4.18 (CH_2) and δ 4.00 (CH_3) ppm. Infrared absorption bands occur at 2958 (s), 2922 (s), 2865 (m), 1405 (m), 1398 (m), 1393 (m), 1262 (m), 1130 (s), 1122 (s), 1112 (s), 1020 (s), 805 (s), 755 (m), 695 (m), 680 (m), and 670 (m) cm^{-1}. With the

TABLE I ^{19}F NMR Data for Lewis Base Adducts

	Chemical Shift[a] (ppm)	$^2J_{F-Cd}$(Hz) ^{111}Cd, ^{113}Cd
$(CF_3)_2Cd$·diglyme	44.1	448, 471
$(CF_3)_2Cd$·2py	46.7	354, 374
$(CF_3)_2Cd$·2THF	44.5	457, 476

[a]Relative to trifluoroacetic acid (ext.)

mass spectrometer operating at ambient temperature, the mass spectrum of $(CF_3)_2Cd \cdot$glyme contains the following *m/e*, ion (ion abundance): 273, $CF_3Cd \cdot g^+$ (65%); 223, $CdF \cdot g^+$ (70%); 204, $Cd \cdot g^+$ (10%); 202, CF_3CdF^+ (7%); 183, CF_3Cd^+ (25%); 129, $CdCH_3^+$ (20%); 114, Cd^+ (100%) (where g is glyme).

The compound $(CF_3)_2Cd \cdot$glyme is soluble in ethers and haloalkanes, sparingly soluble in arenes, and insoluble in alkanes. The product is best stored under N_2 or Ar at low temperatures, since thermal decomposition, ~5%/day, occurs at ambient temperature.

References

1. R. E. Banks, ed., *Organofluorine Chemicals and Their Industrial Applications,* Halstead Press, New York, 1979.
2. C. Heidelberger, *Cancer Res.*, **30,** 1549 (1970).
3. Y. Kobayashi, I. Kumadaki, and K. Yamamoto, *J. Chem. Soc., Chem. Commun.*, **1977,** 536.
4. J. A. Morrison, L. L. Gerchman, R. Eujen, and R. J. Lagow, *J. Fluorine Chem.*, **10,** 333 (1977).
5. R. J. Lagow, R. Eujen, L. L. Gerchman, and J. A. Morrison, *J. Am. Chem. Soc.*, **100,** 1722 (1978).
6. L. J. Krause and J. A. Morrison, *Inorg. Chem.*, **19,** 604 (1980).
7. L. J. Krause and J. A. Morrison, *J. Am. Chem. Soc.*, **103,** 2995 (1981).
8. L. J. Krause and J. A. Morrison, *J. Chem. Soc., Chem. Commun.*, **1981,** 1282.
9. R. Hani and R. A. Geanangel, *Polyhedron,* **1,** 826 (1982).
10. J. A. Morrison, *Adv. Inorg. Chem. Radiochem.*, **27,** 293 (1983).
11. B. L. Dyatkin, B. I. Martynov, I. L. Knunyants, S. R. Sterlin, L. A. Fedorov, and Z. A. Stumbrevichute, *Tetrahedron Lett.*, **1971,** 1345.
12. E. K. Liu and L. B. Asprey, *J. Organometal. Chem.*, **169,** 249 (1979).
13. R. Eujen, *Inorg. Synth.*, **24,** 52 (1986).
14. D. F. Shriver, *The Manipulation of Air-Sensitive Compounds,* McGraw-Hill, New York, 1969.

18. TRIFLUOROMETHYL HYPOCHLORITE AND PERFLUORO-*tert*-BUTYL HYPOCHLORITE (2,2,2-TRIFLUORO-1,1-BIS(TRIFLUOROMETHYL)ETHYL HYPOCHLORITE)

Submitted by FRITZ HASPEL-HENTRICH* and JEAN'NE M. SHREEVE*
Checked by JO ANN M. CANICH† and GARY L. GARD†

Hypochlorites are useful reagents for the introduction of fluorinated alkoxy groups and/or chlorine into a variety of inorganic and organic molecules.[1] Trifluoromethyl hypochlorite has been prepared by the reaction of dichlorine oxide or

*Department of Chemistry, University of Idaho, Moscow, ID 83843.
†Department of Chemistry, Portland State University, Portland, OR 97207.

chlorine monofluoride with carbonyl fluoride. Perfluoro-*tert*-butyl hypochlorite is obtained by the reaction of perfluoro-*tert*-butyl alcohol with chlorine monofluoride.[2–4]

In the following procedures, chlorine monofluoride is used as the source of positive chlorine to convert carbonyl fluoride in the presence of cesium fluoride as well as perfluoro-*tert*-butyl alcohol into the corresponding hypochlorites, because it is much more convenient to handle than the unstable chlorine monoxide.

Gases are transferred in a standard Pyrex glass vacuum line equipped with high-vacuum stopcocks (lubricated with a fluorocarbon grease [Halocarbon]) to which is attached a four-trap system used for low-temperature trap-to-trap distillation (Fig. 1). Because of the reactivity of the compounds, a Heise-Bourdon tube gauge [Dresser Ind.], is used for PVT measurements.

■ **Caution.** *Extreme care should be exercised in handling chlorine monofluoride, carbonyl fluoride, and the hypochlorite products because of their very high reactivities and hazardous properties. Safety shielding and leather gloves*

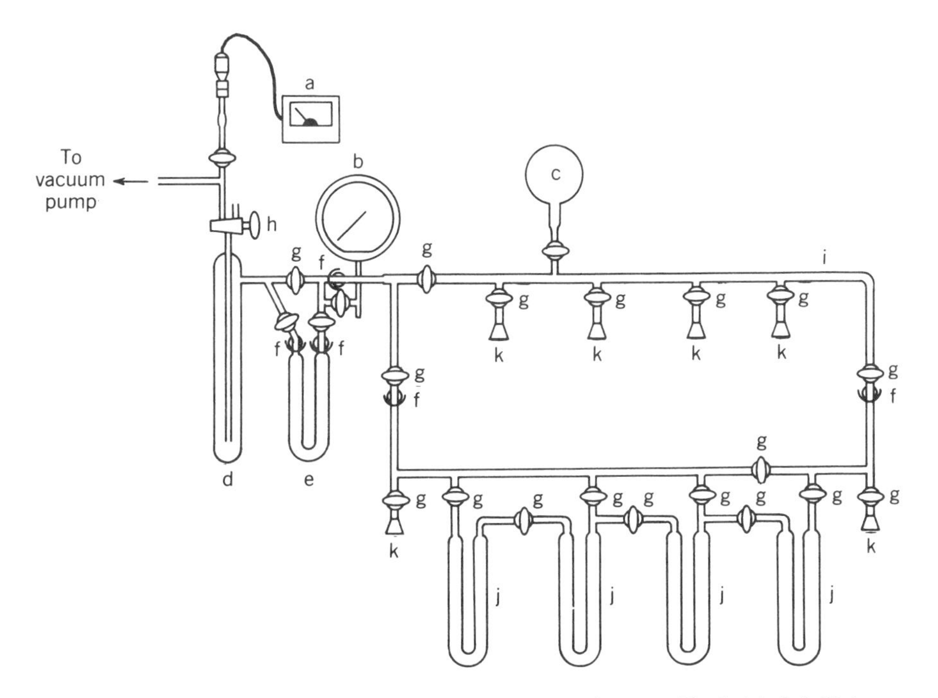

Fig. 1. Standard Pyrex glass vacuum line: a, thermocouple gauge [Fredericks]; *b,* Heise-Bourdon tube gauge [Dresser]; *c,* reservoir, 500 mL; *d,* removable cold trap; *e,* removable U-trap; *f,* 18/9 ball-and-socket joint; *g,* 2-mm Pyrex high-vacuum stopcock [Kontes]; *h,* three-way high-vacuum stopcock; *i,* Pyrex glass manifold, 10–12-mm diam; *j,* Pyrex glass traps to fit into a standard Dewar flask; *k,* 10/30 outer joint.

should be used. All apparatus should be clean and free of organic materials. Liquid nitrogen should be used for condensing reagents.

A. TRIFLUOROMETHYL HYPOCHLORITE

$$OCF_2 + ClF \xrightarrow[-196 \text{ to } 25^\circ]{CsF} CF_3OCl$$

Procedure

A 75-mL spun stainless steel Hoke vessel (test pressure 1300 psi) that is fitted with a stainless steel needle valve [Hoke] is charged with 20 mmol (3.1 g) cesium fluoride that has been dried in an oven at 150–200° for several hours. The cesium fluoride should be ground repeatedly until it remains a finely divided powder at this temperature. The transfer is made within the drying oven itself into the metal vessel, which has been dried for several hours at 150° and is still very hot (use fiber glass gloves), or the cesium fluoride may be dried and placed in an inert atmosphere box, where it may be transferred to the metal vessel. The valve should be screwed into place immediately and the vessel attached to the vacuum line via a Pyrex glass 10/30 inner standard taper joint attached to the valve by means of a standard Gyrolok fitting [Hoke] with Teflon ferrules. The vessel is evacuated at once to 10^{-3} torr. It is then cooled to -196° (liquid nitrogen), and 5 mmol of chlorine monofluoride (measured by expanding the ClF [Ozark-Mahoning] from the cylinder into a known volume to a predetermined pressure, assuming ideal gas behavior) are added. The ClF is used in order to passivate the reactor walls and to assure the completely anhydrous nature of the CsF. The vessel is removed from the vacuum line and is kept at 150° for 24 hr. It is then allowed to cool and is reattached to the vacuum line, where all of the volatile materials are removed. These should be destroyed by passing them through a soda-lime trap. This passivating process needs to be carried out only once and is not necessary again until the metal vessel is cleaned. The cesium fluoride may be used repeatedly with little change in its catalytic role, with the exception that its efficiency may improve with use (probably due to change in available surface area).

The vessel is then cooled to -196°, and 20 mmol of carbonyl fluoride [PCR, Inc.] is condensed into the vessel.

■ **Caution.** *Carbonyl fluoride is a highly poisonous gas.*

To this is added 21 mmol (1 mmol excess) of chlorine monofluoride by condensing the premeasured gas into the metal vessel. The reaction valve is closed, and the vessel is removed from the vacuum line and placed behind a shield. The vessel is allowed to warm slowly to and to remain at 25° for 12 hr.

The product is purified by trap-to-trap distillation using traps at −137° and −196°. The nearly pure CF_3OCl condenses in the trap at −137° (30°/60° petroleum ether/liquid N_2 slush) while any unreacted ClF or COF_2 passes into the trap at −196°. The yield is 19.8 mmol (99%). The product contains traces of chlorine, which is an impurity in the commercially available chlorine monofluoride.

Properties

Trifluoromethyl hypochlorite is a pale yellow liquid at −137° and a colorless gas at 25°. The vapor pressure curve is given by the equation $\log P_{mm} = -1023/T + 7.413$, and the normal boiling point is −47°.[2] The ^{19}F NMR spectrum shows a singlet at ϕ −72 ppm relative to an internal CCl_3F reference. The gas-phase IR spectrum measured by using a Pyrex glass cell with NaCl windows consists of the following absorption bands at 1262 (s), 1220 (sh), 1205 (s), 925 (mw), and 650 (mw), cm^{-1}. The compound can be stored without degradation for extended periods at 25° in a passivated stainless steel vessel equipped with a stainless steel valve.

B. PERFLUORO-*tert*-BUTYL HYPOCHLORITE

$$(CF_3)_3COH + ClF \rightarrow (CF_3)_3COCl + HF$$

■ **Caution.** *The toxicological properties of perfluoro-*tert*-butyl alcohol [PCR] are not fully known. The material should be handled only in a well-ventilated hood, using gloves. Chlorine monofluoride is strongly oxidizing, and all organic materials should be absent. Hydrogen fluoride is a dangerous reagent which upon contact produces slow-healing burns. Hands, arms, and face should be protected with gloves, lab coat, and safety shield.*

Procedure

A spun stainless steel Hoke vessel (test pressure 1300 psi) that is fitted with a stainless steel needle valve [Hoke] is passivated by condensing into it 2 mmol of chlorine monofluoride and then heating the vessel in an oven at 150° for 24 hr. Any residual gas that remains in the vessel is removed by passing the gas slowly through a tube containing soda lime. Twenty millimoles (4.72 g) of $(CF_3)_3COH$, which is weighed into a vessel that can be attached to the vacuum line, is transferred into the evacuated metal vessel, which is then cooled to −196°. Then 21 mmol (1 mmol excess) of chlorine monofluoride is condensed into the vessel at −196°. The valve is closed, and the vessel is placed behind a shield, where it is allowed to warm to and remain at 0° for 12 hr. All of the volatile materials are transferred via the vacuum line to a second Hoke vessel, which contains 40 mmol of powdered anhydrous CsF maintained at $-196°$. The vessel and contents are allowed to warm to 0° while being shaken vigorously with a mechanical shaker. After 40 min, all of the HF will have been taken up by the CsF. The remaining volatile materials are then separated by trap-to-trap distillation by using traps cooled to $-78°$ (Dry Ice/ethanol) and $-196°$ (liquid N_2). While any unchanged chlorine monofluoride passes into the trap at $-196°$, pure perfluoro-*tert*-butyl hypochlorite is retained at $-78°$. The yield is ~100%. If any unchanged alcohol remains, it will also be found in the trap at $-78°$. These compounds are very difficult to separate. Therefore, the mixture should be returned to the Hoke vessel, chlorine monofluoride added (~5 mmol), and the reaction process repeated.

Anal. Calcd. for $(CF_3)_3COCl$: C, 17.74; F, 63.22; Cl, 13.12. Found: C, 17.40; F, 64.91; Cl, 13.84.

Properties

Perfluoro-*tert*-butyl hypochlorite is a pale yellow liquid at 25°. The ^{19}F NMR spectrum shows a singlet at ϕ -70.1 ppm, using CCl_3F as an internal reference. The IR spectrum shows the following absorption bands at 1282 (vs), 1232 (ms), 1190 (mw), 1108 (s), 1003 (s), 983 (s), 788 (w), 758 (m), and 732 (s) cm^{-1}. The compound slowly decomposes at 25° to $(CF_3)_2CO$ and CF_3Cl. Thus, it should be stored at $-78°$, where it is stable for long periods.

References

1. J. M. Shreeve, *Adv. Inorg. Chem. Radiochem.*, **26,** 119 (1983).
2. D. E. Gould, L. R. Anderson, D. E. Young, and W. B. Fox, *J. Am. Chem. Soc.*, **91,** 1310 (1969).
3. C. J. Schack and W. Maya, *J. Am. Chem. Soc.*, **91,** 2902 (1969).
4. D. E. Young, L. R. Anderson, D. E. Gould, and W. B. Fox, *J. Am. Chem. Soc.*, **92,** 2313 (1970).

19. DIFLUOROPHOSPHORANES, DIETHYL PHOSPHOROFLUORIDATE, FLUOROTRIPHENYLMETHANE, AND *N*-FLUORODIMETHYLAMINE

Submitted by STANLEY M. WILLIAMSON,* O. D. GUPTA,† and JEAN'NE M. SHREEVE†
Checked by MEGAN E. LERCHEN‡ and GARY L. GARD‡

*Division of Natural Sciences, University of California-Santa Cruz, Santa Cruz, CA 95064.
†Department of Chemistry, University of Idaho, Moscow, ID 83843.
‡Department of Chemistry, Portland State University, Portland, OR 97207.

Carbonyl fluoride is a gentle and versatile fluorinating reagent.[1] It behaves as an oxidative fluorinating agent toward phosphorus(III)- and arsenic(III)-containing compounds to form the respective phosphorus(V) and arsenic(V) materials. In addition, when COF_2 is used, fluorine displaces hydrogen from C—H, N—H, B—H, and P—H bonds. These reactions occur under mild conditions without need for special apparatus. To demonstrate the wide general applicability of this method for introducing fluorine into compounds, we have selected a few representative examples that illustrate the syntheses of useful compounds.

A. OXIDATIVE FLUORINATION

The same method is used for the preparation of difluorotris(2,2,2-trifluoroethoxy)phosphorane, $(CF_3CH_2O)_3PF_2$, and difluorotriphenylphosphorane, $(C_6H_5)_3PF_2$. The reaction vessel is a 50- or 100-mL Pyrex glass, round-bottomed flask that is equipped with a Teflon-coated magnetic stirring bar and an inner 14/20 standard taper joint. The flask is fitted with a 2-mm Kontes Teflon stopcock to which are attached outer 14/20 and inner 10/30 standard taper joints. The reactions are carried out at less than 1 atm pressure.

■ **Caution.** *Carbonyl fluoride is a poisonous gas. Carbon monoxide, which is highly poisonous, is a product of the reaction. Phosphorus-containing compounds are evil-smelling and harmful. These reactions should be carried out in a well-ventilated fume hood using safety shields. Skin should be protected from contact with reactants or products.*

1. Difluorotris(2,2,2-trifluoroethoxy)phosphorane

$$(CF_3CH_2O)_3P + COF_2 \rightarrow (CF_3CH_2O)_3PF_2 + CO$$

Procedure

Using vacuum transfer techniques,[2] 0.227 g (0.69 mmol) of $(CF_3CH_2O)_3P$ [Aldrich] and 0.046 g (0.69 mmol) of carbonyl fluoride, COF_2[3], [PCR] are placed in the Pyrex glass flask. The stopcock is closed, the vessel is warmed to ambient temperature, and the mixture is stirred overnight. Fractionation through traps at -15, -116, and $-196°$ under dynamic vacuum gives $(CF_3CH_2O)_3PF_2$ in the trap at $-15°$.

■ **Caution.** *Carbon monoxide does not remain in the trap at −196° under dynamic vacuum. The vacuum pump should be vented into a well-ventilated fume hood.*

The yield of $(CF_3CH_2O)_3PF_2$ (mp 21.2–22.1°) is 0.252 g (0.69 mmol; ~100%).[4]

The compound is confirmed by stoichiometric gain in weight, ^{19}F, ^{1}H, and ^{31}P NMR and mass spectral measurements.

Anal.[4] Calcd. for $C_6H_6F_{11}O_3P$: C, 19.68; H, 1.65; F, 57.08; P, 8.46. Found: C, 19.50; H, 1.73; F, 56.82; P, 8.33.

Properties

The colorless compound is stable at room temperature under anhydrous conditions for extended periods. The mass spectrum obtained at 17 eV contains peaks at *m/e* 347 $[FP(OCH_2CF_3)_3]^+$ and at *m/e* 267 $[F_2P(OCH_2CF_3)_2]^+$. The ^{31}P (H_3PO_4) and ^{19}F (CCl_3F) NMR spectra obtained in $CDCl_3$ have peaks at δ −78.4 (tr, J_{P-F} = 749 Hz) and ϕ −77.4 (CF_3, tr, J_{H-F} = 7.8 Hz) and ϕ −58.9 (PF_2, d), respectively. The IR spectrum of $(CF_3CH_2O)_3PF_2$ contains bands at 1287, 1172 (ν_{C-F}), 1120 (ν_{P-O-C}), 960 (ν_{C-CF_3}), 905, 858, and 827 (ν_{P-F}) cm^{-1}.

2. *trans*-Difluorotriphenylphosphorane

$$(C_6H_5)_3P + COF_2 \xrightarrow{CH_2Cl_2} (C_6H_5)_3PF_2 + CO$$

Procedure

Into the Pyrex glass flask are placed 0.27 g (1 mmol) of $(C_6H_5)_3P$ [Aldrich] and 10 mL of dry CH_2Cl_2.

■ **Caution.** *Triphenylphosphine is an irritant. It should be handled in a well-ventilated area, and gloves should be worn.*

Using vacuum transfer techniques,[2] 0.06 g (1 mmol) of COF_2[3] [PCR] is added, the stopcock is closed, and the reaction mixture is stirred for 12 hr at 25°. The volatile materials (CO and CH_2Cl_2) are removed under dynamic vacuum.

■ **Caution.** *Carbon monoxide does not remain in a trap at −196° under dynamic vacuum. The vacuum pump should be vented into a well-ventilated fume hood.*

Remaining is a white solid residue, which is purified by bulb-to-bulb distillation at 100°/1 torr to give 0.24 g (0.8 mmol; 80% yield) of $(C_6H_5)_3PF_2$.[5] Nuclear magnetic resonance and mass spectral measurements are used to confirm the formation of $(C_6H_5)_3PF_2$.

Anal. Calcd. for $C_{18}H_{15}F_2P$: C, 71.99; H, 5.04; F, 12.65; P, 10.32. Found: C, 73.21; H, 5.32; F, 12.25; P, 10.74.

Properties

The hygroscopic white solid $(C_6H_5)_3PF_2$ melts at 159–160°C. It is stable for extended periods under anhydrous conditions. The mass spectrum at 17 eV shows a molecular ion at *m/e* 300. The IR spectrum has a characteristic band at 620 cm^{-1} (ν_{P-F}). The ^{31}P (H_3PO_4) and ^{19}F (CCl_3F) NMR spectra recorded in $CDCl_3$ confirm the synthesis of *trans*-difluorotriphenylphosphorane. ^{31}P: δ -54.14 (tr, J_{P-F} = 666 Hz); ^{19}F: ϕ -37.3 (d). This compound is highly soluble in benzene, dichloromethane, and acetonitrile.

B. DISPLACEMENT OF HYDROGEN FROM P—H, C—H, AND N—H BONDS BY FLUORINE

The same method is used for the preparation of diethyl phosphorofluoridate, fluorotriphenylmethane, and *N*-fluorodimethylamine. The reaction vessel is a 50-mL Pyrex glass, round-bottomed flask that is equipped with a magnetic stirring bar and an inner 14/20 standard taper joint. The flask is fitted with a 2-mm Kontes Teflon stopcock to which are attached outer 14/20 and inner 10/30 standard taper joints. The reactions are carried out at less than 1 atm pressure.

■ **Caution.** *Carbonyl fluoride is a poisonous gas. Carbon monoxide, which is highly poisonous, is a product of the reaction. The triethylammonium fluoride that is formed in the reaction should be placed in the solid waste container. The phosphorus compounds and amines are evil-smelling and harmful. These reactions should be carried out in a well-ventilated fume hood using safety shields. Gloves and other protective clothing should be worn.*

1. Diethyl Phosphorofluoridate

$$(CH_3CH_2O)_2P(O)H + COF_2 + Et_3N \xrightarrow{CH_2Cl_2} (CH_3CH_2O)_2P(O)F + CO + Et_3N\cdot HF$$

In an anhydrous atmosphere, a solution of 0.28 g (2 mmol) of $(CH_3CH_2O)_2P(O)H$ [Aldrich], 25 mL of anhydrous reagent grade dichloromethane, and 0.2 g (2 mmol) of triethylamine is prepared in the Pyrex glass flask. By using vacuum transfer techniques,[2] 0.12 g (2 mmol) COF_2[3] [PCR] is added. The mixture is stirred at room temperature for 12 hr. The volatile materials are transferred into a trap at $-196°$ under dynamic vacuum. Prolonged pumping at 25° will result in loss of product.

■ **Caution.** *Carbon monoxide does not remain in the trap at −196° under dynamic vacuum. The vacuum pump should be vented into a well-ventilated fume hood.*

Remaining in the flask is a solid residue ($Et_3N\cdot HF$) and a nonvolatile liquid. The colorless liquid that is obtained after bulb-to-bulb distillation at 35°/1 torr is $(CH_3CH_2O)_2P(O)F$ (0.2 g; 1.2 mmol; 65%).[6]

Anal. Calcd. for $C_4H_{10}O_3FP$: F, 12.17. Found: F, 12.2.

Properties

Diethyl phosphorofluoridate is a colorless liquid that can be readily distilled under reduced pressure. It is very soluble in benzene, hexane, and diethyl ether. It is stable for long periods under anhydrous conditions. The mass spectrum gives a molecular ion at *m/e* 155. ^{31}P NMR (H_3PO_4): δ −9.52 (d, J_{P-F} = 964 Hz); ^{19}F NMR (CCl_3F): ϕ −81.46 (d).

2. Fluorotriphenylmethane

$$(C_6H_5)_3CH + COF_2 + Et_3N \xrightarrow{CH_2Cl_2} (C_6H_5)_3CF + CO + Et_3N \cdot HF$$

To the Pyrex glass flask are added 0.49 g (2 mmol) of $(C_6H_5)_3CH$ [Aldrich], 20.0 mL of anhydrous reagent grade CH_2Cl_2, and 0.20 g (2 mmol) of triethylamine. By using vacuum transfer techniques,[2] 0.12 g (2 mmol) of COF_2[3] [PCR] is added to the thoroughly stirred mixture, and stirring is continued at room temperature for 12 hr. The color of the solution changes from colorless to yellow. The solvent and other volatile materials are transferred under dynamic vacuum into a trap cooled to −196°.

■ **Caution.** *Carbon monoxide does not remain in the trap at −196° under dynamic vacuum. The vacuum pump should be vented into a well-ventilated fume hood.*

The $(C_6H_5)_3CF$ and triethylammonium fluoride remain in the flask. The latter is removed by bulb-to-bulb distillation at 50°/1 torr. The residue, which is crude $(C_6H_5)_3CF$,[7] is purified by recrystallizing from benzene to give 0.32 g (1.25 mmol; 60%).

Properties

Fluorotriphenylmethane is a white solid that decomposes slowly, gradually becoming brown. It melts at 103–104° with decomposition. ^{19}F NMR (CCl_3F): ϕ −126.5.

3. *N*-Fluorodimethylamine

$$(CH_3)_2NH + COF_2 + Et_3N \xrightarrow{CH_2Cl_2} (CH_3)_2NF + CO + Et_3N \cdot HF$$

In the Pyrex glass flask are mixed 0.09 g (2 mmol) of dimethylamine, 0.20 g (2 mmol) of triethylamine, and 20.0 mL of anhydrous reagent grade dichloromethane. Using standard vacuum transfer techniques,[2] 0.12 g (2 mmol) of COF_2[3]

[PCR] is added. The mixture is stirred for 12 hr at room temperature. The volume of the solvent is reduced to 4 mL, and bulb-to-bulb distillation is performed at room temperature to leave the solid $Et_3N \cdot HF$ in the reaction flask. Final purification is accomplished by trap-to-trap distillation, which separates CH_2Cl_2 (trap at −98°) from $(CH_3)_2NF$ (trap at −126°). Fluorodimethylamine is obtained as a colorless liquid (0.056 g, 0.89 mmol, 45%).

Anal. Calcd. for C_2H_6NF[8]: C, 38.1; H, 9.6; N, 22.2. Found: C, 37.9; H, 9.5; N, 21.9.

Properties

N-Fluorodimethylamine is a colorless volatile liquid that melts at −113° and is not stable at ambient temperature. At 0°, it has a vapor pressure of 221.0 torr.[8] The mass spectrum gives a molecular ion at *m/e* 63; ^{19}F NMR (CCl_3F) ϕ −24.5.

References

1. O. D. Gupta and J. M. Shreeve, *J. Chem. Soc., Chem. Commun., 1984*, 416.
2. W. L. Jolly, *The Synthesis and Characterization of Inorganic Compounds*, Prentice-Hall, Englewood Cliffs, NJ, 1970, Chapter 8.
3. M. W. Farlow, E. H. Man, and C. W. Tullock, *Inorg. Synth.*, **6**, 155 (1960). F. S. Fawcett, C. W. Tullock, and D. D. Coffman, *J. Am. Chem. Soc.*, **84**, 4275 (1962). (The method described in the second reference is much superior.)
4. F. Jeanneaux and J. Riess, *Nouvelle J. Chim.*, **3**, 263 (1979). (This compound $(CF_3CH_2O)_3PF_2$ was one in the mixture of all possible from the fluorolysis of $Me_3SiOCH_2CF_3$ by PF_5 to give $PF_{5-x}(OCH_2CF_3)_x + xMe_3SiF$.)
5. R. Appel and A. Gilak, *Chem. Ber.*, **107**, 2169 (1974); W. C. Smith, *J. Am. Chem. Soc.*, **82**, 6176 (1960); E. L. Muetterties, W. Mahler, and R. Schmutzler, *Inorg. Chem.*, **2**, 613 (1963).
6. N. B. Chapman and B. C. Saunders, *J. Chem. Soc.*, **1948**, 1010; A. Lopusinski and J. Michalski, *J. Am. Chem. Soc.*, **104**, 290 (1982).
7. Y. A. Olah, J. T. Welch, Y. D. VanKar, M. Jogima, J. Kerkes, and J. A. Olah, *J. Org. Chem.*, **44**, 3872 (1979).
8. R. E. Wiesboeck and J. K. Ruff, *Inorg. Chem.*, **5**, 1629 (1966).

20. CHROMIUM DIFLUORIDE DIOXIDE (CHROMYL FLUORIDE)

$$CrO_3 + COF_2 \xrightarrow{185^\circ} CrO_2F_2 + CO_2$$

Submitted by GARY L. GARD*
Checked by STANLEY M. WILLIAMSON†

*Department of Chemistry, Portland State University, Portland, OR 97207.
†Department of Chemistry, University of California at Santa Cruz, Santa Cruz, CA 95064.

Although the literature describes a number of methods for preparing CrO_2F_2, the above reaction represents a convenient, quantitative, and facile synthesis.[1] Chromyl fluoride converts hydrocarbons to ketones and organic acids and is unique in providing easy routes to other chromyl compounds; for example, chromyl nitrate is easily prepared from CrO_2F_2 and $NaNO_3$.[2] Previously, it was necessary to prepare and handle N_2O_5 in order to prepare chromyl nitrate.[3]

Procedure

■ **Caution.** *Carbonyl fluoride is a water-sensitive and highly poisonous gas. It should be handled in a well-ventilated hood and in a well-constructed vacuum line. Chromium difluoride dioxide is a strong oxidizer. Organic materials must be absent.*

Chromium(VI) oxide [American Scientific] (0.90 g, 9 mmol) that has been dried at 110° for 24 hr is added to a 100-mL Hoke stainless steel vessel in a dry box. The vessel is fitted with a Hoke stainless steel needle-nose valve, attached to a standard Pyrex glass vacuum line, and evacuated. Carbonyl fluoride[4] [PCR] (1.58 g, 24 mmol) is measured by means of PVT techniques and transferred into the Hoke vessel cooled to −196°. After the vessel is heated at 185° for 12 hr, it is cooled to −78°, and any materials that are volatile at that temperature (CO_2, COF_2) are removed under dynamic vacuum.

The violet-red solid remaining in the vessel is transferred under vacuum into another Hoke stainless steel vessel. It is pure CrO_2F_2 (1.90 g, >99%).

Anal. Calcd. for CrO_2F_2: Cr, 42.62. Found: Cr, 42.71.

Properties

Chromium difluoride dioxide is a violet-red crystalline solid that at 29.6° has a vapor pressure of 760 mm. It melts to an orange-red liquid at 31.6°, and its vapor pressure at the triple point is 885 mm.[5] Although the fluoride is thermally stable,[6] care should be exercised in its handling because it is water-sensitive and is also a very strong oxidizing agent.

The IR spectrum of CrO_2F_2 has very strong absorption bands at 1016, 1006, 789, and 727 cm^{-1}; strong bands also appear at 304 and 274 cm^{-1}.[7] The gas-phase Raman spectrum[8] contains the symmetric stretching vibrations at 1007 and 728 cm^{-1}; these are found at 995 and 708 cm^{-1} in the liquid state.[9] In the solid state, the symmetric terminal Cr—F stretch at 708 cm^{-1} is replaced by a new peak at 540 cm^{-1}, which is assigned to a bridging fluorine stretching mode (Cr—F—Cr).[9] The positive and negative ion mass spectra for CrO_2F_2 have been determined; singly charged positive ions found were: $CrO_2F_2^+$, $CrOF_2^+$, CrF_2^+, CrO_2F^+, $CrOF^+$, CrF^+, CrO_2^+, CrO^+, and Cr^+.[10] More recently, the molecular structure of chromyl fluoride has been reinvestigated by electron diffraction of the gas.[11]

References

1. P. J. Green and G. L. Gard, *Inorg. Chem.*, **16,** 1243 (1977), and references therein.
2. S. D. Brown and G. L. Gard, *Inorg. Chem.*, **12,** 483 (1973).
3. M. Schmeisser and D. Lutzow, *Angew. Chem.*, **66,** 230 (1954).
4. M. W. Farlow, E. H. Man, and C. W. Tullock, *Inorg. Synth.*, **6,** 155 (1960).
5. A. Engelbrecht and A. V. Grosse, *J. Am. Chem. Soc.*, **74,** 5262 (1952).
6. W. V. Rochat, J. N. Gerlach, and G. L. Gard, *Inorg. Chem.*, **9,** 998 (1970).
7. W. E. Hobbs, *J. Chem. Phys.*, **28,** 1220 (1958).
8. I. R. Beattie, C. J. Marsden, and J. S. Ogden, *J. Chem. Soc., Dalton Trans.*, **1980,** 535.
9. S. D. Brown, G. L. Gard, and T. M. Loehr, *J. Chem. Phys.*, **64,** 1219 (1976).
10. G. D. Flesch and H. J. Svec, *Int. J. Mass. Spectrom. Ion Phys.*, **3**(5), 339 (1969).
11. R. J. French, L. Hedberg, K. Hedberg, G. L. Gard, and B. M. Johnson, *Inorg. Chem.*, **22,** 892 (1983).

21. NITRYL HEXAFLUOROARSENATE

$$2NO_2 + F_2 + 2AsF_5 \rightarrow 2[NO_2][AsF_6]$$

Submitted by MICHAEL J. MORAN,* JOANN MILLIKEN,† and RONALD A. DE MARCO†
Checked by SCOTT A. KINKEAD‡

The recent interest in nitryl hexafluoroarsenate, $[NO_2][AsF_6]$, as an oxidizing agent has emphasized the need for a simple, one-step, high-yield synthesis of this compound. Previous syntheses have involved the initial preparation of NO_2F[1,2] and subsequent reaction with AsF_5; the use of HF[3,4] with HNO_3, $ClNO_2$, or nitrate esters; the reaction of NO_2, BrF_3, and As_2O_5[5]; the use of FNO_3[6]; or metathetical reactions[7] from other $[AsF_6]^-$ salts. These reactions generally are conducted in metal cylinders or quartz vessels. The method reported here involves the direct reaction of NO_2, F_2, and AsF_5 in a Pyrex vessel and provides a pure product.

Procedure

■ **Caution.** *Although this reaction has been conducted repeatedly without incident, extreme care should be exercised at all times, and adequate shielding and protective clothing must be used. Attempts to conduct this reaction at a scale larger than 6 mmol have not been made. Fluorine is highly oxidizing and*

*West Chester State College, West Chester, PA 19380.
†Naval Research Laboratory, Washington, DC 20375.
‡Department of Chemistry, University of Idaho, Moscow, ID 83843.

very reactive. Easily oxidized materials and all organic compounds should be absent. Nitrogen oxides and arsenic pentafluoride are highly poisonous. The reaction should be conducted in a well-ventilated fume hood.

Fluorine [Matheson] is pretreated to remove HF and SiF_4 impurities and is handled in a passivated copper vacuum line designed for fluorine use. Approximately 1 g of NaF and four stainless steel balls are placed in a 200-mL high-pressure stainless steel Hoke cylinder [Koch Associates] and a high-pressure stainless steel needle valve [Koch Associates] is attached to the cylinder. The cylinder is evacuated on the copper line and cooled to $-196°$. A total of 50 mmol of fluorine is carefully condensed into the cylinder. The cylinder valve is closed. The cylinder is allowed to warm slowly to ambient temperature in an empty, precooled Dewar flask ($-196°$). It is removed from the vacuum line, shaken to loosen and spread the NaF, placed on its side for 2 days, and shaken periodically. The cylinder is reattached to the vacuum line, and the space between the valves (interspace) is evacuated. A stainless steel infrared cell (8.0 cm) with AgCl windows is attached to the vacuum line and evacuated. A high-pressure infrared spectrum (2 atm) indicates that traces of CF_4 are present, but HF and SiF_4 are absent.*

The arsenic pentafluoride [Ozark-Mahoning] and NO_2 [Matheson] must be carefully purified by trap-to-trap fractional condensation on a standard Pyrex vacuum line (10^{-6} torr) equipped with Teflon/glass valves. To purify the AsF_5, the less volatile impurities (HF and AsF_3) are condensed at $-95°$ (toluene/liquid nitrogen), the AsF_5 is condensed at $-126°$ (methylcyclohexane/liquid nitrogen), and the more volatile SiF_4 is condensed at $-196°$.† The NO_2 is purified by collecting the NO_2 in a bath at $-78°$ (acetone/Dry Ice) and passing the more volatile NO into a bath at $-196°$. The trap-to-trap distillations are repeated at least three times in each case. The removal of all traces of NO is important. We believe that traces of NO form side products, most likely NOF, that contaminate the product.‡

A Pyrex reaction vessel with a volume of approximately 390 mL is constructed from a 350-mL round-bottomed Pyrex flask with a 29/42 standard taper male joint connected to the neck, and a Teflon/glass stopcock that has a 29/42 standard taper outer joint. The standard taper joints are lubricated with Fluorolube grease

*The checker reports that F_2 is sufficiently purified by passing the F_2 through a NaF scrubber.

†The checker reports that HF and SiF_4 are removed from the AsF_5 by exposing the AsF_5 to NaF for approximately 5 min. The AsF_3 is removed by distilling the AsF_5 through a bath at $-98°$ (methanol/liquid nitrogen).

‡The checker also reports that NO_2 is purified in a modified method. The NO_2 is expanded into the vacuum line, then condensed into a trap and isolated. Pure O_2 is admitted into the vacuum line and the NO_2 allowed to warm up and mix with the O_2. The NO_2 is recondensed into the trap and the O_2 is pumped from the line. The process is repeated until the blue N_2O_3 color is no longer visible in the condensed NO_2.

[Fisher Scientific]. The vessel is attached to the glass vacuum line with Swagelok connectors [Potomac Valve & Fitting], evacuated under high vacuum, and dried under dynamic vacuum by heating with a flame from a torch. Care should be exercised during the heating to prevent the glass from collapsing. When the vessel cools, the valve is closed and the reactor is transferred to the copper line and connected with Swagelok connectors. The interspace is evacuated, and approximately 100 torr of F_2 is added to the vessel to passivate the surface. Then the stopcock is closed. The vessel is removed from the vacuum line and placed in the sunlight for approximately 1 hr. The vessel is reattached to the copper vacuum line, and the interspace is evacuated. The F_2 is removed and passed through a soda lime trap, and the stopcock is closed.

The reactor is taken from the copper line and attached to the glass vacuum line with Swagelok connectors. The interspace is evacuated, and the base of the reactor is cooled to $-196°$. The stopcock valve is opened, and the NO_2 and AsF_5 are added in layers. First, the NO_2 (5.84 mmol) is slowly condensed onto the bottom of the flask, followed by the AsF_5 (5.84 mmol). The reactor stopcock is closed, and the flask, while being maintained at $-196°$, is removed from the glass line and reconnected to the copper line. The interspace is evacuated, and the entire bulb of the reactor is cooled to $-196°$. Excess F_2 (6.0 mmol) is slowly and carefully distilled into the vessel. With the temperature of the vessel maintained at $-196°$, the stopcock on the vessel is closed and the vessel is removed from the copper line. The liquid nitrogen coolant is discarded, and the cold Dewar flask placed around the reaction vessel. The Dewar flask and the vessel are then placed behind a shield and allowed to warm slowly to ambient temperature by placing towels around the neck of the Dewar flask to reduce the rate of warming. As the flask warms to room temperature ($\sim$2 hr), a copious amount of white solid forms. The vessel is attached to the copper line, and the interspace is evacuated. The excess F_2 is removed at $-196°$ and passed through a soda-lime trap, and any unreacted AsF_5 and NO_2 are removed under dynamic vacuum, condensed in a waste trap, and hydrolyzed. After pumping for approximately 15 min, the stopcock valve is closed, and the vessel is disconnected from the vacuum line and transferred into a dry box. The stopcock on the vessel is opened, the standard taper joints are disconnected, and the grease is removed from the standard taper joint. Approximately 1.01 g (75% yield) of product is scraped into a tared Kel-F vessel equipped with a stainless steel Swagelok cap and Teflon ferrules.

Properties

Nitryl hexafluoroarsenate is a white crystalline solid that is stable at room temperature and sensitive to moisture. The product is identified and its purity established (extraneous peaks are not observed) by infrared spectra[8,9] (Nujol and

Fluorolube mulls) 395 (vs), 605 (m), 700 (vs), and 2380 (m) cm^{-1}; a Raman spectrum[8] 373 (vw), 385 (m), 586 (w), 696 (vs), and 1406 (s) cm^{-1}; and Debye-Scherrer X-ray powder pattern (d, Å) 5.05 (s), 4.96 (vs), 4.55 (vs), 3.90 (w), 3.55 (vs), 3.40 (vs), 2.92 (m), 2.68 (w), 2.45 (m), 2.38 (m), 2.21 (s), 2.06 (w), 1.92 (w), 1.85 (w), 1.77 (m,bd), 1.69 (m), 1.64 (w), 1.56 (w), and 1.52 (w).

The salt is soluble in nitromethane, and, like other nitryl salts (e.g., $[NO_2][PF_6]$, $[NO_2][SbF_6]$), it is an excellent nitrating agent for aromatics[10,11] and a good oxidizing and intercalating agent for graphite[12–14] and polyacetylene.[15]

References

1. E. E. Aynsley, G. Hetherington, and P. L. Robinson, *J. Chem. Soc.*, **1954,** 1119.
2. J. E. Griffiths and W. A. Sunder, *J. Fluorine Chem.*, **6,** 533 (1975).
3. S. J. Kuhn, *Can. J. Chem.*, **40,** 1660 (1962).
4. S. J. Kuhn, *Can. J. Chem.*, **45,** 3207 (1967).
5. A. A. Woolf and H. J. Emeléus, *J. Chem. Soc.*, **1950,** 1050.
6. J. Shamir and D. Yellin, *Israel J. Chem.*, **7,** 421 (1969).
7. R. Schmutzler, *Angew. Chem. Int. Ed.*, **7,** 440 (1968) (Ref. 146, K. Christe, private communication).
8. A. M. Qureshi and F. Aubke, *Can. J. Chem.*, **48,** 3117 (1970).
9. R. D. Peacock and I. L. Wilson, *J. Chem. Soc.*, **1969,** 2030.
10. G. Olah, S. Kuhn, and A. Mlinko, *J. Chem. Soc.*, **1956,** 4257.
11. S. J. Kuhn and G. A. Olah, *J. Am. Chem. Soc.*, **83,** 4564 (1961).
12. A. Pron, Ph.D. dissertation, University of Pennsylvania, 1980.
13. D. Billaud, P. J. Flanders, J. E. Fischer, and A. Pron, *Mat. Sci. Eng.*, **54,** 31 (1982).
14. M. Moran, G. R. Miller, R. A. De Marco, and H. A. Resing, *J. Phys. Chem.*, **88,** 1580 (1984).
15. J. Milliken, M. J. Moran, D. C. Weber, and H. A. Resing, 16th ACS Middle Atlantic Regional Meeting, Newark, DE, April 21–23, 1982.

22. SILVER HEXAFLUOROARSENATE AND BIS(*cyclo*-OCTASULFUR)SILVER(1+) HEXAFLUOROARSENATE(1−)

$$2Ag + 3AsF_5 \rightarrow 2Ag[AsF_6] + AsF_3$$
$$2S_8 + Ag(AsF_6) \rightarrow [Ag(S_8)_2][AsF_6]$$

Submitted by HERBERT W. ROESKY* and MICHAEL WITT*
Checked by K. C. MALHOTRA† and R. J. GILLESPIE†

*Institut für Anorganische Chemie der Universität Göttingen, Tammannstrasse 4, D-3400 Göttingen, Federal Republic of Germany.

†Department of Chemistry, McMaster University, Hamilton, Ontario, Canada, L8S 4M1.

Although charge transfer complexes with S_8,[1] metal complexes with ionic sulfur moieties,[2] and several covalent compounds containing homocyclic sulfur rings[3] are well established, the reaction of *cyclo*-octasulfur with silver hexafluoroarsenate in liquid sulfur dioxide provides a new synthetic pathway to a coordination compound of a neutral S_8 ring.[4] Earlier work has shown that metal hexafluoroarsenates form complexes with very weak O, N, and S donors, using liquid sulfur dioxide as solvent.[5]

General Procedure

■ **Caution.** *Arsenic pentafluoride and its derivatives are highly toxic. Liquid sulfur dioxide can generate up to 3 atm of pressure at room temperature. The handling of such solutions requires reinforced glass equipment, suitable shields, and heavy gloves. The reaction should be carried out in a well-ventilated hood. Sulfur dioxide should not be cooled below −78° because solid SO_2 (mp −78°) may cause glass vessels to crack on warming up.*

Solids are transferred into or out of the reaction vessel, a modified Schlenk apparatus shown in Fig. 1,[6] in a nitrogen-purged glove box. Solids and a magnetic

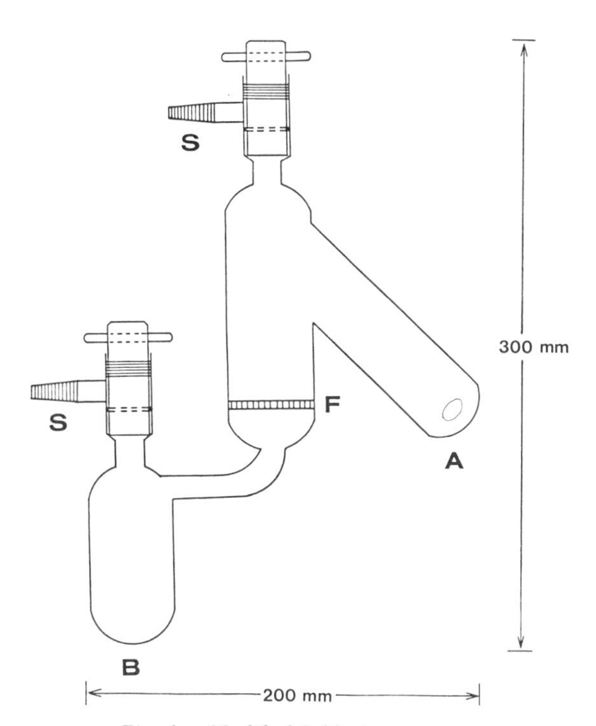

Fig. 1. Modified Schlenk apparatus.

stirring bar are filled into part *A* of the vessel, and the whole apparatus is carefully evacuated on a suitable vacuum line with a crude manometer ranging from 0 to 1000 torr. Anhydrous sulfur dioxide is dried by storing it over phosphorus pentoxide in a metal or reinforced glass cylinder with Teflon valve. The approximate amount of SO_2 is condensed into the flask at $-78°$ (Dry Ice/acetone) by closing the vacuum line to the pump and opening the storage tank. The total pressure should not exceed 300 torr.

A. SILVER HEXAFLUOROARSENATE

Procedure

The use and handling of arsenic fluorides and liquid SO_2 have been described.[7] Powdered silver metal (5.4 g, 50 mmol) is filled into part *A* of the flask, and about 25 mL of SO_2 is condensed as described above. Similarly, arsenic pentafluoride (13.6 g, 80 mmol, 5.4 mL) is condensed from the storage tank into a small metered trap at $-78°$ ($d^{-79°} = 2.47$ g/mL) and then retransferred into the reaction vessel. The system is closed and then slowly warmed with stirring. The reaction is maintained at room temperature until the silver is totally dissolved (about 2 hr).* The solution is filtered through frit *F* into part *B*, and most of the volatile materials are recondensed by cooling to 0° into part *A*. After a short period, white crystals of $Ag[AsF_6]$ begin to precipitate and the volatile materials are pumped away under vacuum. The salt is formed in quantitative yield and can be used without further purification.

Properties

Silver hexafluoroarsenate is a white crystalline solid, sensitive to moisture and light. Therefore, it should be stored in a dark sealed tube. It is very soluble in SO_2, insoluble in nonpolar solvents such as dichloromethane, and soluble in complexing solvents such as acetonitrile and dioxane with complex formation. Contact with such solvents should be avoided, since the solvent complexes are very stable and the salt is no longer useful for complexation.

B. BIS(*cyclo*-OCTASULFUR)SILVER(1+) HEXAFLUOROARSENATE(1−)

Procedure

Commercially available sublimed sulfur is dried under vacuum. Because the product is sensitive to daylight, the reaction is carried out with the flask wrapped in aluminum foil. A mixture of 1.11 g (3.7 mmol) of freshly prepared $Ag[AsF_6]$* and 1.92 g (7.5 mmol) of S_8 is filled into part *A* of the Schlenk apparatus. After evacuation, 20 mL of sulfur dioxide is condensed as described above. After warming up, the solution is stirred at room temperature for 2.5 hr. The solution is filtered into part *B* of the flask, and the solid residue is extracted carefully several times by condensing SO_2 from part *B* into part *A* and pouring it back into part *B* of the vessel. After removal of SO_2 through one of the stopcocks and drying under vacuum, $[Ag(S_8)_2][AsF_6]$, mp 155° (dec.), is obtained in pure form in 90% yield.†

Properties

Bis(*cyclo*-octasulfur)silver(1+) hexafluoroarsenate(1−) is a moisture-sensitive pale yellow crystalline solid that decomposes in the daylight to yield black Ag_2S.[4] It can be stored indefinitely in a dark sealed tube under nitrogen. The compound is soluble only in SO_2, insoluble in nondonor solvents, and decomposes in donor solvents, yielding the solvent complex and elemental sulfur.

An IR spectrum (Nujol mull) with absorption bands at 712 (s), 696 (s), 676 (s), and 395 (s) cm^{-1}, and a Raman spectrum with bands at 675 (w), 470 (m), 465 (m), 437 (w), 263 (w), 223 (s), 162 (m), and 155 (s) cm^{-1} have been recorded. The structure has been established by X-ray diffraction. The unit cell consists of four ion pairs in the space group *C2/c*. The silver atom is distorted 4-coordinate with 1,3 linkage of the two sulfur rings, which exhibit only a slight change in their geometry compared to free *cyclo*-octasulfur.

References

1. (a) T. Bjorvatten, *Acta Chem. Scand.*, **16,** 749 (1962); (b) T. Bjorvatten, O. Hassel, and A. Lindheim, *Acta Chem. Scand.*, **17,** 689 (1963); (c) R. Laitinen, J. Steidel, and R. Steudel, *Acta Chem. Scand.*, **A34,** 687 (1980).
2. (a) A. Müller, E. Diemann, R. Jostes, and H. Bögge, *Angew. Chem. Int. Ed. Engl.*, **20,** 934 (1981); (b) T. W. Wolff, P. P. Power, R. B. Frankel, and R. H. Holm, *J. Am. Chem. Soc.*, **102,** 4694 (1980); (c) G. Christen, C. D. Garner, R. M. Miller, C. E. Hohnsen, and J. D. Rush, *J. Chem. Soc., Dalton Trans.*, **1980,** 2363.
3. (a) R. Steudel and M. Rebsch, *Z. Allg. Anorg. Chem.*, **413,** 252 (1975); (b) R. Steudel and T. Sandow, *Angew. Chem. Int. Ed. Engl.*, **15,** 722 (1976); (c) J. Passmore, P. Taylor, T. K. Whidden, and P. White, *J. Chem. Soc., Chem. Commun.*, **1976,** 689.

*Checkers using 0.897 g (8.3 mmol) of purified granular silver and 2.72 g (13.3 mmol) of AsF_5 report that ~8 hr is necessary for complete dissolution of the silver metal.

†Checkers report that the yield depends on the age of the $Ag[AsF_6]$. If the $Ag[AsF_6]$ is kept 1 day the yield is only 50%, and if it is kept 1 week the yield is only 30%. The $Ag[AsF_6]$ should be freshly prepared and should be used immediately.

4. H. W. Roesky, M. Thomas, J. Schimkowiak, P. G. Jones, W. Pinkert, and G. M. Sheldrick, *J. Chem. Soc., Chem. Commun.*, **1982,** 895.
5. (a) R. Mews, *J. Chem. Soc., Chem. Commun.*, **1979,** 278; (b) H. W. Roesky, M. Thomas, J. Schimkowiak, M. Schmidt, M. Noltemeyer, and G. M. Sheldrick, *J. Chem. Soc., Chem. Commun.*, **1982,** 790; (c) H. W. Roesky, H. Hofmann, P. G. Jones, W. Pinkert, and G. M. Sheldrick, *J. Chem. Soc., Dalton Trans.*, **1983,** 1215.
6. G. Hartmann, R. Froböse, R. Mews, and G. M. Sheldrick, *Z. Naturforsch.*, **37b,** 1234 (1982).
7. B. D. Cutworth and R. J. Gillespie, *Inorg. Synth.*, **19,** 22 (1977).

23. TRIBROMOSULFUR(IV) HEXAFLUOROARSENATE(V)

$$S_8 + 12Br_2 + 12AsF_5 \xrightarrow{SO_2} 8[SBr_3][AsF_6] + 4AsF_3$$

Submitted by MIKE MURCHIE* and JACK PASSMORE*
Checked by ROBERT C. THOMPSON†

Based on the extrapolation of the properties of SF_4 (stable) and SCl_4 (dissociates at $-31°$), sulfur tetrabromide, if it can be prepared, is likely to be very unstable. However, the tribromosulfur(IV) cation[1] can be prepared quantitatively as the salt of the very weakly basic $[AsF_6]^-$ and $[SbF_6]^-$ anions. The advantage of $[AsF_6]^-$ as the anion is that AsF_3 and the solvent SO_2 are readily removed to give the desired $[SBr_3][AsF_6]$ as the sole solid product. The disadvantage is that $[SBr_3][AsF_6]$ is less stable to decomposition than is $[SBr_3][SbF_6]$, but synthesis of the latter results in the formation of antimony(III)/antimony(V) fluoride complex salts, and $[Sb_2F_{11}]^-$ anions if excess SbF_5 is added. Although only the synthesis of $[SBr_3][AsF_6]$ is described here, $[SBr_3][SbF_6]$ is synthesized in essentially the same manner.

Procedure

■ **Caution.** *All reagents should be used in a very efficient fume hood with appropriate precautions, especially AsF_5.[2,3] Arsenic pentafluoride is very poisonous and hydrolyzes readily to form HF. Sulfur dioxide is poisonous and boils at $-10°$. Well constructed glass vessels or metal systems must be employed to*

*Department of Chemistry, University of New Brunswick, Fredericton, New Brunswick, Canada E3B 6E2.

†Department of Chemistry, University of British Columbia, Vancouver, British Columbia, Canada V6T 1Y6.

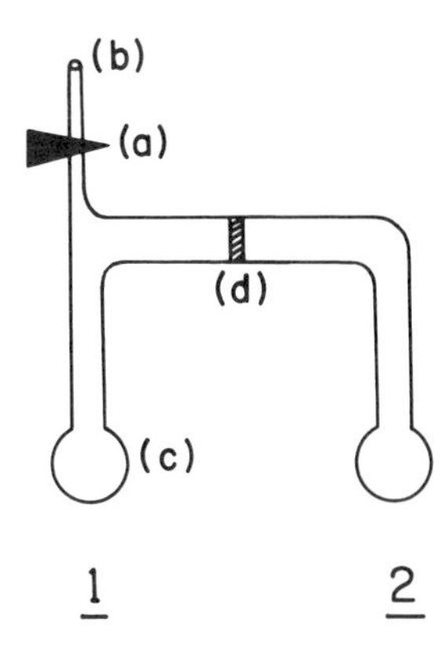

Fig. 1. Two-bulbed glass reaction vessel. (*a*) Teflon-stemmed Pyrex valve [Kontes]; (*b*) ¼-in. o.d. glass tubing joined to the manifold via a Hoke or Whitey (1KS4) [Whitey Tool] metal valve and Swagelok Teflon compression fittings [Crawford Fitting]; (*c*) thick-walled glass bulbs (8–15 mL); and (*d*) coarse sintered glass frit.

prevent pressure bursts. Bromine is corrosive and is harmful to skin and mucous membranes. Gloves must be worn. The $[SBr_3]^+$ *salts react very readily with moisture and should be manipulated in a rigorously moisture-free environment.* Sulfur (0.24 g, 0.9 mmol) is loaded into bulb 1 of the reaction vessel by removing the Teflon valve stem (Fig. 1). The assembled vessel is connected to the vacuum line (Fig. 2) and evacuated for at least an hour while the bulb is heated gently with a Bunsen flame or hot air gun. Prior to use, the metal line is evacuated with gentle heating to remove traces of water and is prefluorinated with elemental fluorine [Matheson], sulfur tetrafluoride [Matheson], or arsenic pentafluoride [Ozark-Mahoning]. Sulfur dioxide (3.66 g, 57.1 mmol) is condensed onto the sulfur with a slow rate of transfer in order to prevent bumping of the liquid SO_2 in the reservoir. The liquid sulfur dioxide is stored over CaH_2. Arsenic pentafluoride [Ozark-Mahoning] (2.09 g, 12.3 mmol) is transferred onto the mixture in aliquots from a premeasured volume at a known pressure.[4,5] On warming to room temperature the solution becomes red-brown and, after 5 min, dark blue, which indicates the formation of $[S_8][AsF_6]_2$.

■ **Caution.** *The pressure in the reaction vessel will be greater than 1 atm. Face shield, heavy gloves, and safety shields should be used.*

Bromine (1.86 g, 11.6 mmol), which is stored in a glass vessel over P_4O_{10}, is condensed onto the mixture at −78°. After it is warmed to room temperature and agitated for 5 min, a yellow-brown solution over bright yellow crystals is obtained. The solution is transferred through the sintered glass frit into bulb 2. The SO_2 is condensed onto the solid by cooling bulb 1 at −78°. On warming to 22°, some of the solid dissolves in the liquid SO_2, and the solution is again

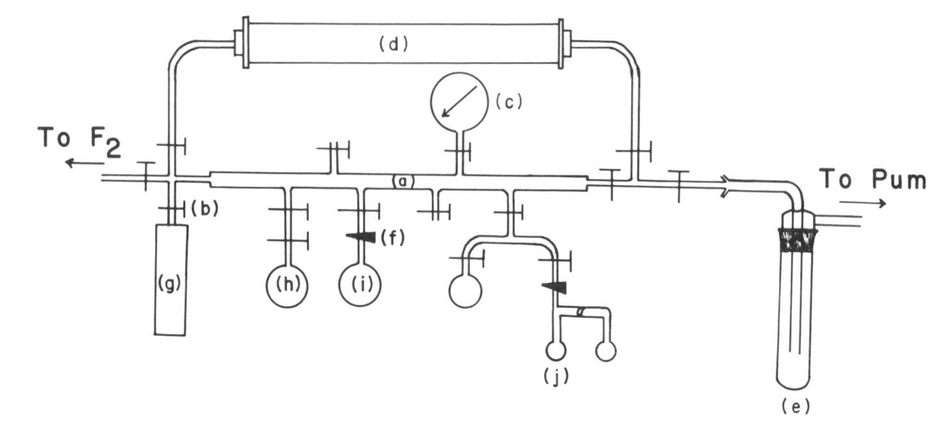

Fig. 2. Metal vacuum line. (*a*) ¼-in. Monel tubing silver soldered to ½-in. Monel tubing to form the manifold; (*b*) stainless steel Hoke or Whitey (1KS4) valve joined to the manifold by Swagelok Teflon compression fittings [Crawford Fitting]; (*c*) stainless steel gauge (30 in. Hg vacuum to 60 psi) [Dresser Industries]; (*d*) copper soda lime trap; (*e*) Pyrex glass cold trap (−196°C); (*f*) Teflon-stemmed Pyrex valves [Kontes]; (*g*) metal lecture bottle of AsF_5; (*h*) SO_2 reservoir; (*i*) Br_2 reservoir; and (*j*) reaction vessel. *Note:* All glass-to-metal seals (round-bottomed flasks to Hoke or Whitey (1KS4) valves) are formed by ¼-in. o.d. glass tubing attached to the Hoke or Whitey (1KS4) valve by Teflon compression fittings [Whitey Tool].

poured through the frit into bulb 2. This process is repeated until all soluble material has been transferred to bulb 2. Highly crystalline material is obtained by condensing the volatile SO_2 and AsF_3 into bulb 1 by cooling it with tap water (~11°) while bulb 2, which contains the solvent/solute mixture, is held at room temperature overnight. The volatile materials are then removed under vacuum (~0.5 − 1.5 hr).

Calcd. weight of $[SBr_3][AsF_6]$ based on sulfur taken: 3.46 g. Found: 3.47 g.

If less highly crystalline material is satisfactory, the reaction vessel (Fig. 1) can be replaced by a single thick-walled tube and the volatile materials removed immediately after the reaction is complete.

Properties

Solid $[SBr_3][AsF_6]$[1] decomposes slowly at room temperature in a sealed glass tube under an atmosphere of dry nitrogen. The decomposition is not sufficiently rapid to prevent spectroscopic investigation and handling over a time period of hours. Samples have been stored at −20° for periods of 6 months without noticeable signs of decomposition. The compound $[SBr_3][AsF_6]$ is best identified by its Raman spectrum: 685 (sh), 674 (m), 573 (w), 562 (w), 429 (mw)†, 414 (m)†, 391 (w), 375 (vvs)†, 367.5 (sh), 175 (w)†, and 128 (w)† cm^{-1} († identifies vibrations due to $[SBr_3]^+$).

References

1. J. Passmore, E. K. Richardson, and P. Taylor, *Inorg. Chem.*, **17**, 1681 (1978).
2. F. A. Hohorst and J. M. Shreeve, *Inorg. Synth.*, **11**, 143 (1968).
3. R. Y. Eagers, *Toxic Properties of Inorganic Fluorine Compounds*, Elsevier, Amsterdam, 1969.
4. P. A. W. Dean, R. J. Gillespie, and P. K. Ummat, *Inorg. Synth.*, **15**, 213 (1974).
5. C. L. Chernick, in *Noble-Gas Compounds*, H. H. Hyman (ed.), The University of Chicago Press, Chicago, 1963, p. 35.

24. OSMIUM(VI) FLUORIDE

$$Os + 3F_2 \rightarrow OsF_6$$

Submitted by ROBERT C. BURNS* and THOMAS A. O'DONNELL*
Checked by ROLAND BOUGON†

In earlier reports of its formation, osmium(VI) fluoride was prepared in a flow system by passing fluorine gas over hot osmium metal. The volatile OsF_6 was collected in a cooled trap.[1] However, it is more convenient to prepare OsF_6 and similar higher fluorides in a closed system.[2] This approach is now used routinely for a wide range of penta-, hexa-, and heptafluorides, for example, VF_5, MoF_6, WF_6, UF_6, ReF_6, IrF_6, and ReF_7, that are stable at 25°. Osmium(VI) fluoride is a useful precursor to $OsCl_5$ by using a halogen exchange reaction with BCl_3 at 25°.[2]

■ **Caution.** *There are several potentially hazardous aspects to this synthesis. Fluorine is a highly oxidizing gas. Contact with any combustible material, such as, hydrocarbon grease, causes a local hot spot, and metal components normally resistant because of formation of a passive layer of fluoride can then burn uncontrollably. All metal components, especially valves, must be thoroughly degreased before use with fluorine. All organic materials must be absent. Skin and clothing should be protected with gloves and a lab coat. All hydrolyzable fluorides, including* OsF_6, *form hydrogen fluoride on contact with moisture or tissue. Hydrogen fluoride causes severe, slowly healing burns. Oxidative hydrolysis of* OsF_6 *can form* OsO_4, *which is very volatile and very toxic. Contact with eyes must be avoided.*

*Department of Inorganic Chemistry, University of Melbourne, Parkville, Victoria, 3052, Australia.
†Centre d'Études Nucléaires de Saclay, 91191 Gif sur Yvette, Cédex, France.

Information is available on safe handling of fluorine and volatile fluorides and on the treatment in the event of personal exposure to these materials.[3]

Procedure

A metal (stainless steel, nickel or Monel) vacuum line is constructed in a well-ventilated hood.[4,5] This line is equipped with a Bourdon tube gauge [Dresser Industries] with a range of at least 0–1500 torr. A small cylinder (0.5 lb) of fluorine [Matheson] is stored behind a suitable barricade [Matheson]. The cylinder is equipped with a standard two-stage regulator [Matheson] and then connected to the metal line via ¼-in. copper, nickel, or Monel tubing.

■ **Caution.** *The vacuum forepump is protected from the corrosive, reactive fluorine gas by pumping all gases from the line through a trap containing granules of activated alumina or soda lime.*

The fluorination of osmium can be carried out in a reaction vessel welded from nickel, Monel, or stainless steel tubing with end plates and flanges of similar metal. A typical vessel is a cylinder 20 cm in height, 6 cm in diameter, and of wall thickness 3 mm or greater. It has an upper flange (12 mm thick) welded to it, and this is bolted to a top plate (12 mm thick) into which is welded 15 cm of ¼-in. o.d. tubing to which is fitted a Whitey valve (SS-IKS4) [Seattle Valve and Fitting]. All dimensions are approximate except the outside diameter of the connecting tube, which must match Swagelok connections [Seattle Valve and Fitting]. The top plate and ¼-in. tube are fitted with a metal jacket through which cooling water circulates to protect the gasket and the valve from thermal damage. An essential feature of the vessel is the shear seal gasket used to seal the top plate to the reactor flange. This type of seal can be constructed from annealed copper or nickel and is particularly useful in the preparation of metal fluorides.[4]

Metallic osmium (10 mmol, 1.9 g) is loaded into the metal vessel, and the top plate is bolted into place. The vessel is attached to the vacuum line and evacuated. An excess of fluorine (~32 mmol) is metered into the vessel. This can be done assuming ideal gas behavior; for example, in a vessel such as the one described, where the volume is ~565 mL, a pressure of 975 torr in the evacuated vessel will assure the addition of ~32 mmol of F_2. The valve is closed. The line is evacuated slowly through the soda lime or alumina trap.

■ **Caution.** *The reaction between fluorine and soda lime or alumina is very exothermic; therefore, the fluorine must be removed slowly.*

The vessel is maintained at 350–400° for about 3 hr. The reactor is reattached to the vacuum line and cooled to −78° (Dry Ice/acetone slush). The excess fluorine is removed by way of the soda lime or alumina trap. Then the reaction vessel is allowed to warm to room temperature and the volatile OsF_6 is transferred into a smaller metal vessel for storage. This vessel should contain a small amount of NaF to act as a "getter" for any HF that may be present. The yield of OsF_6 is ~3 g (9.9 mmol; ~100%).

Anal.[6] Calcd. for OsF_6: Os, 62.5; F, 37.5. Found: Os, 62.7; F, 38.2.

Larger scale preparations are possible. Higher pressures of fluorine gas are required.

Properties

Osmium(VI) fluoride is a water-sensitive yellow solid that melts at 32.1°. The gas-phase IR spectrum recorded in a nickel cell with AgCl windows has bands at 1453 (m), 1352 (w), 969 (w), and 720 (vs) cm^{-1}.

References

1. J. H. Canterford and R. Colton, *Halides of the Second and Third Row Transition Metals,* Wiley-Interscience, New York, 1968, pp. 2–3, 324.
2. R. C. Burns and T. A. O'Donnell, *Inorg. Chem.,* **18,** 3081 (1979).
3. T. A. O'Donnell, in *Comprehensive Inorganic Chemistry,* Vol. II, Trotman-Dickenson, A. F., ed., Pergamon Press, 1973, pp. 1014–1019, 1028–1030.
4. J. H. Canterford and T. A. O'Donnell, *Tech. Inorg. Chem.,* **7,** 273 (1968).
5. D. F. Shriver, *The Manipulation of Air-Sensitive Compounds,* McGraw-Hill, New York, 1969, p. 136.
6. B. Weinstock, H. H. Claassen, and J. G. Malm, *J. Chem. Phys.,* **32,** 181 (1960).

Chapter Two

MAIN GROUP COMPOUNDS

25. ETHYLBORONIC ACID AND TETRAETHYLDIBOROXANE/TRIETHYLBOROXIN (3:1)

Submitted by ROLAND KÖSTER* and PETER IDELMANN*
Checked by GARY M. EDVENSON† and DONALD F. GAINES†

A. ETHYLBOROINIC ACID (ETHYLDIHYDROXYBORANE)

$$(C_2H_5BO)_3 + 3H_20 \xrightarrow{\text{hexane}} 3C_2H_5B(OH)_2$$

Ethylboronic acid, $C_2H_5B(OH)_2$, is a colorless crystalline product that can be used for the preparation of various *O*-substituted ethylboron compounds, such as triethylboroxin and 2-ethyl-1,3,4-dioxaborolanes or -borinanes:

$$C_2H_5B(OH)_2 + \begin{matrix}HO\\ \\HO\end{matrix}\!\!>\!(CH_2)_n \xrightarrow[-2H_2O]{} C_2H_5B\!\!<\!\begin{matrix}O\\ \\O\end{matrix}\!\!>\!(CH_2)_n \qquad n = 2,3$$

*Max-Planck-Institut für Kohlenforschung, Kaiser-Wilhelm-Platz 1, D 4330 Mülheim an der Ruhr, West Germany.
†Department of Chemistry, University of Wisconsin-Madison, Madison, WI 53706.

The 3:1 reagent mixture of tetraethyldiboroxane and triethylboroxin is important for preparative purposes and is prepared from ethylboronic acid. Ethylboronic acid influences the reaction course of carbohydrates.[1]

Ethylboronic acid can be prepared from chloroethylphenylborane,[2] diethyl ethylboronate,[3] or triethylboroxin.[4] Very impure products resulted from the hydrolysis of the products obtained when trimethyl borate was reacted with bromoethylmagnesium.[5,6] Ethylboronic acid is best prepared by the reaction of triethylboroxin with water. Triethylboroxin is easily obtained from diboron trioxide with triethylborane[7,8] or from tetraethyldiboroxane either thermally or by >BH catalysis.[9]

Triethylboroxin[9] may be prepared as follows. Ethyldiboranes(6) [~ 0.05 g; 10.1% H^- ≃ 0.05 mmol >BH][10] are added to 6 g (39 mmol) of tetraethyldiboroxane[11] at ~0°, and the stirred mixture is heated to 130–140° (bath temperature) for 30 min, during which time triethylborane distills off. The solution is then cooled to room temperature and concentrated under vacuum (12 torr) to remove the last traces of triethylborane. Triethylboroxin is obtained as a colorless liquid residue (2.2 g; 100%).

Procedure

A 50-mL three-necked flask is fitted with a stirrer, a 25-mL dropping funnel, and an argon bubbler. The apparatus is evacuated while being heated with a gas flame and is then filled with N_2 or Ar. To a solution of 5 g (29.8 mmol) of triethylboroxin in hexane (30 mL) at ~20°, 1.5 g (83.3 mmol) of water is added dropwise with stirring over a period of about 20 min. The temperature rises to ~30°, and a voluminous precipitate forms. The residual water is flushed from the dropping funnel into the reaction mixture with a small amount of hexane. The mixture is then stirred for ~3 hr at ~20°, filtered (D-3 glassfrit), and washed four times with cold pentane (~5-mL portions) to remove excess triethylboroxin. Air is strictly excluded. The product is dried (max. 5 min!) under vacuum (10^{-1} torr and bath temperature ~30°) to avoid loss due to sublimation. Colorless crystalline ethylboronic acid (6.2 g; 100%)* with mp 90° is obtained.

Anal. Calcd. for $C_2H_7BO_2$: C, 32.51; H, 9.55; B, 14.63; H^+, 2.73. Found: C, 32.59; H, 9.61; B, 14.48; H^+, 2.70.

With activated triethylborane,[12–14] the compound gives 2 equivalents of ethane [volumetric determination of the purity by the ethane number (EZ)[15,16] = 2].

Properties

Ethylboronic acid has been reported several times since 1862.[3] The previously described melting points (e.g., 166–167°[5]; 161–166°[6]) are those of impure boric

*Checkers obtained 32% when using 8 mmole of triethylboroxin.

acid [mp 169°[17] (dehydration)], which forms from ethylboronic acid by moist air oxidation. The oxidation of an aqueous solution of ethylboronic acid in air gives boric acid and ethanol:

$$2C_2H_5B(OH)_2 + O_2 + 2H_2O \rightarrow 2B(OH)_3 + 2C_2H_5OH$$

Above its melting point ethylboronic acid decomposes to give water and triethylboroxin:

$$3C_2H_5B(OH)_2 \xrightarrow[-\ 3H_2O]{>90^\circ} (C_2H_5BO)_3$$

With strict exclusion of air, the very water-soluble ethylboronic acid is stable. Spectral data include the following: Infrared (KBr): $\nu_{OH} \approx 3400\ cm^{-1}$; $\nu_{BO} \approx 1390\ cm^{-1}$. Mass (70 eV): *m/e* 74 (M^+, 4%), 57 (M−OH, 3%), 45 (M−Et, 100%). ^1H NMR [60 MHz, di(methyl-d_3) sulfoxide]: δ 7.04 ppm (broad, OH), 0.9 ppm (m, >BEt) in the ratio 2:5. ^{11}B NMR [32.1 MHz, di(methyl-d_3) sulfoxide]: δ 32.5 ppm (Hwb = 525 Hz). ^{17}O NMR (27.1 MHz, tetrahydrofuran-d_8): δ 94.5 ppm (Hwb ~ 270 Hz).

B. TETRAETHYLDIBOROXANE/TRIETHYLBOROXIN (3:1)

$$3C_2H_5B(OH)_2 + 6B(C_2H_5)_3 \xrightarrow[-\ 6EtH]{cat.} \{3[(C_2H_5)_2B]_2O + (C_2H_5BO)_3\}$$

Mixtures of tetraethyldiboroxane and triethylboroxin are suited for the combined introduction of *O*-diethylboryl and *O*-ethyl boranediyl groups into various hydroxy types. With the help of boroxin/boroxane mixtures, carbohydrates can be transformed to *O*-substituted ethylboron derivatives. Tetraethyldiboroxane/triethylboroxin mixtures can also be used as modifying reagents for the determination of characteristic numbers of oligo- and polyhydroxy compounds with different organoboron reagents.

All sorts of tetraethyldiboroxane/triethylboroxine mixtures can be obtained by mixing the components. Without using tetraethyldiboroxane, a defined 3:1 mixture of tetraethyldiboroxane and triethylboroxin with a constant boiling point can be prepared from triethylboroxin and triethylborane. The first step is the reaction of triethylboroxin with water. In a second step the ethylboronic acid is reacted with activated triethylborane [~1 mole % [(2,2-dimethylpropanoyl)oxy]diethylborane[18] is added]. After separation from excess triethylborane, the distillable 3:1 mixture with constant composition (bp_{765} 152°) is obtained. The expected bis[(diethylboryl)oxy]ethylborane (pentaethyltriboradioxane) cannot be isolated. The decomposition of this compound to 3 parts tetraethyldiboroxane and 1 part triethylboroxin occurs during its formation.

Procedure

A 500-mL three-necked flask is fitted with a stirrer, a 25-mL dropping funnel, and a reflux condenser capped with an argon bubbler. Water (13.3 g; 732.7 mmol) is added quickly with stirring ~20° to 41.8 g (249.4 mmol) of triethylboroxin[9] in 200 mL of hexane. A voluminous precipitate forms, and the flask temperature rises to ~30°. After 4 hr at ~20°, 162.9 g (1.662 moles) of triethylborane that contains 300 mg of [(2,2-dimethylpropanoyl)oxy]diethylborane[18] is added dropwise with stirring over ~8 hr. Ethane (31.2 L, ~95%) is evolved in an exothermic reaction (t_{max} ~70°). The distillation through a 60-cm fractionating column filled with 0.4-cm glass Raschig rings (bath temperature 120–170°) gives, after hexane (bp_{765} 69°) and triethylborane (bp_{765} 94–95°), 142.9 g (93%)* of a colorless 3:1 mixture of tetraethyldiboroxane and triethylboroxin (bp_{765} 152°). A yellow liquid residue (20 g) [boroxane/boroxine mixture including [(2,2-dimethylpropanoyl)oxy]diethylborane] remains. After addition of tetraethyldiboroxane to the 3:1 mixture, the excess boroxane (bp_{765} 143°) can be quantitatively distilled out of the mixture. Excess triethylboroxin (bp_{765} 154°) cannot be removed by distillation because of the small boiling point difference.

Properties

The 3:1 tetraethyldiboroxane/triethylboroxin mixture has a constant boiling point at atmospheric pressure; bp_{765} 152°; n_D^{20} = 1.4015 [tetraethyldiboroxane: n_D^{20} = 1.4032; triethylboroxin: n_D^{20} = 1.3965]. Spectral data are as follows: Infrared (neat): ν_{BO} 1400 cm^{-1}; ν_{CO} 1100 cm^{-1}; mass (70 eV): *m/e* 168 (M^+ from triethylboroxin, 13%); 139 (M^+ − Et; 18%); 125 (M^+ − Et from tetraethyldiboroxane; 100%); ^{11}B NMR (32.1 MHz, neat); δ 52.8 ppm (tetraethyldiboroxane) and 33.5 ppm (triethylboroxin) in the ratio 2:1 (Hwb 125 and 180 Hz). ^{17}O NMR (27.1 MHz, toluene-d_8): δ 224 ppm (tetraethyldiboroxane) and 145 ppm (triethylboroxin) in the ratio 1:1.

*Checkers obtain a yield of 57% when using 10 mmol of triethylboroxin. This synthesis does not adapt well to a small-scale preparation due to the difficulty of performing the small-scale final distillation.

References

1. R. Köster, P. Idelmann, and W. V. Dahlhoff, *Synthesis,* **1982,** 650.
2. B. M. Mikhailov and P. M. Aronovich, *Zh. Obshch. Khim.,* **29,** 1257 (1959); *Chem. Abstr.,* **54,** 8684 (1960).
3. E. Frankland, *J. Chem. Soc.,* **15,** 363 (1862).
4. A. D. McElroy and R. M. Hunt, U.S. Patent 2,996,539 (1959); *Chem. Abstr.,* **56,** 8742 (1962).
5. P. A. McCusker, E. C. Ashby, and H. S. Makowski, *J. Am. Chem. Soc.,* **79,** 5179 (1957).
6. J. P. Laurent, *Compt. Rend.,* **253,** 1812 (1961).
7. J. Goubeau and H. Keller, *Z. Anorg. Allgem. Chem.,* **267,** 1 (1951).
8. R. Köster, *Liebigs Ann. Chem.,* **618**, 31 (1958).
9. R. Köster, *Houben-Weyl,* Vol. XIII/3a, 1982, p. 838, Thieme-Verlag, Stuttgart.
10. R. Köster and P. Binger, *Inorg. Synth.,* **15,** 141 (1974).
11. W. Fenzl and R. Köster, *Inorg. Synth.,* **22,** 188 (1983).
12. R. Köster, K.-L. Amen, and W. V. Dahlhoff, *Liebigs Ann. Chem.,* **1975,** 752.
13. R. Köster and W. Fenzl, *Liebigs Ann. Chem.,* **1974,** 69.
14. R. Köster, H. Bellut, and W. Fenzl, *Liebigs Ann. Chem.,* **1974,** 54.
15. R. Köster and W. Schüssler, unpublished, Mülheim an der Ruhr 1972–1982; see W. V. Dahlhoff and R. Köster, *J. Org. Chem.,* **42,** 3151 (1977).
16. R. Köster and L. Synoradzki, *Chem. Ber.,* **117,** 2850 (1984).
17. *Handbook of Chemistry and Physics,* 58th ed., CRC Press, Boca Raton, FL 1977–1978, B 96.
18. R. Köster and G. Seidel, *Inorg. Synth.,* **22,** 185 (1983).

26. INDIUM(III) IODIDE

$$2\text{In} + 3\text{I}_2 \xrightarrow{\text{Et}_2\text{O}} 2\text{InI}_3$$

Submitted by J. P. KOPASZ,* R. B. HALLOCK,* and O. T. BEACHLEY, JR.*
Checked by W. RODGER NUTT†

Indium(III) iodide is an important starting material for the preparation of a great variety of organoindium compounds. The synthesis described herein is a modification of a published procedure.[1] Since the title compound is hygroscopic, a combination of Schlenk and high vacuum techniques is employed.[2]

Procedure

A 6.00-g (52.3-mmol) sample of indium foil cut into small strips and 20.0 g (78.8 mmol) of iodine are placed in a 250-mL flask equipped with a magnetic

*Department of Chemistry, State University of New York, Buffalo, NY 14214.
†Department of Chemistry, Davidson College, Davidson, NC 28036.

stirring bar and an efficient reflux condenser. The flask is attached to the vacuum line[2] by a Teflon valve [Kontes] and standard taper adapter connected at the top of the condenser, the apparatus is evacuated, and 100 mL of dry diethyl ether, distilled from Na/benzophenone ketyl, is vacuum-distilled into the flask at $-196°$. The flask, with the Teflon valve closed, is warmed to room temperature, and the contents are stirred. As the contents of the flask warm, the reaction becomes more vigorous. The last amount of indium is slow to react, so the solution is stirred at room temperature for 12 hr. The resulting solution without further purification can be used for the syntheses of organoindium compounds[3] such as $In(CH_2SiMe_3)_3$. If indium(III) iodide is to be isolated, the condenser is replaced with a bent sintered-glass filter stick with attached flask (Fig. 1) under an inert (argon or nitrogen) atmosphere, and the reaction mixture is filtered. The diethyl ether is completely removed by vacuum distillation to yield yellow crystalline indium(III) iodide. The last traces of any excess iodine can be removed by washing the product three times with 20 ml of cold benzene to yield 24 g (95%) of InI_3. Further purification can be achieved by vacuum sublimation at 120–130°.

Properties

Indium(III) iodide is a yellow hygroscopic crystalline solid, mp 210°. The compound exists in the solid state as iodine bridged dimers[4] ($I_2InI_2InI_2$) and is readily soluble in organic solvents such as benzene, chloroform, and diethyl ether. The vibrational spectrum has been reported,[5] but the observed infrared and Raman frequencies occur in the far-infrared region, below 250 cm^{-1}.

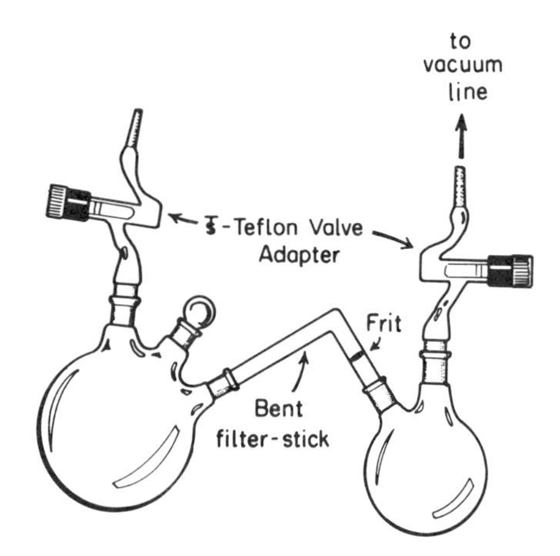

Fig. 1. Bent filter-stick filtration apparatus.

References

1. M. J. S. Gynane, M. Wilkinson, and I. J. Worrall, *Inorg. Nucl. Chem. Lett.*, **9,** 765 (1973).
2. D. F. Shriver, *The Manipulation of Air-Sensitive Compounds*, McGraw-Hill, New York, 1969.
3. J. P. Kopasz, R. B. Hallock, and O. T. Beachley, Jr., *Inorg. Synth*, **24,** 89 (1986).
4. J. D. Forrester, A. Zalkin, and D. H. Templeton, *Inorg. Chem.*, **3,** 63 (1964).
5. (a) I. R. Beattie, T. Gilson, and G. A. Ozin, *J. Chem. Soc. A*, **1968,** 813; (b) N. N. Greenwood, D. J. Prince, and B. P. Straughan, *J. Chem. Soc. A*, **1968,** 1694.

27. TRIS[(TRIMETHYLSILYL)METHYL]INDIUM

$$InI_3 + 3[(Me_3Si)CH_2]MgCl \xrightarrow{Et_2O} In[CH_2(SiMe_3)]_3 + 3\ MgClI$$

Submitted by J. P. KOPASZ,* R. B. HALLOCK,* and O. T. BEACHLEY, JR.*
Checked by R. A. ANDERSEN†

Tris[(trimethylsilyl)methyl]indium[1] and its gallium[2] analog are readily prepared from an appropriate metal halide [Alfa Products] and the (trimethylsilyl)methyl Grignard reagent in diethyl ether. The apparent steric bulk of the three (trimethylsilyl)methyl groups reduces the Lewis acidity of the alkyl indium and gallium sufficiently to enable the diethyl ether to be readily removed at room temperature by vacuum distillation. The combination of the ease of preparation of the Grignard reagent and removal of the diethyl ether from the product made the following procedure preferable to one using the lithium alkyl in a hydrocarbon solvent. The air and moisture sensitivities of the compounds involved in this preparation require the use of inert-atmosphere (argon or nitrogen) and dry-box techniques.

■ **Caution.** *The compound $In[CH_2(SiMe_3)]_3$ burns spontaneously in air, and indium compounds are toxic. The reaction should be carried out in a good hood, and adequate protection should be employed.*

The following procedure can be readily adapted for the preparation of $Ga[CH_2(SiMe_3)]_3$ from $GaCl_3$ and $[(Me_3Si)CH_2]MgCl$ in diethyl ether.[2]

Procedure

Diethyl ether should be dried over Na/benzophenone ketyl and distilled under an inert atmosphere prior to use. The Grignard reagent is prepared from 4.83 g

*Department of Chemistry, State University of New York at Buffalo, Buffalo, NY 14214.
†Department of Chemistry, University of California, Berkeley, CA 94720.

(0.199 mole) of Mg and 23.6 g (0.193 mole) of $(Me_3Si)CH_2Cl$ [PCR] in diethyl ether (total volume of 125 mL) in a 250-mL three-necked round-bottomed flask by means of standard procedures.[3] If the magnesium does not react with the $(Me_3Si)CH_2Cl$, a few crystals of iodine should be added to the magnesium, without stirring. After the iodine color has disappeared, the reaction should begin, as evidenced by bubbling in the solution. The diethyl ether solution of $(Me_3Si)CH_2Cl$ should be added at a rate that maintains a slow but steady reflux. When the reaction is complete and the solution has cooled to room temperature, the solution of Grignard reagent can be transferred to an addition funnel under argon flush and cover.

A diethyl ether solution of InI_3 (25 g, 50 mmol, 100 mL Et_2O) is contained in a 500-mL flask equipped with a magnetic stirring bar, reflux condenser, and a 150-mL pressure-equalizing addition funnel and is purged with argon gas. The Grignard reagent solution is transferred to the addition funnel under argon by using either a syringe or a pipette. The Grignard reagent is then added dropwise to the InI_3/diethyl ether solution over a period of 1 hr. A precipitate of magnesium halide forms but redissolves as more Grignard reagent is added. After the addition is complete, the reaction mixture is stirred for about 8 hr at room temperature. The reflux condenser, stirrer, and dropping funnel are replaced with stoppers and a standard taper Teflon valve adapter. The diethyl ether is then removed by vacuum distillation.

■ **Caution.** *At this stage, the reaction flask contains the alkyl indium compound which burns spontaneously in air and reacts violently with water.*

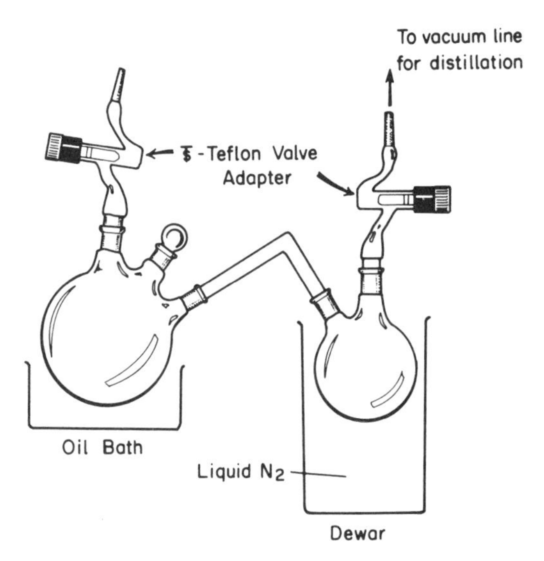

Fig. 1. Apparatus for isolation of alkyl metal compound by distillation.

In the dry box, the reaction vessel is fitted with a purged bent tube connected to a 100-mL two-necked flask equipped with a vacuum adapter (Fig. 1). The apparatus is now attached to a vacuum line through the vacuum adapter on the small flask. Using an oil bath at 110°, the volatile components are distilled into the small flask, which is cooled to − 196°. After about 4 hr, the distillation is usually complete and heating is discontinued. The receiving flask is warmed to room temperature, and small amounts of diethyl ether are removed by pumping off the components that are volatile at room temperature into a trap at − 196°. In order to ensure that all traces of diethyl ether are removed, the apparatus is usually pumped on for 8–12 hr at room temperature to yield 18.1 g (91.8%) of $In[CH_2(SiMe_3)]_3$.*

Anal.[1] Calcd. for $C_{12}H_{33}Si_3In$: In, 30.5. Found: 30.4.

A ^1H NMR spectrum of the product with a small quantity of benzene as a reference can be used to identify the product and ensure the complete removal of diethyl ether.

Properties

At room temperature $In[CH_2(SiMe_3)]_3$ is a colorless pyrophoric liquid. The ^1H NMR spectrum of the compound in benzene (reference δ 7.13) consists of two lines in area ratio 9:2, with the larger methyl line at δ +0.13 and the smaller methylene line at δ −0.05.† The IR spectrum[1] of the neat liquid has bands at 2935 (vs), 2885 (s,sh), 1440 (w), 1400 (w), 1350 (w), 1291 (w,sh), 1244 (vs), 920 (s), 825 (vs), 757 (vs), 720 (vs), 692 (s,sh), 580 (m), and 490 (m) cm^{-1}. The compound can be used to prepare [(trimethylsilyl)methyl]indium halogen compounds by appropriate stoichiometric exchange reactions[1] with $InCl_3$, $InBr_3$, and InI_3. Properties of many of these organoindium halogen compounds have been reported.[1] The experimental procedure is related to that described for $Al(CH_2SiMe_3)_2Br$,[4] except that solvents are used for the reactions.

References

1. O. T. Beachley, Jr., and R. N. Rusinko, *Inorg. Chem.*, **18,** 1966 (1979).
2. O. T. Beachley, Jr., and R. G. Simmons, *Inorg. Chem.*, **19,** 1021 (1980).
3. R. A. Andersen and G. Wilkinson, *Inorg. Synth.*, **19,** 262 (1979).
4. C. Tessier-Youngs and O. T. Beachley, Jr., *Inorg. Synth.*, **24,** 94 (1986).

*The checker used $[(Me_3Si)CH_2]_2Mg$ in diethyl ether rather than $[(Me_3Si)CH_2]MgCl$. Starting with 1.5 g of $InCl_3$, 1.0 g (88%) of $In[CH_2(SiMe_3)]_3$ was obtained.

†The checker reports that the ^1H NMR spectrum in C_6D_6 at 28° shows two resonances at δ 0.42 and δ −0.89 in area ratio 9:2, respectively. The ^{13}C NMR spectrum consists of a triplet centered at δ 12.0 ($^1J_{CH}$ = 118 Hz) and a quartet centered at δ 2.98 ($^1J_{CH}$ = 118 Hz).

28. TRIS[(TRIMETHYLSILYL)METHYL]ALUMINUM

$$3Li[CH_2(SiMe_3)] + AlBr_3 \xrightarrow{\text{Hexane}} Al[CH_2(SiMe_3)]_3 + 3LiBr$$

Submitted by CLAIRE TESSIER-YOUNGS* and O. T. BEACHLEY, JR.*
Checked by JOHN P. OLIVER† and KATHALEEN BUTCHER†

The title compound, a very useful precursor to many organoaluminum compounds,[1–4] was first prepared from the reaction of the dialkylmercury compound and aluminum metal in refluxing toluene.[5] The toxicity of the organomercury reagent prompted us to attempt the synthesis of $Al[CH_2(SiMe_3)]_3$ using the alkyllithium reagent.[1] The present synthesis is a modification of this latter work. The air and moisture sensitivities of the reagents and products involved in this preparation require the use of inert atmosphere and dry-box techniques and/or the use of high vacuum.[6]

■ **Caution.** *The compounds $Li[CH_2(SiMe_3)]$ and $Al[CH_2(SiMe_3)]_3$ burn spontaneously in air and react violently with water. The reaction should be carried out in a good hood, and adequate protection should be employed.*

Procedure

In the dry box, 6.74 g (71.7 mmol) of diethyl ether–free $Li[CH_2(SiMe_3)]$[7] and 70 mL of dry hexane are combined in a 100-mL pressure-equalizing funnel equipped with a high-vacuum Teflon valve [Ace Glass] and a stirring bar. Also in the dry box, a 250-mL three-necked flask equipped with a stirring bar and a standard taper Teflon valve vacuum adapter is loaded with 6.39 g (23.9 mmol) of $AlBr_3$ (previously sublimed at 60°C) and 100 mL of hexane. A reflux condenser and the addition funnel are placed on the flask under an argon flow. The complete dissolution of the $Li[CH_2(SiMe_3)]$ in the addition funnel is ensured by agitation with the stirring bar and gentle heating with a hot air gun. The $Li[CH_2(SiMe_3)]$ solution is added slowly to the stirred $AlBr_3$ suspension. An exothermic reaction takes place immediately, and LiBr precipitates. After the $Li[CH_2(SiMe_3)]$ addition is complete (about 1 hr), the reaction mixture is heated at reflux for 12 hr. The reflux period is necessary to avoid the isolation of products that contain bromine. Heating is stopped, and the reaction vessel is cooled under argon flow. The addition funnel and reflux condenser are replaced by glass stoppers, and the

*Department of Chemistry, State University of New York at Buffalo, Buffalo, NY 14214.
†Department of Chemistry, Wayne State University, Detroit, MI 48202.

hexane is removed by vacuum distillation. The reaction vessel is then transferred to the dry box, and a stopper is replaced by a U-tube connected to a 50- or 100-mL two-necked flask with a standard taper Teflon valve vacuum adapter. The apparatus[8] is now attached to a vacuum line through the vacuum adapter on the small flask and, by using an oil bath at 70°, 5.9 g (85%)* of $Al[CH_2(SiMe_3)]_3$ is distilled from the reaction vessel under high vacuum. The product is pure at this stage. Redistillation under high vacuum using a microdistillation apparatus [Kontes] modified by the addition of a Teflon valve gives one fraction boiling at 48–49° (0.01 torr) (lit.[5] bp 51°/0.08 torr).

■ **Caution.** *The compound $Al[CH_2(SiMe_3)]_3$ is pyrophoric. All manipulations of $Al[CH_2(SiMe_3)]_3$ are best performed in the dry box.*

Properties

The compound $Al[CH_2(SiMe_3)]_3$ is a colorless pyrophoric liquid at room temperature, and it reacts violently with water. The ^1H NMR spectrum of the compound in CH_2Cl_2 exhibits two lines in area ratio 9:2, with the larger methyl line at δ 0.07 and the smaller methylene line at δ -0.24. The IR spectrum of a Nujol solution has bands at 1265 (m,sh), 1253 (vs), 958 (s), 925 (s), 860 (vs), 832 (vs), 764 (vs), 736 (s), 690 (m), 662 (m), 590 (m) cm^{-1}. The compound exists in benzene solution as a monomer-dimer equilibrium mixture.[5]

References

1. O. T. Beachley, Jr., C. Tessier-Youngs, R. G. Simmons, and R. B. Hallock, *Inorg. Chem.*, **21,** 1970 (1982).
2. O. T. Beachley, Jr., and C. Tessier-Youngs, *Organometallics,* **2,** 796 (1983).
3. C. Tessier-Youngs, W. J. Youngs, O. T. Beachley, Jr., and M. R. Churchill, *Organometallics,* **2,** 1128 (1983).
4. C. Tessier-Youngs, C. Bueno, O. T. Beachley, Jr., and M. R. Churchill, *Inorg. Chem.*, **22,** 1054 (1983).
5. J. Z. Nyathi, J. M. Ressner, and J. D. Smith, *J. Organometal. Chem.*, **70,** 35 (1974).
6. D. F. Shriver, *The Manipulation of Air-Sensitive Compounds,* McGraw-Hill, New York, 1969.
7. C. Tessier-Youngs and O. T. Beachley, Jr., *Inorg. Synth.*, **24,** 95 (1986).
8. J. P. Kopasz, R. B. Hallock, and O. T. Beachley, Jr., *Inorg. Synth.*, **24,** 87 (1986).

*Checkers ran the reaction on a one-half scale and obtained the same percentage yield.

29. BROMOBIS[(TRIMETHYLSILYL)METHYL]ALUMINUM

$$2Al[CH_2(SiMe_3)]_3 + AlBr_3 \rightarrow 3Al[CH_2(SiMe_3)]_2Br$$

Submitted by CLAIRE TESSIER-YOUNGS* and O. T. BEACHLEY, Jr.*
Checked by JOHN P. OLIVER† and KATHALEEN BUTCHER†

The title compound is obtained by a stoichiometric exchange reaction between 2 moles of the trialkylaluminum compound and 1 mole of $AlBr_3$. This procedure is based on the original preparation[1] and can be easily adapted to the synthesis of $AlEt_2Br$ (bp 54°/0.01 torr). In general, the synthesis of bromo-substituted organoaluminum compounds by exchange reactions is easier than that for the corresponding chloro derivatives. The higher lattice energy of $AlCl_3$ requires higher reaction temperatures, and the use of sealed tubes is frequently necessary. The compound chlorobis[(trimethylsilyl)methyl]aluminum has been synthesized[2] from $AlCl_3$ and 2 moles of $Al[CH_2(SiMe_3)]_3$. The air and moisture sensitivity of the reagents and products involved in this preparation require the use of inert atmosphere and dry-box techniques or the use of high vacuum.[3]

■ **Caution.** *The compounds $Al[CH_2(SiMe_3)]_3$ and $Al[CH_2(SiMe_3)]_2Br$ burn spontaneously in air and react violently with water. Adequate protection should be used.*

Procedure

In the dry box, 1.75 mL (1.40 g, 4.87 mmol) of $Al[CH_2(SiMe_3)]_3$[4] and 0.650 g (2.43 mmol) of freshly sublimed $AlBr_3$ are combined in a 10-mL flask. Aluminum(III) bromide sublimes at 60° under high vacuum. Heat is evolved rapidly and a colorless liquid is produced. (■ **Caution.** *If the reaction quantities are increased, the reagents should be combined slowly to prevent excessive heating.*) The flask is attached to a microdistillation apparatus [Kontes] modified by the addition of a Teflon valve using a minimum of Dow Corning silicone grease or, preferably, Halocarbon grease 25-5S [VWR Scientific] (Apiezon L or N greases are unsatisfactory for the distillation). The title compound reacts with all greases tested. At 60–62° (0.01 torr), $Al[CH_2(SiMe_3)]_2Br$ (1.926 g, 94%) distills as a colorless pyrophoric liquid.

Anal. Calcd. for $AlBrC_8H_{22}Si_2$: Al, 9.59; Br, 28.41; $SiMe_4$, 2.00 mole/mole. Found: Al, 9.45; Br, 27.73, $SiMe_4$, 1.98 mole/mole.

*Department of Chemistry, State University of New York at Buffalo, Buffalo, NY 14214.
†Department of Chemistry, Wayne State University, Detroit, MI 48202.

Properties

The compound $Al[CH_2(SiMe_3)]_2Br$ exists as an equilibrium mixture of monomeric and dimeric species in benzene solution. The ^1H NMR spectrum of a benzene solution (reference δ 7.13) of the compound exhibits two lines in area ratio 9:2, with the larger methyl line at δ 0.13 and the smaller methylene line at δ −0.30. The IR spectrum of a Nujol solution has bands at 1259 (s), 975 (s,b), 946 (m), 858 (vs,b), 763 (s), 734 (s), 698 (m), 668 (m), 636 (vs), 577 (s), and 315 (w) cm^{-1}.

References

1. O. T. Beachley, Jr., and C. Tessier-Youngs, *Organometallics*, **2**, 796 (1983).
2. S. Al-Hashimi and J. D. Smith, *J. Organometal. Chem.*, **153**, 253 (1978).
3. D. F. Shriver, *The Manipulation of Air-Sensitive Compounds*, McGraw-Hill, New York, 1969.
4. C. Tessier-Youngs and O. T. Beachley, Jr., *Inorg. Synth.*, **24**, 92 (1986).

30. [(TRIMETHYLSILYL)METHYL]LITHIUM

$$2Li + (Me_3Si)CH_2Cl \rightarrow Li[CH_2(SiMe_3)] + LiCl$$

Submitted by CLAIRE TESSIER-YOUNGS* and O. T. BEACHLEY, JR.*
Checked by JOHN P. OLIVER† and KATHALEEN BUTCHER†

Organolithium compounds are excellent alkylating reagents for the formation of organometallic compounds in hydrocarbon solvents, specifically in the absence of ethers. The syntheses of $Li[CH_2(SiMe_3)]$ described herein are modifications of published procedures.[1,2] The preparation in diethyl ether is reproducible, but care must be taken to remove all the diethyl ether from the product if it is to be used in the synthesis of $Al[CH_2(SiMe_3)]_3$. The preparative method in hexane is highly dependent on the purity of the lithium dispersion and the alkyl halide. For this reason it is advised that the lithium dispersion be purchased in small quantities or be repackaged in ampules in an argon-filled dry box. The air and moisture sensitivity of the reagents and products involved in this preparation require the use of inert atmosphere (argon) and dry-box techniques or the use of high vacuum.[3]

*Department of Chemistry, State University of New York at Buffalo, Buffalo, NY 14214.
†Department of Chemistry, Wayne State University, Detroit, MI 48202.

■ **Caution.** *Lithium dispersion is very reactive with air and water. The compound Li[$CH_2(SiMe_3)$] burns spontaneously in air, but solutions are less reactive. The reaction should be carried out in a good hood, and adequate protection should be provided.*

Procedure in Diethyl Ether

Oil is removed from commercially available lithium dispersion [Aldrich Chemical] in paraffin oil by washing with hexane under vacuum. In an argon-filled dry box, a three-necked 250-mL flask equipped with a stirring bar is loaded with 3.0 g (0.43 mole) of lithium dispersion. Dry diethyl ether from Na/benzophenone ketyl is vacuum-distilled into the flask. Under argon flush, a 100-mL pressure-equalizing addition funnel equipped with a high-vacuum Teflon valve [Ace Glass] and a *very efficient* reflux condenser‡ are attached to the flask. Then $(Me_3Si)CH_2Cl$ [PCR] (17 mL, 15 g, 0.12 mole) is placed in the funnel. The alkyl halide is added slowly to the stirred lithium suspension. A vigorous exothermic reaction occurs, precipitating LiCl, and the reaction mixture takes on a reddish hue due to by-products. The reaction mixture is stirred for 12 hr and then is filtered under vacuum. The apparatus for filtering is set up by replacing the reflux condenser and funnel with a stopper and a bent coarse 30-mm fritted filter stick attached to a two- or three-necked 250-mL flask.[4] A bent filter stick is much easier to use than the commercially available straight filter sticks, because a portion of the reaction mixture can be decanted, resulting in less clogging of the frit and thereby speeding the filtration process. After the filtration is complete, the filter stick is disconnected under an argon flush, and the solvent is removed by vacuum distillation. The resultant light brown sticky material contains appreciable quantities of diethyl ether. Under an argon flush, the flask is fitted with a sublimation cold finger. It is then evacuated and is heated with an oil bath. Although it would appear preferable to fit the flask with a sublimation cold finger in the dry box, the vapor pressure of the diethyl ether makes it difficult to subject the flask to a pump-down cycle for entering the dry box without securely fastening all standard taper joints. Diethyl ether is completely removed by heating the flask to about 50°. At 90°, 9.0 g (78%) of colorless Li[$CH_2(SiMe_3)$] is collected (mp 106–108°, lit mp[1] 112° from hexane; ^1H NMR: C_5H_{10}, δ −1.88). Toward the end of the sublimation, the red-brown viscous residue is stirred to avoid splashing it onto the sublimed Li[$CH_2(SiMe_3)$]. Frequently, a second sublimation is required for complete separation of pure Li[$CH_2(SiMe_3)$] from the red residue.

Procedure in Hexane

The following procedure is useful for the preparation of small quantities of Li[$CH_2(SiMe_3)$] in the absence of diethyl ether. As in the previous procedure,

‡Catalog No. 07-736B, 300 mm jacket length Allihu condenser [Fisher Scientific].

1.83 g (0.264 mole) of lithium dispersion from a freshly opened container is placed in a three-necked flask equipped with a stirring bar. Hexane is vacuum-distilled into the flask at −196°. Under argon flush, while the hexane is still frozen, the flask is equipped with an *efficient* reflux condenser. When the hexane is partially melted, 11.9 mL (10.5 g, 0.0860 mole) of $(Me_3Si)CH_2Cl$ (purified by distillation) is added by syringe, and stirring is initiated. A vigorous reaction takes place while the mixture is still slightly below room temperature. After being stirred for 12 hr, the reaction mixture is worked up as in the previous procedure. The yield of $Li[CH_2(SiMe_3)]$ is 7.3 g (90%).

The reaction is very dependent on the purity of the lithium dispersion surface. As the lithium dispersion (which is being stored in the dry box) ages, the lithium has to be activated by adding 50 mL of diethyl ether, and later the reaction has to be initiated by using a warm water bath. These steps cause a decrease in the yield of $Li[CH_2(SiMe_3)]$ and complicate the sublimation of pure $Li[CH_2(SiMe_3)]$ because larger amounts of the red by-products are obtained.

■ **Caution.** *This preparative procedure in hexane should not be used for the preparation of larger quantities of* $Li[CH_2(SiMe_3)]$*, because the reaction might be too vigorous to control.*

References

1. J. W. Connolly and G. Urry, *Inorg. Chem.*, **2,** 645 (1963).
2. H. L. Lewis and T. L. Brown, *J. Am. Chem. Soc.*, **92,** 4664 (1970).
3. D. F. Shriver, *The Manipulation of Air-Sensitive Compounds*, McGraw-Hill, New York, 1969.
4. J. P. Kopasz, R. B. Hallock, and O. T. Beachley, Jr., *Inorg. Synth.*, **24,** 87 (1986).

31. CYCLOPENTADIENYLTHALLIUM (THALLIUM CYCLOPENTADIENIDE)

$$Tl_2SO_4 + 2NaOH \rightarrow 2TlOH + Na_2SO_4$$
$$2TlOH + 2C_5H_6 \rightarrow 2TlC_5H_5 + 2H_2O$$

Submitted by A. J. NIELSON,* C. E. F. RICKARD,* and J. M. SMITH*
Checked by BRUCE N. DIEL†

Reagents used for introducing the cyclopentadienyl ligand (Cp) into metal complexes[1] often give low yields of product or suffer from difficulties in handling.

*Department of Chemistry, University of Auckland, Private Bag, Auckland, New Zealand.
†Department of Chemistry, University of Idaho, Moscow, ID 83843.

The frequently used and easily prepared NaCp in tetrahydrofuran solution[2] is extremely sensitive to traces of oxygen and does not store well. Product yields are usually lower with CpMgBr[3] or Cp_2Mg,[4] both of which are air-sensitive, with the latter involving a tedious preparation. Dicyclopentadienylmercury[5] has low thermal stability and does not store well.

Cyclopentadienylthallium[6] is a superior reagent for many syntheses. It is easily prepared and stored, it may be handled in air, and the insolubility of thallium halide leads to high yields of product that may be filtered easily and worked up. The reagent has been used for preparing a variety of actinide,[7] lanthanide,[8] and transition metal[9] cyclopentadiene complexes.

■ **Caution.** *Thallium compounds are extremely toxic. Avoid inhalation of dust and contact with skin. All operations should be carried out in a fume hood with the use of neoprene gloves. Wastes should be stored in bottles or disposed of by thoroughly mixing with a large amount of sand and burying the mixture in a safe, open area.*

Procedure

Thallium(I) sulfate (25 g, 0.0495 mole) and sodium hydroxide (8 g, 0.2 mole) are dissolved in 250 mL of water in a 500-mL round-bottomed flask. Freshly cracked cyclopentadiene[2,10] (8.2 mL) is added, the flask is stoppered, and the solution is stirred magnetically for 12 hr. The brown precipitate is removed by filtering, washed with 50 mL of water followed by 50 mL of methanol, and dried under vacuum for 2 hr. The solid is transferred to a water-cooled sublimation apparatus. A plug of glass wool is placed on top, and the TlCp is sublimed at 90–100° at 10^{-3} torr. The sublimation should be continued until no further product sublimes onto the cold finger after the bulk of the TlCp has been scraped away. A brown powdery residue remains after the sublimation is completed. Yield: 24.4 g, (91%*).

Anal. Calcd. for C_5H_5Tl: C, 22.3; H, 1.9. Found: C, 22.5; H, 2.2.

Properties

Cyclopentadienylthallium is a light yellow solid that decomposes at 230°.[6] It is stable for several months if stored under N_2 in a Schlenk flask kept in the dark. Slow decomposition takes place in the air and light, whereupon the solid turns brown, but pure product may be sublimed away from this material. The IR spectrum shows absorption bands at 3022, 1584, 995, and 725 cm^{-1}. Cyclopentadienylthallium is soluble in most organic solvents. It may be handled in

*Checker obtained 94.4% yield.

air, and it is normally heated at reflux in THF solution with the substrate for about 2 hr under an inert gas atmosphere. When allowed to react with metal halides, the insoluble thallium halide produced may be filtered from the reaction solution. Most cyclopentadienyl complexes are air- and moisture-sensitive and should be handled with appropriate techniques.[11]

References

1. J. M. Birmingham, *Adv. Organometal. Chem.*, **2,** 365 (1964).
2. G. Wilkinson, *Org. Synth.*, **36,** 31 (1956).
3. G. Wilkinson, F. A. Cotton, and J. M. Birmingham, *J. Inorg. Nucl. Chem.*, **2,** 95 (1956).
4. W. A. Barber, *Inorg. Synth.*, **6,** 11 (1960).
5. G. Wilkinson and T. S. Piper, *J. Inorg. Nucl. Chem.*, **2,** 32 (1956).
6. H. Meister, *Angew. Chem.*, **69,** 533 (1957).
7. P. Zanella, S. Faleschini, L. Doretti, and G. Faraglia, *J. Organometal. Chem.*, **26,** 353, (1971); **43,** 339 (1972).
8. E. O. Fischer and H. Fischer, *J. Organometal. Chem.*, **3,** 181 (1965).
9. L. E. Manzer, *Inorg. Synth.*, **21,** 84 (1982).
10. R. B. King and F. G. A. Stone, *Inorg. Synth.*, **7,** 99 (1963).
11. D. F. Shriver, *The Manipulation of Air-Sensitive Compounds*, McGraw-Hill, New York, 1969.

32. TETRAISOCYANATOSILANE

$$SiCl_4 + 4KOCN \xrightarrow[-10^\circ]{liq.SO_2} Si(NCO)_4 + 4KCl$$

Submitted by JOSEPH S. THRASHER*
Checked by DIETER LENTZ†

Even though the preparation of tetraisocyanatosilane has been previously described in *Inorganic Syntheses*,[1] there exists a less time-consuming and far less expensive patented procedure[2] that has been largely overlooked. This procedure involves the reaction of silicon tetrachloride with sodium or potassium cyanate in place of the silver or lead salt, which first must be freshly prepared. The use of liquid sulfur dioxide as the solvent, instead of benzene, offers no real increase in handling problems. The details of this alternate method adapted to a laboratory

*Department of Chemistry, The University of Alabama, University, AL 35486.
†Institut für Anorganische und Analytische Chemie, Freie Universität Berlin, Fabeckstrasse, 34–36, D-1000 Berlin 33.

scale are given below. It should be added that compounds such as $R_nSi(NCO)_{4-n}$, where R = CH_3, OCH_3, or OC_6H_5, can also be easily prepared by this method.[2]

Tetraisocyanatosilane has been found to be a useful starting material in the field of sulfur-nitrogen-fluorine chemistry as shown, for example, in the preparation of (fluorocarbonyl)imidosulfurous difluoride, FC(O)N=SF_2.[3]

Procedure

■ **Caution.** *Sulfur dioxide is a poisonous gas; therefore, this reaction must be carried out in an efficient fume hood.*

Anhydrous powdered potassium cyanate (405 g, 5.0 moles) is placed in a 2-L four-necked, round-bottomed Pyrex flask fitted with a low-temperature thermometer, a gas inlet valve, a sealed mechanical stirrer, and a Dry Ice–cooled condenser topped by a drying tube. The flask is flushed with dry argon or nitrogen and cooled to −70° (Dry Ice/2-propanol) before approximately 500 mL of sulfur dioxide is introduced through the gas inlet valve. Stirring is begun, and the temperature of the mixture is raised to −30° by warming the cooling bath by the addition of 2-propanol. The gas inlet valve is then replaced with a pressure-equalizing addition funnel containing silicon tetrachloride (115 mL, 1 mole). Over the next hour the silicon tetrachloride is added very slowly. The reaction temperature must not exceed −10°, and the initial reaction is strongly exothermic. The temperature is maintained between −30 and −10° by adding increments of Dry Ice to the cooling bath as needed. After completing the addition of the silicon tetrachloride, the reaction mixture is allowed to stir at −10° for an additional 4 hr.

At this time, the sulfur dioxide is stripped into another trap cooled by Dry Ice by allowing the reaction mixture and reflux condenser to warm to room temperature. This trap is then disconnected and maintained at −78° until the sulfur dioxide can be transferred to an appropriate storage vessel for reuse in this procedure. To carry out this transfer, the trap is chilled to liquid nitrogen temperature and attached to a standard vacuum line where the noncondensible materials are removed by pumping. It is often necessary to complete several freeze-thaw cycles to totally degas the sulfur dioxide, which is then transferred in the vacuum system with careful warming to the storage vessel held at −196°. The pressure in the system must be monitored throughout this transfer.

Meanwhile, the thermometer and stirrer of the reaction flask are replaced with stoppers and the condenser with a drying tube while the system is purged with argon or nitrogen. The reaction flask is then filled with 400 mL of dry diethyl ether and shaken to wash the product from the potassium salts. In order to filter the resulting solution, the drying tube is replaced with a fritted funnel (50-mm diameter coarse glass frit, porosity 3), which is connected to a vertical vacuum adapter topped by a 1-L two-necked flask. The apparatus is then inverted as the flow of argon is replaced with vacuum aspiration to aid in the filtration. The potassium salts are washed with several 50–100-mL quantities of diethyl ether. After the diethyl ether is distilled off, the product is collected as the fraction boiling between 185 and 190°. The product can also be distilled at reduced pressure after first removing the diethyl ether. The yield of tetraisocyanatosilane by this method is generally 75–80%. This reaction has also been carried out on a one-quarter scale with similar results.

Properties

Tetraisocyanatosilane is a moisture-sensitive, white crystalline solid, mp $26.0 \pm 0.5°$, bp $185.6 \pm 0.3°$.[4] It dissolves not only in diethyl ether but also in benzene, chloroform, carbon tetrachloride, acetone, and petroleum naphtha.[1] Its reaction with most primary and secondary amines followed by hydrolysis is convenient for syntheses of substituted ureas.[5]

References

1. R. G. Neville and J. J. McGee, *Inorg. Synth.*, **8,** 23 (1966).
2. K. Matterstock, German Patent 1,205,099; *Chem. Abstr.*, **64,** P8236h (1966).
3. J. S. Thrasher, *Inorg. Synth.*, **24,** 10 (1986).
4. G. S. Forbes and H. H. Anderson, *J. Am. Chem. Soc.*, **62,** 761 (1940).
5. R. G. Neville and J. J. McGee, *Org. Synth.*, **45,** 69 (1965).

33. DIALKYL [(*N*,*N*-DIETHYLCARBAMOYL)METHYL]PHOSPHONATES

Submitted by S. M. BOWEN* and R. T. PAINE*
Checked by LOUIS KAPLAN†

Several bifunctional organophosphorus ligands have been found to be efficient liquid-liquid extractants for lanthanide and actinide ions present in highly acidic aqueous nuclear waste solutions. The (carbamoylmethyl)phosphonates, $(RO)_2P(O)CH_2C(O)NR'_2$, and in particular dihexyl [(*N*,*N*-diethylcarbamoyl)methyl] phosphonate (DHDECMP) have attracted the most attention because of the low aqueous-phase solubilities and superior radiolytic stabilities of the ligands.[1–10]

*Department of Chemistry, University of New Mexico, Albuquerque, NM 87131.
†Chemistry Division, Argonne National Laboratory, Argonne, IL 60439.

Syntheses for (carbamoylmethyl)phosphonates based upon the Arbusov and Michaelis reactions have been described briefly by Siddall.[11] DHDECMP (~60% purity) is available from a commercial source [Wateree Chemical]. The ligands are obtained typically with acidic phosphonate impurities, which have a deleterious effect on extraction selectivity. In order to overcome this problem, McIsaac and coworkers designed a purification procedure for DHDECMP that provides samples with 98–99% purity.[12]

The increasing interest in (carbamoylmethyl)phosphonates, the fact that only one of these ligands is commercially available, the lack of adequate detail for their syntheses, and the relative inaccessibility of the purification scheme require descriptions of general procedures for the syntheses and purification of the ligands. The procedures are outlined here for R = R′ = Et. However, the same preparations have been used for R = Me, *i*-Pr, Bu, Hex with R′ = Et.

Procedure

Although the (carbamoylmethyl)phosphonates are not particularly air-sensitive, it is found that the purities of the products are enhanced by performing the following reactions under dry nitrogen.

1. Arbusov Reaction

$$(RO)_3P + ClCH_2C(O)N(C_2H_5)_2 \xrightarrow{150^\circ} (RO)_2P(O)CH_2C(O)N(C_2H_5)_2 + RCl$$

$$R = CH_3,\ C_2H_5,\ i\text{-}C_3H_7,\ n\text{-}C_4H_9,\ n\text{-}C_6H_{13}$$

The reaction apparatus consists of a 1-L three-necked round-bottomed flask equipped with a magnetic stirring bar, a water-cooled condenser, a nitrogen gas inlet tube, and a Y-tube. The open end of the condenser (top) is attached to a Hg-filled pressure relief bubbler. One arm of the Y-tube is fitted with a sliding O-ring seal thermometer adapter and thermometer [Kontes Glass], and the second arm is fitted with a 250-mL pressure-equalizing dropping funnel. The thermometer is inserted through the sliding O-ring seal so that the thermometer bulb is immersed in the stirred reaction mixture. The entire apparatus is purged with dry nitrogen, and 220 g (1.32 moles) of triethyl phosphite (99%) [Aldrich Chemical] is added to the 1-L flask. 2-Chloro-*N*,*N*-diethylacetamide [Fairfield Chemical] 189 g (1.26 moles) is added to the dropping funnel under steady nitrogen purge or in a nitrogen-filled glove bag. The phosphite is then heated with stirring to 150° under nitrogen purge. The temperature of the phosphite is measured directly by the thermometer held in place in the Y-tube. At 150°, the acetamide is slowly dripped into the phosphite. (■ **Caution.** *The addition should be slow due to* *the exothermic nature of the reaction.*) Chloroethane is evolved rapidly and is swept from the reaction vessel by the nitrogen stream. After addition of the acetamide is complete (~2 hr), the temperature of the reaction pot is maintained at 150° for another hour. The resulting yellow mixture is vacuum-distilled, and the $(C_2H_5O)_2P(O)CH_2C(O)N(C_2H_5)_2$ is collected at 122–130° (<1 torr). The yield is about 60%, and the purity is better than 95%. Similar yields and purity levels are obtained for $(CH_3O)_2P(O)CH_2C(O)N(C_2H_5)_2$ (86–120°/<1 torr) and (i-$C_3H_7O)_2P(O)CH_2C(O)N(C_2H_5)_2$ (112–128°/<1 torr). Similar yields are obtained for $(C_4H_9O)_2P(O)CH_2C(O)N(C_2H_5)_2$ (112–130°/<1 torr) and $(C_6H_{13}O)_2P(O)CH_2C(O)N(C_2H_5)_2$ (105–125°/<1 torr), but there are several impurities that cannot be removed by simple distillation.*

2. Michaelis Reaction

$$(RO)_2P(O)H + NaH \xrightarrow{\text{THF}} [(RO)_2P(O)]Na + H_2$$

$$[(RO)_2P(O)]Na + ClCH_2C(O)N(C_2H_5)_2 \xrightarrow{\text{Et}_2\text{O}} (RO)_2P(O)CH_2C(O)N(C_2H_5)_2 + NaCl$$

$$R = C_2H_5,\ n\text{-}C_4H_9,\ n\text{-}C_6H_{13}$$

Diethyl phosphonate (98%) [Aldrich Chemical], 215 g (1.55 moles), is added to a 250-mL addition funnel under dry nitrogen. A slight excess of NaH [Aldrich Chemical], 37.5 g (1.56 moles), is placed in a nitrogen-filled 1-L three-necked flask containing a stirring bar. (■ **Caution.** *Sodium hydride is extremely pyrophoric.*) The flask is evacuated, and approximately 200 mL of dry tetrahydrofuran (THF) (distilled from sodium-benzophenone ketal) is vacuum-transferred from a storage vessel into the flask at −196°, using standard Schlenk techniques.[13] The contents of the flask are warmed to room temperature, the flask is back-filled with dry nitrogen, and the flask is fitted with a water-cooled condenser, a dropping funnel that contains the diethyl phosphonate, and a nitrogen gas inlet tube. The diethyl phosphonate is added slowly over 1 hr to the stirred slurry of NaH/THF. A vigorous evolution of hydrogen occurs, and this gas is swept out of the vessel with the nitrogen stream. Vacuum evaporation of

*The same procedure, with minor modifications, was used by the checker to prepare the (carbamoylmethyl)phosphonate compounds with R = n-hexyl, R′ = isobutyl and R = 2-ethylbutyl, R′ = ethyl. Running the reactions at somewhat reduced pressure (~90 torr) facilitates the removal of these higher-boiling alkyl halides, minimizing by-products from their participation in the Arbusov reactions. Crude yields of ~95% pure material are almost quantitative; yields of >99% pure material obtained by mercury purification are 85–90%.

all the volatile components leaves a gummy white solid, $[(C_2H_5O)_2P(O)]Na$. Yield: 209 g (85%). The yields for the other salts are similar.

Sodium diethyl phosphonate, 255 g (1.6 moles), is loaded into a 2-L three-necked flask under dry nitrogen. The flask is evacuated, and about 1 L of anhydrous diethyl ether is vacuum-transferred into the flask. 2-Chloro-*N,N*-diethylacetamide [Fairfield Chemical], 201 g (1.3 moles), is added to a 500-mL dropping funnel under nitrogen. The flask containing the frozen $NaP(O)(OC_2H_5)_2$/Et_2O solution is back-filled with nitrogen, and a water-cooled condenser, a Hg bubbler, a dropping funnel that contains the $ClCH_2C(O)N(C_2H_5)_2$, and a nitrogen gas inlet tube are put in place. While the diethyl ether is still frozen, the $ClCH_2C(O)N(C_2H_5)_2$ is added slowly. The entire contents are then allowed to warm to 25°. Heat is evolved, and a fine precipitate of NaCl forms. The mixture is heated at reflux for 1 hr. The condenser and dropping funnel are then removed, and the ether is evaporated under vacuum. Benzene (500 mL, dried over Na) is transferred under nitrogen to the flask, and this mixture is stirred for 1 hr. The benzene, containing the (carbamoylmethyl)phosphonate, is filtered in air several times through a medium-porosity fritted funnel to remove NaCl. Vacuum evaporation of the benzene produces $(C_2H_5O)_2P(O)CH_2C(O)N(C_2H_5)_2$ in about 65% yield. For R = n-C_4H_9 and n-C_6H_{13}, the extraction is more efficient with hexane. The crude products are purified by distillation as described in Section 1.*

3. Purification

$$2(RO)_2P(O)CH_2C(O)N(C_2H_5)_2 + 2Hg(NO_3)_2 \xrightarrow{HNO_3} \{Hg[(RO)_2P(O)CHC(O)N(C_2H_5)_2]NO_3\}_2$$

$$\{Hg[(RO)_2P(O)CHC(O)N(C_2H_5)_2]NO_3\}_2 + 8KCN \xrightarrow[H_2O]{Na_2CO_3} 2Hg(CN)_4^{2-} + 8K^+ + 2(RO)_2P(O)CH_2C(O)N(C_2H_5)_2$$

$$R = n\text{-}C_4H_9,\ n\text{-}C_6H_{13}$$

*The same procedure was used by the checker to prepare the (carbamoylmethyl)phosphonate compound with R = isopropyl, R′ = ethyl in a yield of 72% of distilled product with purity >99%. Some simplification is achieved by adding the tetrahydrofuran [Aldrich, 99.5+% Gold Label] to the NaH [Aldrich] through the dropping funnel instead of distilling from sodium benzophenone, and by adding anhydrous ether through the funnel to the sodium salt of the dialkyl phosphite (without pumping off the THF) in the same vessel in which the latter is prepared (using a flask of appropriate size for the scale of reaction). The 2-chloro-*N,N*-dialkylacetamide is then added through the funnel so as to maintain spontaneous reflux.

The (carbamoylmethyl)phosphonates, with R = CH_3, C_2H_5, and i-C_3H_7, are typically obtained with 95% or better purity after one distillation. If higher purities are required, the following purification procedure can be used, but there is significant loss of the desired ligands to the aqueous phase. The ligands with R = *n*-C_4H_9 and *n*-C_6H_{13} are typically obtained with 60–80% purity after one distillation. Subsequent distillations result in significant product degradation. On the other hand, these two ligands can be easily purified using the following scheme given for R = *n*-C_4H_9.

In a 1-L polyethylene bottle, crude dibutyl [(*N*,*N*-diethylcarbamoyl)methyl]phosphonate (56 g, 0.18 mole) is diluted with hexane to give a 20% (v/v) solution of the ligand. To this solution is added 400 mL of a 1 *M* HNO_3 solution that is also 1 *M* in $Hg(NO_3)_2$. (■ **Caution.** *Mercury salts are toxic.*) The two-phase mixture is agitated for 24 hr on a shaker. A white precipitate forms at the interface. The solid is collected by filtration and washed with three 50-mL portions of water and three 50-mL portions of hexane. The complex has been shown to be $\{Hg[(BuO)_2P(O)CHC(O)N(C_2H_5)_2]NO_3\}_2$.[14]†

The ligand is reclaimed by dissolving the mercury complex (85 g, 0.15 mole) in 1 L of water containing 130 g (2 moles) of KCN and 53 g (0.5 mole) of Na_2CO_3, and combining this solution with 200 mL of hexane. (■ **Caution.** *Cyanides are very poisonous.*) If the complexes containing ligands with R = Me, Et, or *i*-Pr are being used, benzene is the preferred solvent. The mixture is shaken for 1 hr. The phases are separated, and the ligand is recovered from the organic phase by vacuum evaporation of the solvent. The ligand is recovered with greater than 99% purity. The success of the above extraction can be checked by removing a small aliquot of hexane from the organic phase and adding a copper bead and a few milliliters of 1 *M* HNO_3 to the aliquot. If mercury deposits on the copper bead, an additional KCN/Na_2CO_3 extraction is required. The aqueous solution containing the $Hg(CN)_2$ is evaporated slowly at 25° in an efficient fume hood, and the solid residue can then be reclaimed or safely disposed.

Properties[14]

The (carbamoylmethyl)phosphonates are colorless to slightly yellow viscous liquids that are stable to water and air. It is advised that for long-term storage the ligands be contained in a vessel with a dry nitrogen atmosphere. The dihexyl (R = C_6H_{13}) and dibutyl (R = C_4H_9) derivatives are soluble in hexane, tetrahydrofuran, benzene, and diethyl ether. They are only very slightly soluble in water. The dimethyl (R = CH_3), diethyl (R = C_2H_5), and diisopropyl (R = *i*-C_3H_7) derivatives are soluble in THF, diethyl ether, and benzene and moderately

†The mercury purification was used by the checker to purify commercial DHDECMP as well as the dihexyl *N*,*N*-diisobutyl and bis(2-ethylhexyl) *N*,*N*-diethyl compounds referred to in Section 1.

soluble in water. They are not appreciably soluble in hexane. The principal infrared bands of a thin film appear in the following regions: ν_{CO} 1648–1638, ν_{PO} 1262–1250, and ν_{POC} 980 cm^{-1}. The following NMR spectral resonances[14] are useful in identifying the products: (CH_2Cl_2 solvent) ^1H δ ($PC\underline{H}_2C$) 2.98–2.86 ppm (doublet J_{PH} = 22 Hz); ^{13}C δ ($P\underline{C}H_2C$) 34.81–32.65 ppm (doublet, J_{PC} = 133.6–131.6 Hz); ^{31}P δ 24.53–22.00 ppm.

References

1. W. W. Schulz and J. D. Navratil, in *Recent Developments in Separation* Science. Vol. VII, N. N. Li (ed.), CRC Press, Boca Raton, FL 1981.
2. L. D. McIsaac, J. D. Baker, J. F. Krupa, R. E. LaPointe, D. H. Meikrantz, and N. C. Schroeder, Allied Chemical Corp.-Idaho Chemical Programs, Idaho Falls, Rept. No. ICP-1180 (1979).
3. J. D. Baker, L. D. McIsaac, J. F. Krupa, D. H. Meikrantz, and N. C. Schroeder, Allied Chemical Corp.-Idaho Chemical Programs, Idaho Falls, Rept. No. ICP-1182 (1979).
4. W. W. Schulz and L. D. McIsaac, Atlantic Richfield Hanford Co., Rept. No. ARH-SA-263.
5. R. R. Shoun and W. J. McDowell, in *Actinide Separations,* ACS Symp. Ser. 117, Navratil, J. D. and Schulz, W. W. (eds.) American Chemical Society, Washington, DC, 1980, p. 71.
6. L. D. McIsaac, J. D. Baker, J. F. Krupa, D. H. Meikrantz, and N. C. Schroeder, in *Actinide Separations,* ACS Symp. Ser. 117, Navratil, J. D. and Schulz, W. W. (eds.) American Chemical Society, Washington, D.C., 1980, p. 395.
7. W. W. Schulz and L. D. McIsaac, in *Transplutonium Elements,* W. Müller and R. Lindner (eds.), North-Holland, Amsterdam, 1976, p. 433.
8. E. P. Horwitz, D. G. Kalina, and A. C. Muscatello, *Sep. Sci. Technol.,* **16,** 403 (1981).
9. D. G. Kalina, E. P. Horwitz, L. Kaplan, and A. C. Muscatello, *Sep. Sci. Technol.,* **16,** 1127 (1981).
10. A. C. Muscatello, E. P. Horwitz, D. G. Kalina, and L. Kaplan, *Sep. Sci. Technol.,* **17,** 859 (1982).
11. T. H. Siddall, *J. Inorg. Nucl. Chem.,* **25,** 883 (1963).
12. N. C. Schroeder, L. D. McIsaac, and J. F. Krupa, Rept. No. ENICO-1026, Exxon Nuclear Idaho Co., Idaho Falls, Jan. 1980.
13. D. F. Shriver, *The Manipulation of Air-Sensitive Compounds,* McGraw-Hill, New York, 1969.
14. S. M. Bowen, E. N. Duesler, R. T. Paine, and C. F. Campana, *Inorg. Chim Acta,* **59,** 53 (1982).

34. TRIPHENYL(TRICHLOROMETHYL)PHOSPHONIUM CHLORIDE AND (DICHLOROMETHYLENE)TRIPHENYLPHOSPHORANE

Submitted by ROLF APPEL*
Checked by FRITZ HASPEL-HENTRICH†

Triphenyl(trichloromethyl)phosphonium chloride and (dichloromethylene)triphenylphosphorane are important intermediates in the reaction system triphenylphosphine/carbon tetrachloride.[1] Solutions of (dichloromethylene)triphenylphosphorane, which is a convenient Wittig reagent, were first prepared by the addition of dichlorocarbene to triphenylphosphine in chloroform.[2,3] However, the ylide could not be separated without degradation caused by the reaction medium.

The method of preparation described here is based on the dechlorination of triphenyl(trichloromethyl)phosphonium chloride in benzene, which makes the isolation of the (dichloromethylene)triphenylphosphorane possible without decomposition.[4]

A. TRIPHENYL(TRICHLOROMETHYL)PHOSPHONIUM CHLORIDE

$$Ph_3P + CCl_4 \rightarrow [Ph_3P{-}CCl_3]Cl$$

■ **Caution.** *Carbon tetrachloride and benzene are suspected carcinogens. The reaction should be carried out in a well-ventilated hood, and gloves should be worn at all times. The reaction should be protected from moisture at all times.*

Procedure

A 1000-mL round-bottomed flask equipped with a side-arm gas inlet and a large magnetic stirring bar is flushed with nitrogen. Triphenylphosphine [Aldrich] (105 g, 0.4 mole), anhydrous carbon tetrachloride (400 mL), and dry acetonitrile (200 mL) are placed in the flask. Carbon tetrachloride and acetonitrile are dried

*Anorganisch-Chemisches Institut der Universität Bonn, Gerhard-Domagk-Strasse, 53 Bonn 1, West Germany.

†Department of Chemistry, University of Idaho, Moscow, ID 83843.

by stirring with P_4O_{10} and are distilled before use. The vessel is sealed with a mercury valve, and the suspension is stirred for 1 hr at room temperature. Approximately one-third of the solvent mixture is removed under vacuum. The vacuum pump is protected with a trap cooled with Dry Ice/ethanol and the solvents are retained in a similarly cooled trap. (■ **Caution.** *Do not use a water aspirator!*) The reaction flask is flushed with nitrogen, and 28.8 g (30 mL, 0.4 mole) of 1,2-epoxybutane [Aldrich] is added to destroy the reaction by-product, dichloro(triphenyl)phosphorane.

After stirring for an additional hour, the solid is collected by filtration through a Schlenk filter tube,[5] washed with 50 mL of dry benzene and 50 mL of dry diethyl ether, and dried under vacuum. Benzene and diethyl ether should be heated at reflux over sodium metal. The yield of crude $[Ph_3P{-}CCl_3]Cl$ is approximately 40 g. The product must be recrystallized before it is used as a starting material for Part B.

For recrystallization, the water-sensitive compound is transferred under anhydrous conditions into a 1000-mL round-bottomed flask and is dissolved in approximately 100 mL of dry dichloromethane. Dichloromethane is dried by stirring with P_4O_{10} and distilling before use. Dry diethyl ether, which is contained in a pressure-equalizing dropping funnel, is added in small portions until the product starts to separate. After 3 hr under an inert atmosphere, recrystallization is complete and the white solid material is separated by filtration through a Schlenk filter tube and dried under vacuum. Yield: 29 g (17.4%); mp: 171–180° (dec.).

Anal. Calcd. for $C_{19}H_{15}Cl_4P$: C, 54.84; H, 3.63; P, 7.44; Cl, 34.1. Found: C, 54.95; H, 3.67; P, 7.53; Cl, 33.5.

The IR spectrum (KBr disk) has bands at 3020 (vw), 1582 (vw), 1480 (w), 1435 (m), 1117 (m), 1060 (vw), 992 (w), 755 (m), 720 (m), 688 (m), 536 (m), and 518 (m) cm^{-1}. $^{31}P\ \{^1H\}$ NMR (CD_2Cl_2): δ 47.5 (s) relative to H_3PO_4.

B. (DICHLOROMETHYLENE)TRIPHENYLPHOSPHORANE

$$[Ph_3P{-}CCl_3]Cl + [(CH_3)_2N]_3P \rightarrow Ph_3P{=}CCl_2 + [(CH_3)_2N]_3PCl_2$$

Procedure

■ **Caution.** *Benzene, hexamethylphosphorous triamide, and dichlorophosphoranetriyltris[dimethylamine] are suspected carcinogens. The reaction should be conducted in a well-ventilated hood, and gloves should be worn at all times. The synthesis must be carried out in the absence of any moisture. Airless Schlenk ware and two-ended filters with side-arm gas inlets for filtration are recommended.*[5] *Solvents should be dried rigorously.*

After flushing with nitrogen, 500 mL of dry benzene (heated at reflux over sodium) and 20.8 g (50 mmol) of triphenyl(trichloromethyl)phosphonium chloride (Part A) are added through a Schlenk tube to a 1000-mL round-bottomed flask equipped with a side-arm gas inlet and a magnetic stirring bar. The suspension is treated with 8.15 g (8.2 mL, 50 mmol) of hexamethylphosphorous triamide[6] [Aldrich] and is stirred for 4 hr at room temperature. The intensely colored yellow-orange solution is separated from the pale brown precipitate, dichlorophosphoranetriyltris[dimethylamine], by filtration through a Schlenk filter tube. The product separates when the filtrate is evaporated under vacuum to a volume of about 50 mL. The vacuum pump is protected by a Dry Ice/ethanol–cooled trap, and the solvent is collected in a trap cooled similarly. (■ **Caution.** *A water aspirator must not be used!*) The yellow solid is collected by filtration through a Schlenk filter tube, washed with 5 mL of benzene and 5 mL diethyl ether, and dried under vacuum. Yield: 14.1 g (82%).

Anal. Calcd. for $C_{19}H_{15}Cl_2P$: C, 66.11; H, 4.38; Cl, 20.54; P, 8.97. Found: C, 67.23; H, 4.48; Cl, 19.22; P, 9.03.

Properties

(Dichloromethylene)triphenylphosphorane, $Ph_3P{=}CCl_2$, is a water-sensitive, bright-yellow crystalline solid that melts at 112° with decomposition. The IR spectrum (KBr disks) has bands at 3055 (w), 1590 (w), 1485 (w), 1440 (s), 1310 (vw), 1190 (s), 1120 (s), 1090 (m), 1072 (w), 1028 (vw), 994 (m), 923 (vw), 860 (vw), 765 (ms), 721 (s), 697 (s), 540 (s), 520 (m), 507 (m), 465 (w), and 292 (w) cm^{-1}. $^{31}P\{^1H\}$ NMR (C_6D_6): δ 18.6 (s) relative to H_3PO_4.

References

1. R. Appel, *Angew. Chem. Int. Ed. Engl.*, **14,** 801 (1975).
2. A. J. Speziale, K. W. Ratts, and D. E. Bissing, *Org. Synth.*, **45,** 33 (1965).
3. A. J. Speziale and K. W. Ratts, *J. Am. Chem. Soc.*, **84,** 854 (1962).
4. R. Appel and H. Veltmann, *Tetrahedron Lett.*, **1977,** 399.
5. D. F. Shriver, *The Manipulation of Air-Sensitive Compounds,* McGraw-Hill, New York, 1969.
6. A. B. Burg and P. J. Slotta, *J. Am. Chem. Soc.*, **80,** 1107 (1958).

35. [PHENYL(TRIMETHYLSILYL)METHYLENE]-PHOSPHINOUS CHLORIDE [CHLORO-[PHENYL(TRIMETHYLSILYL)METHYLENE]PHOSPHINE]

Submitted by ROLF APPEL*
Checked by LI-BEI-LI† and ROBERT H. NEILSON†

[Phenyl(trimethylsilyl)methylene]phosphinous chloride is a compound that is of high value for the syntheses of other phosphaalkenes.[1] Because of its sensitivity toward moisture and oxygen, all reactions are carried out under argon and with dry solvents.

A. *N,N,N′,N′*-TETRAMETHYL-*P*-[PHENYL-(TRIMETHYLSILYL)METHYL]PHOSPHONOUS DIAMIDE

$$PhCH_2Si(CH_3)_3 + nBuLi \xrightarrow{TMEDA} PhCH(Li)Si(CH_3)_3{\cdot}TMEDA + nBuH$$

$$PhCH(Li)Si(CH_3)_3{\cdot}TMEDA + [(CH_3)_2N]_2PCl \rightarrow PhCH[Si(CH_3)_3]P[N(CH_3)_3]_2 + LiCl/TMEDA$$

$$TMEDA = [(CH_3)_2NCH_2]_2$$

Procedure

A solution of 100 g (0.61 mole) of trimethyl(phenylmethyl)silane[2] [Pfalz and Bauer] and 87.5 mL (0.61 mole) of *N,N,N′,N′*-tetramethyl-1,2-ethanediamine (TMEDA) [Aldrich] dissolved in 150 mL of diethyl ether is placed in a 1-L three-necked flask provided with a dropping funnel, a mercury relief valve, and a magnetic stirrer. This solution is cooled to −20°, and 364.3 mL (0.61 mole) of a 1.6 *M* solution of butyllithium in hexane is added over a 90-min period. After being warmed to room temperature, the reaction mixture is stirred for at least 6.5 hr while the orange-yellow lithium complex salt precipitates. The mixture is cooled to −20°, and a solution of 92.7 g (0.61 mole) of *N,N,N′,N′*-tetramethylphosphorodiamidous chloride[3] [Alfa Products] in 80 mL of diethyl ether is added over a period of 90 min. This mixture is allowed to warm to ambient temperature and is stirred for 1 hr. Then the LiCl is filtered and washed

*Anorganisch-Chemisches Institut der Universität, Gerhard-Domagk-Strasse 1, 53 Bonn 1, West Germany.

†Department of Chemistry, Texas Christian University, Fort Worth, TX 76129.

three times with 50-mL portions of anhydrous diethyl ether. The solvent and washings are distilled off, and the product is fractionated through a 30-cm Vigreux column. During distillation, only mild cooling (18–25°) is used because the product may crystallize in the condenser. Yield: 118–127 g (70–75%) bp 103–107° at 0.3 torr. The ^{31}P {^{1}H} NMR shows a singlet at δ 96.2 ppm.

Anal. Calcd. for $C_{14}H_{27}N_2PSi$: C, 39.38; H, 9.56; N, 9.92; P, 10.97; Si, 9.95. Found: C, 39.52; H, 9.51; N, 9.71; P, 10.98; Si, 9.71.

B. [PHENYL(TRIMETHYLSILYL)METHYLENE]PHOSPHINOUS CHLORIDE

$$PhCH[Si(CH_3)_3]P[N(CH_3)_2]_2 + 4HCl \rightarrow PhCH[Si(CH_3)_3]PCl_2 + 2(CH_3)_2NH{\cdot}HCl$$

$$PhCH[Si(CH_3)_3]PCl_2 + DBO \rightarrow Ph[Si(CH_3)_3]C{=}PCl + DBO{\cdot}HCl$$

$$DBO = N(CH_2CH_2)_3N$$

Procedure

In a 2-L three-necked flask, provided with a mechanical stirrer and a mercury relief valve, 120 g (0.425 mole) of *N,N,N′,N′*-tetramethyl-*P*-[phenyl(trimethylsilyl)methyl]phosphonous diamide is dissolved in 1–1.5 L of pentane. With vigorous stirring at 0°, hydrogen chloride is passed into the reaction mixture at a rate such that a small escape of gas through the relief-valve takes place.[4] (■ **Caution.** *The reaction must be carried out in a well-ventilated fume hood.*) The hydrogen chloride is dried by passing through a 70-cm drying tube filled with P_4O_{10}. The reaction is completed when the amine hydrochloride separates at the bottom of the flask. Then the mixture is filtered into a 4-L three-necked flask, and the amine hydrochloride is washed three times with 50-mL portions of pentane. The solvent and washings are removed under vacuum at ambient temperature. The yield of $PhCH[Si(CH_3)_3]PCl_2$ is 101–107 g (90–95%); ^{31}P {^{1}H} NMR: δ 189.2 ppm (s). This crude product can be used without further purification.

To a stirred solution of 105 g (0.396 mole) of [phenyl(trimethylsilyl)methyl]phosphonous dichloride in 2–2.5 L of diethyl ether in a 4-L flask, 178 g (1.584 moles) of 1,4-diazabicyclo[2.2.2]octane (DBO) is added. The mixture is stirred overnight at ambient temperature and then filtered. The solid is washed three times with 100-mL portions of pentane. The collected filtrates are concentrated to 1.5 L at ambient temperature under reduced pressure, and the base and its hydrochloride are separated by filtration. The solid is washed three times with 50-mL portions of pentane. This procedure (concentration and filtration of the precipitate) is repeated twice. After that, all of the solvent is removed, and the resulting red oil is treated with 150 mL of pentane. This solution is cooled to −30° for 12 hr. After the precipitate is filtered and the solid is washed three times with 20-mL portions of pentane, the solvent and washings are removed under reduced pressure. The resulting dark red oil still contains DBO and polymers. The yield of the crude product is 74.6 g (82.4%). The distillation is carried out with maximum amounts of 35 g of crude product in a short-path, large-diameter distillation apparatus as shown in Fig. 1. After evacuation with a mercury diffusion pump, the crude product is stirred for 1 hr

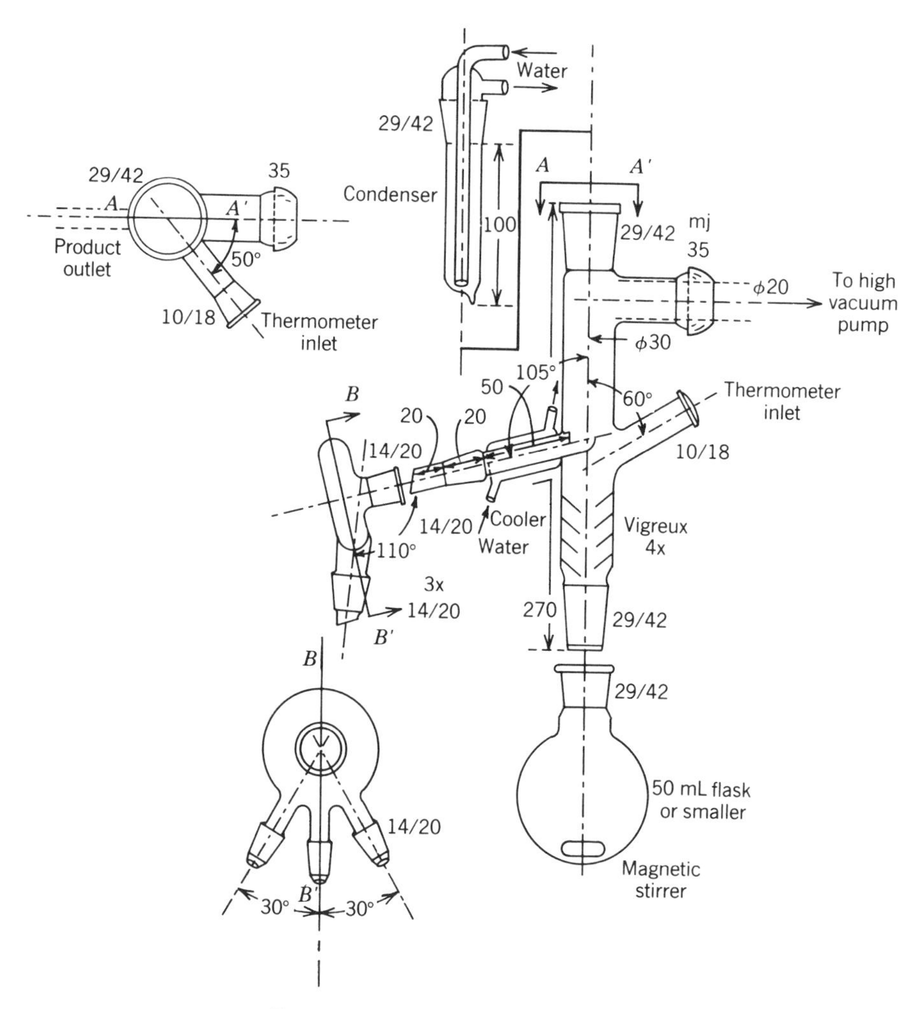

Fig. 1. Short-path distillation apparatus.

at room temperature without cooling the condenser in order to sublime the dissolved DBO into the cooling trap, which is between the distillation apparatus and the vacuum pump. During the distillation, the bath temperature is raised to 50° before the condenser is cooled, and then to 64°. The product distills at 46° as a slightly yellow oil. Overheating results in decomposition of the compound. Yield: 18.5 g (53% relative to 35 g crude $Ph[Si(CH_3)_3]C{=}PCl$; 42.9% relative to $PhCH[Si(CH_3)_3]PCl_2$).

Anal. Calcd. for $C_{10}H_{14}ClPSi$: C, 52.53, H, 6.12, Cl, 15.50, P, 13.54, Si, 12.28. Found, C, 52.86, H, 6.28, Cl, 15.76, P, 12.98, Si, 12.84.

Properties

Phenyl[(trimethylsilyl)methylene]phosphinous chloride is a yellow oil that solidifies at $\sim -10°$ and boils at 46° (10^{-4} torr). The proton-decoupled phosphorus NMR spectrum shows a singlet at 273.4 ppm.

In analogy to the chlorophosphines, chloro(methylene)phosphine undergoes substitution reactions via HCl elimination that result in P-N, P-O, P-S, and P-P linkages.[5] Additionally, a scrambling reaction takes place with aminophosphines similar to the scrambling reactions of the chlorophosphines in which the phosphorus has coordination number 3.[3,5]

References

1. R. Appel and A. Westerhaus, *Angew. Chem.*, **93,** 215 (1981).
2. H. Gilmann, *J. Am. Chem. Soc.*, **71,** 2066 (1949).
3. H. Nöth, *Chem. Ber.*, **96,** 1109 (1963).
4. A. Burg, *J. Am. Chem. Soc.*, **80,** 1107 (1958).
5. R. Appel and U. Kündgen, *Angew Chem.*, **94,** 227 (1982).

36. METHANETETRAYLBIS(PHOSPHORANES) (CARBODIPHOSPHORANES)

Submitted by ROLF APPEL*
Checked by JOSEPH G. MORSE†

Twenty years after the discovery of the methanetetraylbis(phosphoranes) (carbodiphosphoranes),[1] preparative chemists have an interest in this class of compounds because of its potential as a precursor to other materials.[2] The first of

*Anorganisch-Chemisches Institut der Universität, Gerhard-Domagk-Strasse 1, 53 Bonn, West Germany.
†Department of Chemistry and Biochemistry, Utah State University, Logan, UT 84322.

these was the hexaphenyl compound,[1] for which we have developed a simpler synthesis.[3] Methanetetraylbis(triphenylphosphorane) shows typical phosphorus-ylide reactions with organic compounds.[4] The ambident ligand properties of this ylide make it a valuable reagent for complex formation.[5–7]

A. *P,P'*-METHANETETRAYLBIS[*N,N,N',N',N'',N''*-HEXAMETHYLPHOSPHORANETRIAMINE] [HEXAKIS(DIMETHYLAMINO)CARBODIPHOSPHORANE]

$$[(CH_3)_2N]_2PCH_2P[N(CH_3)_2]_2 + 2CCl_4 + 3(CH_3)_2NH \rightarrow [(CH_3)_2N]_3P{=}CHP[N(CH_3)_2]_3{}^+Cl^- + 2CHCl_3 + [(CH_3)_2NH_2]^+Cl^-$$

$$[(CH_3)_2N]_3P{=}CHP[N(CH_3)_2]_3{}^+Cl^- + NaH \rightarrow [(CH_3)_2N]_3P{=}C{=}P[N(CH_3)_2]_3 + H_2\uparrow + NaCl$$

This molecule, in contrast to other carbodiphosphoranes for which X-ray structural measurements have been made, has a PCP angle of exactly 180°.[8–12]

Procedure

■ **Caution.** *Carbon tetrachloride is a suspected carcinogen. All reactions should be carried out in a well-ventilated hood, and because of the sensitivity of the reactants and products to oxygen and moisture, they should be protected with anhydrous argon or nitrogen.*

To a 250-mL two-necked round-bottomed flask equipped with an argon inlet, a magnetic stirring bar, and a mercury relief valve that contains a solution of 4.5 g (17.9 mmol) of *P,P'*-methylenebis[*N,N,N',N'*-tetramethylphosphonous diamide][13] ($[(CH_3)_2N]_2PCH_2P[N(CH_3)_2]_2$) in 100 mL of tetrahydrofuran (THF) at 0° is added 2.4 g (53.7 mmol) of dimethylamine and 5.5 g (35.6 mmol) of carbon tetrachloride. After stirring for 12 hr at 0°, the reaction mixture is allowed to warm to ambient temperature over several hours. The mixture is filtered to remove $[(CH_3)_2NH_2]^+Cl^-$. The volume of solvent is reduced under vacuum until a precipitate forms. This solid is filtered and dried at 40° (0.3 torr). The yield of tris(dimethylamino)[[tris(dimethylamino)phosphoranylidene]methyl]phosphonium chloride after recrystallization from THF is 3.9 g (58%),‡ mp177°; $^{31}P\{^1H\}$ NMR ($CDCl_3$): δ 54.2.

Anal. Calcd. for $C_{13}H_{37}ClN_6P_2$: C, 41.65; H, 9.95; Cl, 9.46; N, 22.42; P, 16.52. Found: C, 41.59; H, 9.91; Cl, 9.76; N, 22.27; P, 16.8.

‡The checker reports that an oil separates as the solvent is evaporated. Upon chilling overnight, crystals form; these are filtered and recrystallized from tetrahydrofuran/hexane.

To a solution of 8.2 g (22 mmol) of the phosphonium salt dissolved in 100 mL of THF contained in a 250-mL flask equipped as above is added 2.4 g (100 mmol) of sodium hydride.

■ **Caution.** *Sodium hydride is very reactive with easily reducible substances. Hydrogen is a product of the reaction, and a well-ventilated hood must be used. Flames should be kept away.*

The mixture is stirred at ambient temperature until the generation of hydrogen can no longer be observed (~2 days). The sodium chloride and unreacted sodium hydride are removed by filtration, the solvent is removed under vacuum, and the residue is distilled under high vacuum using the short-path distillation apparatus described in this volume.[14] The yield of *P,P'*-methanetetraylbis [*N,N,N',N',N'',N''*-hexamethylphosphoranetriamine] $[[(CH_3)_2N]_3P{=}C{=}P[N(CH_3)_2]_3]$ is 6.4 g (86%),* bp 87° (5×10^{-3} torr). It is a colorless liquid that crystallizes to a solid that melts at 51°; $^{31}P\{^1H\}$ NMR (d-THF): δ 27.7.

Anal. Calcd. for $C_{13}H_{36}N_6P_2$: C, 46.14; H, 10.72; N, 24.83; P, 18.31. Found: C, 45.81; H, 10.67; N, 24.62; P, 18.3.

Properties

P,P'-Methanetetraylbis[*N,N,N',N',N'',N''*-hexamethylphosphoranetriamine] is a colorless water-sensitive solid. In the mass spectrum, a molecular ion at 338 and peaks at 294 $[M\text{-}N(CH_3)_2]^+$ and 250 $[M\text{-}2N(CH_3)_2]^+$ are observed.

B. METHANETETRAYLBIS[TRIPHENYLPHOSPHORANE]

$$3Ph_3P + CCl_4 \rightarrow [Ph_3P{=}C(Cl)PPh_3]^+Cl^- + Ph_3PCl_2$$

$$[Ph_3P{=}C(Cl)PPh_3]^+Cl^- + P[N(CH_3)_2]_3 \rightarrow Ph_3P{=}C{=}PPh_3 + [(CH_3)_2N]_3PCl_2$$

Procedure

■ **Caution.** *Carbon tetrachloride is a known carcinogen. The reactions should be carried out in a well-ventilated hood. Because of the sensitivity of the compounds toward moisture and oxygen, all reactants must be protected by nitrogen or argon, and all solvents must be anhydrous.*

Into a 500-mL three-necked round-bottomed flask equipped with an argon inlet, a magnetic stirring bar, a thermometer, and a mercury relief valve is placed

*The checker finds that upon distillation the product crystallizes directly on the cold finger condenser (sublimation apparatus) and can be removed by scraping under anhydrous conditions.

78.6 g (0.3 mole) of Ph_3P [Strem Chemical] dissolved in 150 mL of CH_2Cl_2. Carbon tetrachloride (31.0 g; 0.2 mole) is added, and the mixture is stirred at ambient temperature. After a few minutes, the solution becomes yellow-brown; after 60 min it is warm and dark brown. Crystalline Ph_3PCl_2 precipitates after 4 hr of stirring. When the mixture has been stirred for an additional 2 hr (7 hr total), 15.0 g (0.2 mole) of 1,2-epoxybutane [Aldrich Chemical] is added slowly while the temperature is not allowed to exceed 40°. (■ **Caution.** *1,2-Epoxybutane is a moisture-sensitive, flammable liquid and must be handled in an inert atmosphere.*) Approximately 80 mL of diethyl ether is added dropwise until a slight turbidity appears. After stirring for another 20 min, the precipitation is complete. Then the mixture is filtered. The solid residue is washed again with CH_2Cl_2/Et_2O. The white solid is dried under vacuum at 80–100°. The yield of [chloro(triphenylphosphoranylidene)methyl]triphenylphosphonium chloride, $[Ph_3P{=}C(Cl)PPh_3)^+Cl^-$, is 45 g (74%)*; mp 258–260°.[15]

Anal. Calcd. for $C_{37}H_{30}Cl_2P_2$: C, 73.15; H, 4.94; P, 10.21; Cl^-, 5.85; total Cl, 11.70. Found: C, 73.30; H, 4.87; P, 10.27; Cl^-, 5.88; total Cl, 11.77.

To 150 mL of benzene in which 40.0 g (0.066 mole of $[Ph_3P{=}C(Cl)PPh_3]^+Cl^-$ is suspended is added 10.8 g (0.066 mole) of hexamethylphosphorous triamide [Aldrich]. After 24 hr the mixture is heated to boiling as rapidly as possible in a preheated oil bath (~100°) and is then immediately filtered through a glass frit. The filtrate is cooled to 25°, and the precipitate that forms on cooling is separated by filtration. The solid is washed twice with 10-mL portions of diethyl ether. The methanetetraylbis[triphenylphosphorane], $Ph_3P{=}C{=}PPh_3$, is dried at 25° under reduced pressure (0.3 torr). The yield is 21.8 g (61%),† mp 213–215°.

Anal. Calcd. for $C_{37}H_{30}P_2$: C, 82.8; H, 5.6; P, 11.5. Found: C, 83.3; H, 5.8; P, 11.6.

Properties

Methanetetraylbis[triphenylphosphorane] is a yellow crystalline solid that is stable for long periods in an anhydrous environment at 25°. In the IR spectrum (Nujol mull), a strong band at 1315 cm^{-1} with a shoulder at 1282 cm^{-1} is a most characteristic feature. In cyclohexane it absorbs in the 275–379-mμ region with λ_{max} 325 mμ (ϵ 0.7 × 10^4), also at λ_{max} 258 mμ (ϵ 0.6 × 10^4) and λ_{max} 225mμ (ϵ 3 × 10^4).[1]

*The checker used approximately 1/10 the quantity of reactants and found that turbidity appeared only when the diethyl ether/CH_2Cl_2 ratio was ~3:4. Crystallization required refrigeration overnight with 85% yield being obtained.

†The checker obtained a 20% yield, which may arise from the smaller scale reaction. In a second attempt, he preheated the filter before filtering the hot solution, but only a small improvement in yield was noted.

References

1. R. Ramirez, H. B. Desai, B. Hansen, and N. McKelvie, *J. Am. Chem. Soc.*, **83**, 3539 (1961).
2. H. J. Bestmann, *Angew. Chem.*, **89**, 361 (1977).
3. R. Appel, F. Knoll, H. Schöler, and H. D. Wihler, *Angew. Chem.*, **88**, 769 (1976).
4. C. N. Matthews and G. H. Birum, *Tetrahedron Lett.*, **46**, 5707 (1966); *J. Am. Chem. Soc.*, **90**, 3842 (1968); *Chem. Commun.*, **1966**, 267.
5. C. N. Matthews and G. H. Birum, *Acc. Chem. Res.*, **2**, 373 (1969).
6. W. C. Kaska, D. K. Mitchell, R. F. Reichelderfer, and W. D. Konte, *J. Am. Chem. Soc.*, **96**, 2847 (1974).
7. W. C. Kaska, D. K. Mitchell, and R. F. Reichelderfer, *J. Organometal. Chem.*, **47**, 391 (1973).
8. A. T. Vincent and P. J. Wheatley, *J. Chem. Soc., Dalton Trans.*, **1972**, 617.
9. G. E. Hardy, J. I. Zink, W. C. Kaska, and J. C. Baldwin, *J. Am. Chem. Soc.*, **100**, 8001 (1978).
10. E. Ebsworth, T. Fraser, D. Rankin, O. Gasser, and H. Schmidbaur, *Chem. Ber.*, **110**, 3508 (1977).
11. U. Schubert, C. Kappenstein, B. Milewski-Mahrla, and H. Schmidbaur, *Chem. Ber.*, **114**, 3070 (1981).
12. R. Appel, U. Baumeister, and F. Knoch, *Chem. Ber.*, **116**, 2275 (1983).
13. Z. S. Novikova, A. A. Prishchenko, and I. F. Lutsenko, *Zh. Obshch. Khim.*, **47**, 775 (1977). *J. Gen. Chem.* (Russ.) *Eng. Trans.*, **47**, 707 (1977).
14. R. Appel, *Inorg. Synth.*, **24**, 110 (1986).
15. R. Appel, F. Knoll, W. Michel, W. Morbach, H.-D. Wihler, and H. Veltmann, *Chem. Ber.*, **109**, 58 (1976).

37. BIS[BIS(TRIMETHYLSILYL)METHYLENE]-CHLOROPHOSPHORANE

Submitted by ROLF APPEL*
Checked by ROBERT T. PAINE†

The interaction of phosphorus trichloride with $LiCCl[Si(CH_3)_3]_2$ yields $[(CH_3)_3Si]_2C{=}PC(Cl)[Si(CH_3)_3]_2$, which rearranges by thermal chlorine migration to the title compound. Bis[bis(trimethylsilyl)methylene]chlorophosphorane, when decomposed at high temperature, is a useful precursor to a longer lived phosphoalkyne, $P{\equiv}CSi(CH_3)_3$ ($t_{1/2} \simeq 50$ min).

*Anorganisch-Chemisches Institut der Universität, Gerhard-Domagk-Strasse 1, 53 Bonn, West Germany.
†Department of Chemistry, University of New Mexico, Albuquerque, NM 87131.

A. (DICHLOROMETHYLENE)BIS[TRIMETHYLSILANE]

$$CH_2Cl_2 + 2(CH_3)_3SiCl \xrightarrow{2BuLi} Cl_2C[Si(CH_3)_3]_2 + 2LiCl + 2C_4H_{10}$$

Procedure

■ **Caution.** *Butyllithium reacts violently with water. It should be handled under an atmosphere of nitrogen or argon, and all solvents should be rigorously anhydrous.*

A rigorously dried 2-L three-necked round-bottomed flask equipped with a mechanical stirrer, a thermometer, and a 1-L jacketed pressure-equalizing addition funnel is charged with 51 g (0.6 mole, 38.5 mL) of dichloromethane, 800 mL of tetrahydrofuran, 80 mL of pentane, 80 mL of diethyl ether, and 160 g (1.47 moles, 187 mL) of freshly distilled chlorotrimethylsilane [Aldrich Chemical]. The vessel is immersed in an ethanol/liquid nitrogen slush bath (−118°). After the vessel is flushed with nitrogen or argon, 1.2 moles of butyllithium dissolved in 750 mL of hexane is transferred to the addition funnel, the apparatus is closed with a mercury relief valve, and the butyllithium solution is cooled with a cryostat to −70°. It is then added dropwise to the reaction vessel with vigorous stirring while the temperature inside the vessel is maintained at $< -105°$. This procedure requires ~4.5 hr. When addition is complete, the reaction mixture is allowed to warm to 25° overnight and the addition funnel is replaced by a distillation bridge. Solvents and excess chlorotrimethylsilane are removed by distillation until the temperature reaches 70°. The lithium chloride is removed by filtration through a fritted filter funnel. The filtrate is collected in a 250-mL round-bottomed flask. Distillation of the remaining liquid by means of a 30-cm Vigreux column yields 80 g (58%) of $Cl_2C[Si(CH_3)_3]_2$, bp 204° (62–65°/0.1 torr).

Anal. Calcd. for $C_7H_{18}Cl_2Si_2$: C, 36.7; H, 7.9; Si, 24.5. Found: C, 37.0; H, 8.2; Si, 24.6.

Properties

(Dichloromethylene)bis[trimethylsilane] is a colorless water-sensitive liquid. ^1H NMR (CCl_4): δ 0.29 and IR spectrum (neat): 1266 (s), 878 (s), 851 (s), 810 (s), 768, 747, 703 (s), 649, and 632 cm^{-1}.

B. [BIS(TRIMETHYLSILYL)METHYLENE][CHLOROBIS-(TRIMETHYLSILYL)METHYL]PHOSPHINE

$$Cl_2C[Si(CH_3)_3]_2 + BuLi \rightarrow LiC(Cl)[Si(CH_3)_3]_2 + BuCl$$

$$3LiC(Cl)[Si(CH_3)_3]_2 + PCl_3 \rightarrow [(CH_3)_3Si]_2C{=}PC(Cl)[Si(CH_3)_3]_2 + 3LiCl + Cl_2C[Si(CH_3)_3]_2$$

Procedure

■ **Caution.** *Butyllithium reacts violently with water. All manipulations should be carried out under anhydrous argon or nitrogen and with anhydrous solvents.*

Into a 1-L three-necked flask equipped with a pressure-equalizing dropping funnel, a mechanical stirrer, a thermometer, and a mercury relief valve are placed 500 mL of tetrahydrofuran, 50 mL of pentane, 50 mL of diethyl ether, and 40 g (0.175 mole) of $Cl_2C[Si(CH_3)_3]_2$. While being flushed with dry nitrogen or argon, the flask and its contents are cooled to and maintained at −105° (ethanol/liquid nitrogen slush). Over a period of 1 hr, 110 mL (0.175 mole) of 1.6 *M* solution of *n*-BuLi in hexane [Aldrich] is added from the dropping funnel. After being stirred at −105° for 4.5 hr, a solution of 9 g (0.065 mole, 10% excess) of PCl_3 dissolved in 25 mL of ether is added over a period of 10 min. (■ **Caution.** *The internal temperature must not exceed −90°.*) The course of the reaction can be followed by the change in color of the solution from yellow to dark green. After warming to 25° overnight, the solvent and the excess PCl_3 are removed under vacuum. The residue is mixed with 50 mL of pentane, and the LiCl is removed by filtering through a fritted funnel. The pentane is removed under vacuum, and the product is distilled through a short-path distillation apparatus.[2] The distillation must be carried out at <120°, above which rearrangement and decomposition begin to occur. The yield of $[(CH_3)_3Si]_2C{=}PC(Cl)[Si(CH_3)_3]_2$ is 12 g (54% based on PCl_3).[1]

Properties

[Bis(trimethylsilyl)methylene][chlorobis(trimethylsilyl)methyl]phosphine is a yellow oil (bp 86°/10^{-3} torr) that crystallizes to a yellow solid (mp 48–50°). $^{31}P\{^1H\}$ NMR δ 173.6 (s).

C. BIS[BIS(TRIMETHYLSILYL)METHYLENE]-CHLOROPHOSPHORANE

$$[(CH_3)_3Si]_2C{=}PC(Cl)[Si(CH_3)_3]_2 \xrightarrow{135^\circ} [(CH_3)_3Si]_2C{=}\overset{\displaystyle Cl}{\overset{|}{P}}{=}C[Si(CH_3)_3]_2$$

Procedure

The thermolysis is carried out under anhydrous conditions. The yellow solid $[(CH_3)_3Si]_2C{=}PC(Cl)[Si(CH_3)_3]_2$ (12 g, 0.031 mole) is heated under reflux in *o*-xylol for 1 hr or neat for 5 hr. The temperature must not exceed 135°, because at higher temperatures dimerization occurs by elimination of chlorosilane. The brown-red product is distilled through a short-path distillation apparatus.[2] The yield of $[(CH_3)_3Si]_2C{=}P(Cl){=}C[Si(CH_3)_3]_2$ is 7 g (58%).[1]

Properties

Bis[bis(trimethylsilyl)methylene]chlorophosphorane is an orange-red oil (bp 56°/10^{-3} torr) that crystallizes to a solid of the same color (mp 58°). $^{31}P\ \{^1H\}$ NMR: δ 136.6. $^{13}C\ \{^1H\}$ NMR: δ 83.3 (d, $J_{P=C}$ = 38.6 Hz).

References

1. R. Appel and A. Westerhaus, *Tetrahedron Lett.*, **23,** 2017 (1982).
2. R. Appel, *Inorg. Synth.*, **24,** 110 (1986).

38. *N,N'*-DIMETHYL-*N,N'*-BIS(TRIMETHYLSILYL)UREA

$$MeN(SiMe_3)_2 + MeNCO \rightarrow [Me_3SiN(Me)]_2CO$$

Submitted by H. W. ROESKY* and J. LUCAS*
Checked by ARLAN D. NORMAN†

N,N'-Dimethyl-*N,N'*-bis(trimethylsilyl)urea[1] is a useful precursor for synthesizing heterocyclic compounds containing the —(Me)NC(O)N(Me)— moiety. The reaction of the disilylated urea with $S_3N_3Cl_3$ results in a bicyclic SN

*Institut für Anorganische Chemie der Universität Göttingen, Tammannstrasse 4, D-3400 Göttingen, West Germany.

†Department of Chemistry, University of Colorado, Boulder, CO 80309.

compound that contains a carbon atom in the ring skeleton.[2] With arsenic trichloride, an eight-membered ring containing As, N, and C atoms is formed.[3] These two examples give an idea of the scope of this reagent and the easy cleavage of the nitrogen-silicon bond.

Procedure

■ **Caution.** *Methyl isocyanate is malodorous and toxic. Avoid skin contact and inhalation. All procedures should be carried out in a well-ventilated hood. Due to the moisture sensitivity of the starting materials and the product, all substance transfers must be carried out under an atmosphere of nitrogen.*

The reaction is carried out in a 100-mL two-necked flask equipped with a magnetic stirring bar and a reflux condenser topped with a nitrogen bypass. The apparatus is purged with nitrogen. Twenty grams (114 mmol) of heptamethyldisilazane[4] and 6.5 g (114 mmol) of freshly distilled methyl isocyanate [Aldrich] are transferred to the flask with a syringe. The reaction mixture is stirred for 24 hr at 35°. Distillation of the moderately viscous colorless liquid at 55–58° (0.1 torr) yields 18.6 g (70%) of $[Me_3SiN(Me)]_2CO$. The pure product solidifies easily in the condenser, so that occasional warming is necessary.

Anal. Calcd. for $C_9H_{24}N_2OSi_2$: C, 46.5; H, 10.4; N, 12.0. Found: C, 46.6; H, 10.4; N, 12.0.

Properties

The silyl-substituted urea is a colorless crystalline solid (mp 22–23°, n_D^{20} 1.4523). It is moisture-sensitive. At −30° it can be stored in a sealed flask for several months. ^1H NMR (CH_2Cl_2): δ 2.75 (NCH_3) and δ 0.21 ($SiCH_3$). IR ν_{CO}: 1625 (s) cm^{-1} (Nujol mull).

References

1. J. F. Klebe, J. B. Bush, Jr., and J. E. Lyons, *J. Am. Chem. Soc.*, **86,** 4400 (1964).
2. H. W. Roesky, T. Müller, and E. Rodek, *J. Chem. Soc., Chem. Commun.*, **10,** 439 (1979).
3. W. S. Sheldrick, H. Zamankhan, and H. W. Roesky, *Chem. Ber.*, **113,** 3821 (1980).
4. R. C. Osthoff and S. W. Kantor, *Inorg. Synth.*, **5,** 55 (1957).

39. 1,3,5,7-TETRAMETHYL-1*H*,5*H*-[1,4,2,3]DIAZADIPHOSPHOLO-[2,3-*b*][1,4,2,3]DIAZADIPHOSPHOLE-2,6(3*H*,7*H*)-DIONE AND A MOLYBDENUM COMPLEX

Submitted by H. W. ROESKY* and J. LUCAS*
Checked by ARLAN D. NORMAN†

Although the syntheses of compounds containing phosphorus-phosphorus bonds are well established,[1] the incorporation of this diatomic unit into polycyclic molecules is not easy to accomplish. The method presented here is a single-step reaction starting with silyl-substituted urea and phosphorus trichloride.[2] Variation of the urea substituents and the use of $RPCl_2$ instead of PCl_3 demonstrate the scope of this reaction, which leads to bicyclic $\lambda^3P—\lambda^5P$ as well as spirobicyclic $\lambda^3P—\lambda^5P$ molecules.[3,4] Reactions of the title compound with metal carbonyls do not result in an oxidative cleavage of the phosphorus-phosphorus bond, since the diphosphane acts as a monodentate and as a bridging ligand.[5]

A. 1,3,5,7-TETRAMETHYL-1*H*,5*H*-[1,4,2,3]DIAZADIPHOSPHOLO[2,3-*b*]-[1,4,2,3]DIAZADIPHOSPHOLE-2,6(3*H*,7*H*)-DIONE

$$[Me_3SiN(Me)]_2CO + 2PCl_3 \xrightarrow{CH_2Cl_2}$$

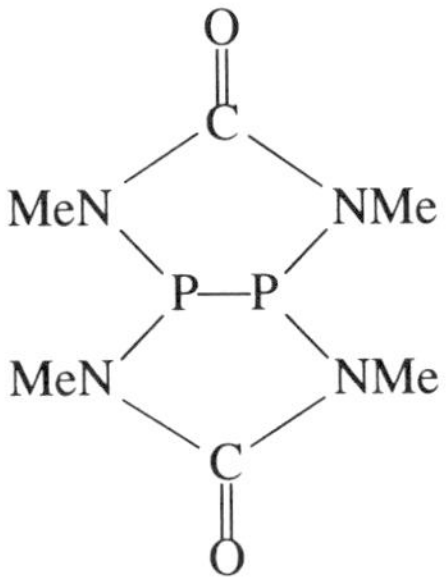

```
            O
            ‖
            C
          /   \
      MeN       NMe
          \   /
           P—P          + 4Me3SiCl + solid[6]
          /   \
      MeN       NMe
          \   /
            C
            ‖
            O
```

*Institut für Anorganische Chemie der Universität Göttingen, Tammannstrasse 4, D-3400 Göttingen, West Germany.
†Department of Chemistry, University of Colorado, Boulder, CO 80309.

Procedure

■ **Caution.** *Phosphorus trichloride is a highly corrosive and easily hydrolyzed compound. It should be handled in a dry nitrogen atmosphere. Gloves should be worn.*

The silyl-substituted urea $[Me_3SiN(Me)]_2CO$ is prepared according to the literature.[7,8] Because of the moisture sensitivity of the starting materials and the product, all compound transfers are carried out in an inert-atmosphere glove box.[9] Solvents are carefully dried. Dichloromethane and CCl_4 are distilled from P_4O_{10}, and tetrahydrofuran (THF) is distilled before use from sodium benzophenone ketyl solution. The reaction is carried out in a 250-mL two-necked flask equipped with a reflux condenser, a dropping funnel, and a nitrogen-inlet bubbler. While a slow stream of nitrogen is passed through the apparatus, a solution of 17.2 g (74 mmol) of $[Me_3SiN(Me)]_2CO$ in 60 mL of CH_2Cl_2 is added dropwise to a stirred solution of 10.1 g (74 mmol) of freshly distilled PCl_3, in 30 mL of CH_2Cl_2. To maintain a constant reaction temperature, the flask is immersed in an ambient temperature water bath during the reaction. After completion of the addition, the solvent Me_3SiCl and other volatile substances are evaporated and trapped at liquid nitrogen temperature by applying oil-pump vacuum and heating the flask at 40°. The sticky white residue is stirred in 50 mL of CCl_4, filtered, and washed three times with 5-mL portions of CCl_4. Heating this residue at 110° (0.01 torr) traps an oily product (bp 95°/0.2 torr), which has been shown to be a six-membered P,N,C heterocyclic compound.[6]

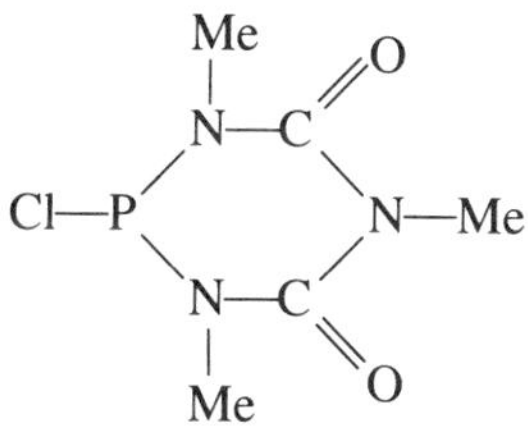

```
          Me
          |      O
          |    //
          N—C
        /     \
  Cl—P          N—Me
        \     /
          N—C
          |    \\
          |      O
          Me
```

At 140–150° (0.01 torr) the title compound sublimes. Further purification by resublimation at 180–190° (760 torr) results in 2.3 g (13%) of pure $P_2[MeNC(O)NMe]_2$.

Anal. Calcd. for $C_6H_{12}N_4O_2P_2$: C, 30.78; H, 5.16; N, 23.93; P, 26.46. Found: C, 30.7; H, 5.1; N, 23.9; P, 26.4.

Properties

The colorless crystalline compound is soluble in CH_2Cl_2 and tetrahydrofuran (THF) and only slightly soluble in aliphatic hydrocarbons. The moisture- and air-sensitive diphosphane can be stored for several weeks in a carefully sealed flask under an atmosphere of dry nitrogen. mp 174–176°; IR: ν_{CO} at 1685 (s) and 1660 (s) cm^{-1} (Nujol mull); ^1H NMR (CH_2Cl_2, 200 MHz): δ_{NCH_3} 3.40 (J_{PNCH} = 5 Hz); ^{31}P NMR (85% H_3PO_4, CH_2Cl_2, 36.43 MHz) δ 27.3. The mass spectrum (70 eV) exhibits a molecular ion (*m/e* 234, 30% relative intensity) and peaks corresponding to M − NCH_3CO (177, 30%), $P_2NCH_3CONCH_3$ (148, 10%), $PNCH_3CO$ (88, 20%), and $PNCH_3$ (60, 100%). The X-ray structure exhibits an "open-book" arrangement of two five-membered P_2N_2C rings joined along the P—P bond. Two molecules form centrosymmetric pairs around the four phosphorus atoms with short internuclear P· · ·P′ distances.

B. HEXACARBONYLTRIS[μ-1,3,5,7-TETRAMETHYL-1*H*,5*H*-[1,4,2,3]DIAZADIPHOSPHOLO[2,3-*b*]-[1,4,2,3]DIAZADIPHOSPHOLE-2,6(3*H*,7*H*)-DIONE]-DIMOLYBDENUM

$$2Mo(C_7H_8)(CO)_3 + 3P_2[MeNC(O)NMe]_2 \xrightarrow{THF} Mo_2(CO)_6\{P_2[MeNC(O)NMe]_2\}_3 + 2C_7H_8$$

Procedure

■ **Caution.** *Metal carbonyls are extremely hazardous and should be handled in a well-ventilated hood.*

In addition to the previously mentioned precautions, the solvents are carefully deoxygenated by refluxing in a stream of nitrogen for ~1 hr. A solution of 1.1 g (4.1 mmol) of $Mo(C_7H_8)(CO)_3$[10,11] in 20 mL of freshly distilled tetrahydrofuran (THF) is added slowly to 1.15 g (4.9 mmol) of $P_2[MeNC(O)NMe]_2$ in 30 mL of THF and is heated to 50° for 2 hr. The light yellow precipitate is filtered off and washed with 10 mL of THF. Recrystallization from CH_2Cl_2 results in 0.5 g of $Mo_2(CO)_6\{P_2[MeNC(O)NMe]_2\}_3$ (28%).

Anal. Calcd. for $C_{24}H_{36}Mo_2N_{12}O_{12}P_6$: C, 27.14; H, 3.39; N, 15.83; P, 17.5. Found: C, 27.7; H, 3.3; N, 16.2; P, 17.7.

Properties

The air- and moisture-sensitive compound can be stored for several weeks in a carefully sealed flask under an atmosphere of dry nitrogen. mp 289°; IR: ν_{CO} at 1980, 1940, 1690 cm^{-1} (Nujol mull); ^1H NMR (CH_2Cl_2, 60 MHz) δ_{NCH_3} 3.0, ^{31}P NMR (85% H_3PO_4, CH_2Cl_2, 36.43 MHz) δ 69.5. The X-ray structure determination indicates that two $(CO)_3Mo$ fragments are bridged by three diphosphane ligands in such a way that the molecule is of approximate D_3 symmetry along the Mo···Mo axis. The Mo···Mo distance of 509.7(1) pm is too long for any bonding interaction.

References

1. E. Fluck, *Prep. Inorg. React.*, **5,** 103 (1968).
2. H. W. Roesky, H. Zamankhan, W. S. Sheldrick, A. H. Cowley, and S. K. Mehrotra, *Inorg. Chem.*, **20,** 2810 (1981).
3. H. W. Roesky, K. Ambrosius, and W. S. Sheldrick, *Chem. Ber.*, **112,** 1365 (1979).
4. H. W. Roesky, K. Ambrosius, M. Banek, and W. S. Sheldrick, *Chem. Ber.*, **113,** 1847 (1980).
5. H. W. Roesky, D. Amirzadeh-Asl, and G. M. Sheldrick, *J. Chem. Soc. Dalton Trans.*, **1983,** 855.
6. G. Sidiropulos, Ph.D. thesis, Universität Frankfurt am Main, West Germany, 1978.
7. J. F. Klebe, J. B. Bush, Jr., and J. E. Lyons, *J. Am. Chem. Soc.*, **86,** 4400 (1961).
8. H. W. Roesky and J. Lucas, *Inorg. Synth.*, **24,** 120 (1986).
9. D. F. Shriver, *The Manipulation of Air-Sensitive Compounds,* McGraw-Hill, New York, 1969.
10. E. W. Abel, J. A. Bennett, R. Burton, and G. Wilkinson, *J. Chem. Soc.*, **1958,** 4559.
11. F. A. Cotton, J. A. McCleverty, and J. W. White, *Inorg. Synth.*, **8,** 121 (1967).

40. SULFUR DICYANIDE

$$SCl_2 + 2(CH_3)_3SiCN \xrightarrow[1\ hr]{0\text{–}25^\circ} S(CN)_2 + 2(CH_3)_3SiCl$$

Submitted by RAMESH C. KUMAR* and JEAN'NE M. SHREEVE*
Checked by JERRY FOROPOULOS, Jr.†

Sulfur dicyanide, $S(CN)_2$, was first synthesized in 1919.[1] It has since been studied vigorously. It is usually prepared by the reaction of silver cyanide dispersed in warm carbon disulfide with sulfur dichloride. However, the yield of $S(CN)_2$ when this method is employed has not been reported.[2] The compound is useful as a precursor to thiols.

The following procedure is a convenient and relatively inexpensive route to sulfur dicyanide by the reaction of sulfur dichloride and trimethylsilyl cyanide. No solvent is required, and thus recovery of $S(CN)_2$ in high yield is very straightforward.[3]

*Department of Chemistry, University of Idaho, Moscow, ID 83843.
†Chemistry Division, Argonne National Laboratory, 9700 S. Cass Ave., Argonne, IL 60439.

Procedure

■ **Caution.** *Sulfur dichloride is an evil-smelling, volatile material that should be handled with gloves and only in well-ventilated areas. Trimethylsilyl cyanide is readily hydrolyzed to form HCN. It should be kept free of moisture and handled only in a fume hood or vacuum line.*

Freshly distilled sulfur dichloride (0.612 g, 6 mmol) and trimethylsilyl cyanide [Aldrich] (1.19 g, 12 mmol) are condensed at − 196° into a 50-mL Pyrex round-bottomed flask that is equipped with a Kontes Teflon stopcock to which is attached a 10/30 inner standard taper joint. A Teflon-coated stirring bar is placed in the vessel. The reaction vessel is connected to the vacuum line by a 10/30 outer joint. The vacuum system is equipped with a Heise Bourdon tube gauge [Dresser Industries] and a Televac thermocouple gauge[4] [Fredericks]. The quantities of SCl_2 and $(CH_3)_3SiCN$ are measured in the vacuum line by means of PVT techniques, assuming ideal gas behavior.

The reactor is warmed to 0° while the mixture is stirred. Warming should continue so that over a 1-hr period the reaction mixture reaches 25°. The vessel is then cooled to −23° (50/50 ethanol water, v/v, and liquid N_2), and all substances that are volatile at this temperature should be transferred out under dynamic vacuum into a U-trap that can be removed subsequently to the fume hood. The contents of the U-trap are destroyed by hydrolysis in aqueous base. An off-white crystalline solid, $S(CN)_2$, remains in the reaction vessel. The pure white crystalline sulfur dicyanide is obtained by subliming under vacuum ($1–5 \times 10^{-3}$ torr) (0.42 g, 80–85% yield). Sulfur dicyanide is stored in a glass vessel equipped with a Teflon stopcock at 0–10°.

Anal. Calcd. for C_2N_2S: C, 28.57, S, 38.90. Found: C, 28.80, S, 38.90.

Properties

Sulfur dicyanide is a white crystalline solid that sublimes under vacuum at 25°. It is hygroscopic but can be stored under anhydrous conditions at 0° for long periods. It is not markedly stable at 25°. On standing at this temperature in the absence of air, it slowly forms a yellow material of unknown composition. It is moderately soluble in water, very soluble in ether, and soluble in warm carbon disulfide. The Raman spectrum of $S(CN)_2$ suggests a bent structure. The IR spectrum of the compound in a Nujol mull has absorption bands at 2190 (m) (C≡N) and 1090 (w) (S—C) cm^{-1}.

References

1. E. Söderbäck, *Ann.*, **419,** 217 (1919).
2. D. A. Long and D. Steele, *Spectrochim. Acta,* **19,** 1731 (1963).
3. R. C. Kumar and J. M. Shreeve, *Z. Naturforsch.*, **36b,** 1407 (1981).
4. F. Haspel-Hentrich and J. M. Shreeve, *Inorg. Synth.*, **24,** 58 (1986).

41. DISILYL SELENIDE (DISILASELENANE)

$$Li[AlH_4] + 4H_2Se \rightarrow Li[Al(SeH)_4]^* + 4H_2$$

$$Li[Al(SeH)_4] + 4H_3SiI \rightarrow 2(H_3Si)_2Se + 2H_2Se + LiI + AlI_3$$

Submitted by JOHN E. DRAKE† and BORIS M. GLAVINCEVSKI†
Checked by E. A. V. EBSWORTH‡ and S. G. HENDERSON‡

Disilyl selenide was first prepared in 1955 by the reaction of iodosilane and silver selenide.[1] More recent methods have involved the exchange of bromosilane with dilithium selenide[2] and the interaction of trisilylphosphine[3] or -amine,[4] or of SiH_4[5] with H_2Se. The former methods require initial syntheses of Ag_2Se and Li_2Se and not only are less convenient but also give lower yields than the interaction of iodosilane with $Li[Al(SeH)_4]$ described herein. This route is suitable for the preparation of the remaining hydromethyl-disilyl or -digermyl selenides and sulfides.[6]

Apparatus

The manipulation of all volatile material is carried out in a Pyrex glass vacuum system of conventional design.[7] The authors' system consists of two manifolds interconnected by four U-traps and a central manifold, which leads to two liquid nitrogen backing traps and mercury diffusion and rotary oil pumps. The vacuum in the system is monitored by a Pirani-type gauge fitted to the central manifold. Pressure readings in excess of about 1 torr are registered by mercury manometers. High-vacuum Teflon-in-glass valves and a silicone-type grease for ground glass joints are preferred because of the marked solubilities of the materials in hydrocarbon grease.

Starting Materials

Iodosilane§ is readily prepared by the reaction of phenylsilane [Petrach Systems] with hydrogen iodide[8,9] or by the reaction of silane [Matheson Gas Products] with hydrogen iodide in the presence of a catalytic amount of aluminum triiodide.[9,10] Hydrogen selenide is conveniently prepared by the hydrolysis of

*Authors do not know exact nature of this species.

†Department of Chemistry, University of Windsor, Windsor, Ontario, Canada N9B 3P4.
‡Department of Chemistry, University of Edinburgh, United Kingdom EH9 3JJ.

§The checkers have prepared iodosilane by the reaction of trisilylamine with hydrogen iodide.

Al_2Se_3.[11] Lithium tetrahydridoaluminate [Alfa Organics] Li[AlH_4], and dimethyl ether [Matheson] are used as supplied.

■ **Caution.** *The selenium compounds in this preparation are toxic and vile-smelling. The nauseating smell usually results from their dissolution in stopcock lubricants. To minimize this, manipulations should be performed under vacuum conditions using greaseless high-vacuum Teflon-in-glass valves and a minimum of silicone-type grease.*† *Rapid decomposition of disilyl selenide will occur on exposure to air or moisture.*

Procedure

The reaction vessel is a bulb (~60 mL) the neck of which is extended into a tube (~6 cm long and ~10 mm o.d.) terminating in a high-vacuum Teflon-in-glass valve and standard taper 18/9 glass joint for attachment to the vacuum line. The vessel is purged with dry nitrogen before the addition of fresh Li[AlH_4] (0.05 g, 1.32 mmol). The connector is packed with glass wool to prevent contamination of the vacuum line, and the system is thoroughly evacuated. Dimethyl ether (~10 mL) is then condensed in at −196° (liquid nitrogen), followed by an excess of hydrogen selenide, H_2Se (6.80 mmol). The reactants are allowed to react with cautious warming while the pressure of the volatile species is monitored.* A vigorous effervescence, with evolution of hydrogen, is apparent as soon as the two phases mix.

■ **Caution.** *It is advisable to incorporate a bubbler manometer into the vacuum line to offset any difficulties arising from sudden expansion of the volatile material.*

When the gas pressure increases too close to atmospheric, the vessel is cooled to −196°, the stopcock is closed, and the noncondensable gas, H_2, is pumped off from the manifold and manometer. After thorough evacuation, the stopcock to the pumping system is closed, the stopcock of the vessel is reopened, and the bath at −196° is removed, again allowing the volatile material to expand against the mercury manometer, with quenching as necessary. To slow the reaction, a −78° bath (Dry Ice/methanol slush) is more convenient than one at −196°. The entire process is repeated until no more hydrogen is liberated. In this manner, over a relatively short reaction time (~30 min), 5.22 mmol of hydrogen is evolved.‡ Iodosilane, H_3SiI (2.39 mmol) is then condensed at −196° into the reaction vessel and allowed to react with the excess selenoaluminate at −45°

†The checkers used conventionally greased fittings on a vacuum line and encountered no difficulties.

*The checkers advocate that the reaction between hydrogen selenide and lithium tetrahydridoaluminate in dimethyl ether is very rapid and care should be taken to avoid evolution of H_2 at too great a rate.

‡The checkers found that in a −78° bath this part took about 40 min.

(chlorobenzene/liquid nitrogen slush) for ~3–4 hr with occasional shaking. After this time the volatile material is fractionated through U-traps held at −78° and −196°. Pure $(H_3Si)_2Se$ (0.15 g, 1.09 mmol, 91%)* is retained in the former trap while Me_2O with traces of H_3SiSeH (identified by its ^1H NMR parameters[12]) is collected in the latter.

■ **Caution.** *The selenium residue in the reaction vessel should be handled in an efficient fume hood. Smell contamination may be considerably reduced by treatment of the residues with a strong bleach solution followed by an acid wash. These solutions should be disposed of in the normal way prescribed for toxic wastes.*

Properties

Disilyl selenide (mp 68.0 ± 0.2°, bp 85.2 ± 1°) is a colorless volatile liquid.[1] Its vapor pressure relationship is given in the range −43 to +74.5° by: $\log_{10} P = 7.894 - 1796/T$, which leads to a ΔH_{vap} of 8219 cal $mole^{-1}$ and a Trouton constant of 22.9.[1] The density at 20° is 1.36.[1] The ^1H NMR spectrum of disilyl selenide recorded in cyclohexane consists of a singlet (δ_{SiH}) at 4.12 ppm downfield of tetramethylsilane ($J_{HH'}$ 0.63 Hz, $J_{^{29}SiH}$ 225 ± 1 Hz).[5,12] The IR spectrum[13] shows prominent bands at 2185 (vs), 1872 (w), 1210 (w), 1000 (ms sh), 932 (vs) [895, 889, 884 (vs) A], 635 (wsh), 597 (ms) C, and 530 (wsh) cm^{-1}. The mass spectrum[6] contains a molecular ion at *m/e* 130–146 $(H_nSi_2Se)^+$. Disilyl selenide appears to be stable if kept in the refrigerator in a break-seal glass ampule. However, decomposition may occur after extended periods of time if it is left at room temperature.

References

1. H. J. Emeléus, A. G. MacDiarmid, and A. G. Maddock, *J. Inorg. Nucl. Chem.*, **1,** 194 (1955).
2. S. Cradock, E. A. V. Ebsworth, and D. W. H. Rankin, *J. Chem. Soc. Ser. A*, **1969,** 1628.
3. E. A. V. Ebsworth, C. Glidewell, and G. M. Sheldrick, *J. Chem. Soc. Ser. A*, **1969,** 352.
4. S. Cradock, E. A. V. Ebsworth, and H. F. Jessep, *J. Chem. Soc., Dalton Trans.*, **1972,** 359.
5. J. E. Drake and C. Riddle, *J. Chem. Soc. Ser. A*, **1970,** 3132.
6. B. M. Glavincevski, Ph.D. thesis, University of Windsor, 1978.
7. D. F. Shriver, *The Manipulation of Air-Sensitive Compounds*, McGraw-Hill, New York, 1969.
8. G. Fritz and D. Kummer, *Z. Anorg. Allgem. Chem.*, **304,** 322 (1960).
9. J. E. Drake, B. M. Glavincevski, R. T. Hemmings, and H. E. Henderson, *Inorg. Synth.*, **19,** 268 (1979).
10. H. J. Emeléus, A. G. Maddock, and C. Reid, *J. Chem. Soc.*, **1941,** 353.
11. G. R. Waitkins and R. Shutt, *Inorg. Synth.*, **2,** 183 (1946).
12. C. Glidewell, D. W. H. Rankin and G. M. Sheldrick, *Trans. Faraday Soc.*, **66,** 1409 (1969).
13. E. A. V. Ebsworth, R. Taylor, and L. A. Woodward, *Trans. Faraday Soc.*, **55,** 211 (1959).

*The checkers report a yield of 85%.

42. ORGANIC SUPERCONDUCTING SOLIDS*

Synthetic metals (synmetals) are of ever-increasing interest because of their novel physical properties such as anisotropic, and sometimes metallic, electrical conductivity. 4,4′,5,5′-Tetramethyl-2,2′-bi-1,3-diselenolylidene (tetramethyltetraselenafulvalene, TMTSF), first synthesized[1,2] in 1974, is such a material which, although it contains no metals, forms radical-cation derivatives that exhibit widely varying electrical properties including insulating [$(TMTSF)_2[SiF_6]$], semiconducting [$(TMTSF)_2[BrO_4]$], and superconducting [$(TMTSF)_2[ClO_4]$] behavior.[3] The 2:1 $(TMTSF)_2X$ salts are most prominent because when X = $[SbF_6]^-$, $[TaF_6]^-$, $[AsF_6]^-$, $[PF_6]^-$, and $[ReO_4]^-$ they become superconducting (SC) under modest (~8–12 kbar) pressure. In the case of X = $[ClO_4]^-$ the SC state is achieved at *ambient pressure* at ~1.2 K. Although $(TMTSF)_2[PF_6]$ was the first true organic superconductor to be discovered, the $[ClO_4]^-$ congener remains the only known ambient pressure organic superconductor in this class of materials.

Recently two more detailed but different procedures for synthesizing deuterated[4] and hydrogenated[5] TMTSF have become available. Certain steps in these procedures are difficult to perform and give low yields. Therefore, we have developed a synthesis that combines and elaborates on the two procedures and is easily performed with good yields.

Materials

The following common laboratory solvents are distilled before use as follows: chloroform (over alumina), diethyl ether (over $FeSO_4$), dichloromethane, benzene, heptane, and 1,1,2-trichloroethane. Methanol and ethyl acetate are used as obtained. Distilled water and aqua regia are used as described herein. It should be noted that *N*-(dichloromethylene)-*N*-methyl methanaminium chloride (phosgene iminium chloride) [Aldrich Chemical], trimethyl phosphite [Strem Chemicals] and hydrogen selenide [Scientific Gas Products] are all air- and/or moisture-sensitive and can be used only in a dry, inert atmosphere.

*Work performed under the auspices of the Office of Basic Energy Sciences, Division of Materials Sciences, of the U.S. Department of Energy under contract W-31-109-ENG-38.

A. SYNTHESIS OF 4,4′,5,5′-TETRAMETHYL-2,2′-BI-1,3-DISELENOLYLIDENE (TETRAMETHYLTETRASELENAFULVALENE, TMTSF)

$$Et_3N + H_2Se \rightarrow Et_3NH^+HSe$$

$$2Et_3NH^+HSe^- + (CH_3)_2\overset{+}{N}{=}CCl_2\ Cl^- \xrightarrow{2Et_3N} (CH_3)_2N{-}C({=}Se){-}Se^-Et_3NH^+ + 3Et_3NH^+Cl^-$$

$$(CH_3)_2N{-}C({=}Se){-}Se^-Et_3NH^+ + CH_3C({=}O)CHBrCH_3 \rightarrow \underset{\textbf{(I)}}{(CH_3)_2NC({=}Se)SeCH(CH_3)C({=}O)CH_3} + Et_3NH^+Br^-$$

$$\textbf{(I)} + H_2SO_4 \rightarrow [CH_3C{=}C(CH_3){-}Se{-}C({=}N(CH_3)_2^+){-}Se]\ HSO_4^- \xrightarrow{NaPF_6}$$

$$[CH_3C{=}C(CH_3){-}Se{-}C({=}N(CH_3)_2^+){-}Se]\ PF_6^- \quad \textbf{(II)}$$

$$\textbf{(II)} + H_2Se \xrightarrow{70\%\ MeOH} \underset{\textbf{(III)}}{[CH_3C{=}C(CH_3){-}Se{-}C({=}Se){-}Se]} + (CH_3)_2NH_2^+\ PF_6^-$$

$$2\textbf{(III)} \xrightarrow[\Delta,\ benzene]{(CH_3O)_3P} \underset{\textbf{(IV)}\ \text{TMTSF}}{[CH_3C{=}C(CH_3){-}Se{-}C({-}Se){=}]\,[{=}C{-}Se{-}C(CH_3){=}C(CH_3){-}Se]}$$

Submitted by JULIE M. BRAAM,* CLARK D. CARLSON,* DENNIS A. STEPHENS,* ANN E. REHAN,* STEVE J. COMPTON,* AND JACK M. WILLIAMS†
Checked by C. F. SIELICKI‡ and F. S. WAGNER‡

Procedure

■ **Caution.** *Hydrogen selenide is a highly toxic gas. Its odor deadens olfactory nerves, which makes it difficult to detect. It can also be absorbed through the skin. Extreme care should be exercised in using H_2Se and other selenium compounds. This synthesis should be done only in a well-ventilated hood, gloves should always be worn, and any excess H_2Se should be trapped in aqueous KOH-H_2O_2.*

1. 1-Methyl-2-oxopropyl dimethylcarbamodiselenoate

■ **Caution.** *Until product is extracted with diethyl ether, the procedure is carried out with total exclusion of oxygen under argon or nitrogen.*[6]
Initially, 5 mL (35.9 mmol) of triethylamine [Baker Chemical] distilled just prior to use, is added to 150 mL of chloroform (distilled over alumina) in a 250-mL three-necked round-bottom flask. Complete exclusion of the ambient atmosphere is required throughout step 1. This solution is degassed with argon or nitrogen (fine-frit gas dispersion tube) for 30–45 min and then maintained at $-10°$ (ice/acetone bath). Under a flow of argon, hydrogen selenide [Scientific Gas Products] (99.9%) is then bubbled through the solution using a gas dispersion tube (fine frit) for approximately 30–60 sec, until the reaction is complete as observed by the detection of excess H_2Se by the formation of a red solid in a trap of aqueous KOH-H_2O_2. The solution is purged with argon for 30–45 min to remove any excess H_2Se, catching any excess in the KOH-H_2O_2 trap. It is essential that the solution be purged completely of H_2Se. A simple test is to insert a microliter syringe filled with triethylamine. A drop on the end of the needle should not "smoke" when held above the reaction mixture.[4]

An additional 5 mL of triethylamine is then added by means of a syringe (stainless steel needle). A tube-shaped flask containing 2.9 g (18 mmol) of *N*-(dichloromethylene)-*N*-methyl methanaminium chloride [Aldrich], which is loaded in a dry box, is attached to the reaction flask under a flow of argon, and the solid is added slowly to the solution over a period of 10–15 min. The reaction

*Research participants sponsored by the Argonne Division of Educational Programs: Julie M. Braam, St. Mary's College, Winona, MN; Clark D. Carlson, University of Minnesota, Morris, MN; Dennis A. Stephens, St. Xavier College, Chicago, IL; Ann E. Rehan, Carroll College, Helena, MT; and Steve J. Compton, Dartmouth College, Hanover, NH.
†Correspondent, Chemistry Division, Argonne National Laboratory, Argonne, IL 60439.
‡Strem Chemicals, Inc., 7 Mulliken Way, Newburyport, MA 01950.

flask is allowed to warm from −10° to room temperature and the solution is stirred for 2 hr, over which time the color changes from orange to very dark brick red. The solution is then cooled to 0° and 1.9 mL (18 mmol) of 3-bromo-2-butanone [Kodak] (prac. grade) is added dropwise (5 min) using a syringe equipped with a stainless steel needle. The mixture is allowed to warm to room temperature and is then stirred for another 2 hr. The solvent is removed under vacuum, and the resulting solid is extracted with six 50-mL aliquots of dry diethyl ether. After the extracts are combined, the solvent is again removed under vacuum to give 3.7–4.4 g (72–86%) of a yellow oil or oily solid of 1-methyl-2-oxopropyl carbamodiselenoate (**I**) that is used without further purification in the next step of the synthesis. Spectroscopic data (IR, NMR, mass spec.) for (**I**) are published.[5]

2. *N*-(4,5-Dimethyl-1,3-diselenol-2-ylidene)-*N*-methylmethanaminium hexafluorophosphate

In a 50-mL three-necked flask, under a flow of argon to ensure a dry atmosphere, 10 mL of H_2SO_4 (conc.) containing a few drops of acetic anhydride to assure dryness, is cooled to −10°. Then, 3.4 g (12 mmol) of 1-methyl-2-oxopropyl carbamodiselenoate (**I**) dissolved in 2–3 mL of acetic anhydride is added dropwise by means of a glass pipet with stirring. The reaction mixture is heated slowly to 58° (internal temperature) and is stirred for 5 min at that temperature. It is cooled in an ice bath, poured onto 25 g of ice (from distilled water), and filtered immediately through a coarse-frit, a medium-frit, and then a fine-frit glass funnel. Next, the filtrate is treated with a filtered solution of 3 g of $Na[PF_6]$ dissolved in 10 mL of distilled water, causing immediate formation of a tan-colored precipitate. The precipitate is separated by filtration and is washed with three 25-mL aliquots of distilled water. The precipitate is then dissolved in 100 mL of dichloromethane, and the solution is dried with $MgSO_4$. It is filtered, and the solvent is removed by vacuum at 25° to give 2.4 g (48%) of the tan-colored salt, *N*-(4,5-dimethyl-1,3-diselenol-2-ylidene)-*N*-methylmethanaminium hexafluorophosphate (**II**). Infrared and UV spectroscopic data for (**II**) are published.[4]

3. 4,5-Dimethyl-1,3-diselenole-2-selone

In a 250-mL three-necked flask, 2.4 g (5.8 mmol) of (**II**) is suspended in 100 mL of 70% aqueous MeOH, and the solution is degassed with argon and then cooled to −30° (xylene/liquid N_2 slush bath). Under a flow of argon, to exclude any outside air, H_2Se is bubbled through the solution, again using a gas dispersion tube (fine frit), for approximately 15–20 sec (until excess is detected in the KOH/H_2O_2 trap) until the suspension turns bright orange. The mixture is warmed to room temperature and stirred for 2 hr with a continuous flow of argon over the solution to remove any excess H_2Se, again trapping it in a KOH/H_2O_2 mixture. During this time the color of the mixture changes from orange to bright red. Argon is bubbled through the solution for 20 min to remove any remaining H_2Se. The product is removed by filtering, washed with 50 mL of distilled water, dissolved in 75 mL of dichloromethane-benzene (3:1 by volume), and dried with $MgSO_4$, after which the solvent is removed. The red-orange crystals are harvested, the container is washed with dichloromethane, and the wash is concentrated by vacuum to dryness and combined with the red-orange crystals of selone to give 1.5 g (83%) of the selone[4] (**III**). The selone, for which IR and UV spectroscopic data are published,[4] is purified by recrystallization in heptane.

4. 4,4′,5,5′-Tetramethyl-2,2′-bi-1,3-diselenolylidene (TMTSF)

Initially, in a 100-mL 3-necked flask equipped with a reflux condenser, 1.4 g (4.6 mmol) of the selone (**III**) is suspended in 5 mL of benzene and refluxed under argon. Then 0.75 mL (6 mmol) of freshly distilled trimethyl phosphite is added with a syringe, and the mixture is refluxed for 90 min. Freshly distilled trimethyl phosphite is crucial to obtaining good yields of TMTSF. The mixture is cooled in an ice bath and then suction filtered. The resulting violet crystals are washed with cold diethyl ether. The mother liquor is concentrated by evaporation under vacuum to give additional product. The products are combined to give 1.8 g (87%) of needle-shaped violet crystals of TMTSF (**IV**).* The product can be further purified by vacuum gradient sublimation onto Teflon (temperature at sample 165° at 10^{-5} torr).[3,4] The vacuum gradient sublimation is accomplished by lining a glass chamber with a Teflon sheet to avoid possible reaction of TMTSF with glass (at 165°). Gradient sublimation of recrystallized TMTSF results in yields of ~50% based on the initial amount of TMTSF used.

Anal. Calcd. for $C_{10}H_{12}Se_4$: C, 26.81; H, 2.70. Found: C, 26.72; H, 2.68.

Properties

4,4′,5,5′-Tetramethyl-2,2′-bi-1,3-diselenolylidene crystallizes as violet needles that decompose at 250°. The ^1H NMR spectrum in $CDCl_3$ (TMS) is a singlet at δ 1.98. The UV/vis spectrum in CH_2Cl_2 is: λ max, 508 ± 5 nm, 299 ± 1 nm.[5] The IR spectrum obtained by subliming TMTSF onto a NaCl plate exhibits bands at 2970 (m), 2902 (s), 2840 (m), 1617 (m), 1434 (vs), 1145 (m), 1062 (s), and 655 (s) cm^{-1}.[4] TMTSF will oxidize slowly if exposed to light and air, but it can be kept many months if it is stored under argon in a laboratory freezer (−15°) in a light-free environment.

*We wish to thank Prof. D. O. Cowan for many helpful suggestions.

B. TETRABUTYLAMMONIUM PERCHLORATE AND BIS(4,4′,5,5′-TETRAMETHYL-2,2′-BI-1,3-DISELENOLYLIDENE) RADICAL ION (1+) PERCHLORATE

$$[CH_3(CH_2)_3]_4NHSO_4 + H[ClO_4] \xrightarrow{H_2O} [CH_3(CH_2)_3]_4N[ClO_4] + H_2SO_4$$

$$[CH_3(CH_2)_3]_4N[ClO_4] + 2TMTSF \xrightarrow{4.5\mu A} (TMTSF)_2[ClO_4]$$

Submitted by DENNIS A. STEPHENS,* ANN E. REHAN,* STEVE J. COMPTON,* ROBERT A. BARKHAU,* and JACK M. WILLIAMS†
Checked by MARSHA M. LEE‡

■ **Caution.** *Perchlorates are potentially shock-sensitive. Appropriate face shields, gloves, and reaction shields should be used at all times. Easily oxidized organic materials must be absent.*

1. Tetrabutylammonium Perchlorate

Procedure

Initially, 5 mL (34.9 mmol) of 70% perchloric acid is added dropwise, with stirring, to a solution of 9.91 g (29.2 mmol) of tetrabutylammonium hydrogen sulfate [Fluka] in 120 mL of distilled water. A white precipitate, $[CH_3(CH_2)_3]_4N[ClO_4]$, forms immediately. The solution is allowed to stir for 5 min and is then cooled at 0° for 40 min and filtered. The solid is washed with distilled water until the pH of wash is 7 and is dried under vacuum. The mother liquor is concentrated under vacuum (25°) to give additional product. The products are combined and recrystallized three times from ethyl acetate; the resulting solid is washed each time with distilled water. The crystals are dried in a vacuum desiccator. This yields 7.95 g (79.7%) of white crystals of tetrabutylammonium perchlorate (mp 208–210°).

*Research participants sponsored by the Argonne Division of Educational programs: Dennis A. Stephens, St. Xavier College, Chicago, IL; Ann E. Rehan, Carroll College, Helena, MT; Steven J. Compton, Dartmouth College, Hanover, NH; Robert A. Barkhau, Carthage College, Kenosha, WI.

†Chemistry Division, Argonne National Laboratory, Argonne, IL 60439.

‡Department of Chemistry, The Johns Hopkins University, Baltimore, MD 21218 (electrolytic synthesis only).

Anal. Calcd. for $[CH_3(CH_2)_3]_4N[ClO_4]$: C, 56.20; H, 10.61; N, 4.10; Cl, 10.37; O, 18.72. Found: C, 56.01; H, 10.37; N, 4.16; Cl, 10.59; O, 18.91.

2. Bis(4,4′,5,5′-tetramethyl-2,2′-bi-1,3-diselenolylidene) Radical Ion (1+) Perchlorate

The crystal-growing apparatus consists of a glass H-cell (15 mL capacity) with a fine-porosity glass frit and two platinum wire electrodes (0.367 cm^2) as shown in Fig. 1.[7] The power supply consists of a variable dc constant-current source with a range of 0–5 μA. The crystal-growing apparatus is covered with aluminum foil to keep out light during crystal growth. Solutions of TMTSF in the organic solvents described herein have very low electrical conductivity. Therefore, the tetrabutylammonium cation and a correspondingly monovalent anion are added, both to increase solubility and conductivity and to promote crystal growth of the desired anionic derivative by electrolytic oxidation at the anode.

The cell is cleaned and thoroughly dried before use as follows: Fresh aqua regia is drawn through the frit three times in each direction, using a water aspirator

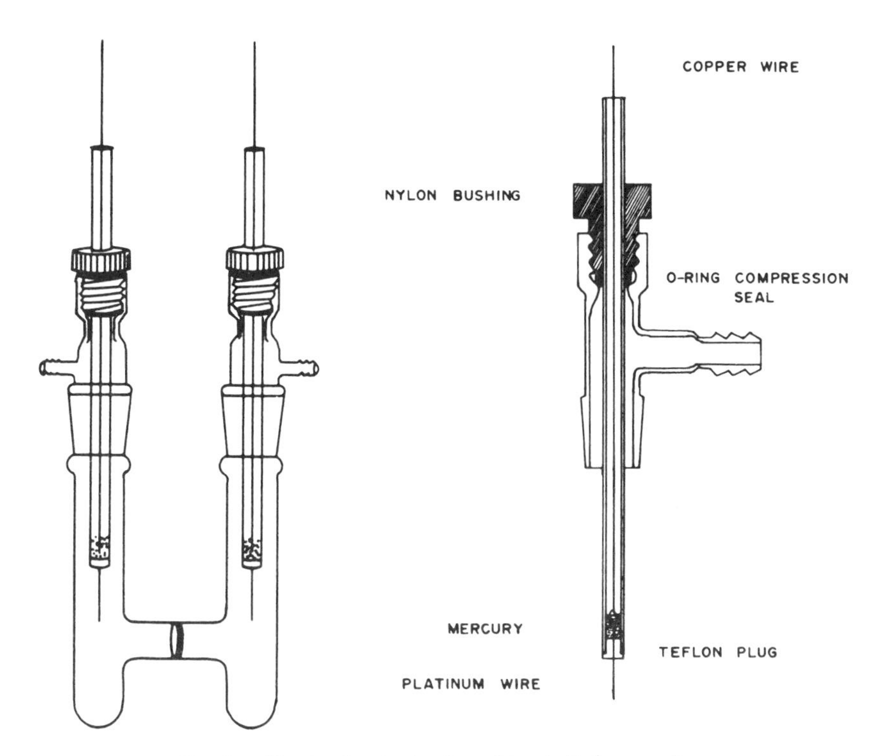

Fig. 1. Electrochemical cell and inset. Reference 7.

and changing the acid each time. Tap water, followed by distilled water, is then drawn through the frit in exactly the same manner. Finally, the cell is rinsed three times with anhydrous methanol and then dried in an oven at 125°.

In order to prepare the cell for crystal growth, 10 mL of a 0.15 *M* tetrabutylammonium perchlorate salt solution [0.513 g, 1.5 mmol in 10 mL of dry 1,1,2-trichloroethane (TCE)] in TCE and 5 mL of a 1.2×10^{-2} *M* TMTSF solution (0.027 g, 0.06 mmol in 5 mL of dry TCE) in TCE are needed to produce final concentrations of 0.1 and $4 \times 10^{-3}M$, respectively. The 1,1,2-trichloroethane should be distilled and dried using an alumina column prior to use. First 5 mL of the TMTSF solution is added to the anode compartment, and then 5 mL of the tetrabutylammonium salt solution is added to the cathode compartment. The remaining 5 mL of the tetrabutylammonium salt solution is added in equal amounts to the anode and cathode compartments to produce equal solution heights on both sides. The solution in each compartment (cathode compartment first) is purged for 30 sec with argon. The precleaned and dried platinum electrodes are placed in their respective compartments and adjusted in height so that they are immersed in the solutions but are not touching the cell bottom. The cell pressure vents are covered with parafilm. The anode and cathode leads from the constant-current regulator are connected and the current is adjusted to the desired setting (~0.5–5 μA). The cell is covered with aluminum foil to exclude light that may cause premature oxidation of the TMTSF, which is indicated by the formation of an orange or green color in the solution.

The crystals are harvested at the desired stage of crystal growth by carefully lifting the anode from the solution and then lightly brushing the crystals from the platinum electrode into a suction filter. After the cathode has been removed, the remaining contents of the anode cell are also poured into the filter. The cell is rinsed several times with TCE to remove any remaining crystals. Finally, the crystals are washed with 10 mL of TCE, dried for 10 min on the suction filter, and stored in a glass sample vial. If crystals are harvested when the TMTSF solution has lost its pink color, a yield of nearly 100%, or ~30 mg, of metallic black crystals of bis(4,4′,5,5′-tetramethyl-2,2′-bi-1,3-diselenolylidene) perchlorate $(TMTSF)_2[ClO_4]$ is obtained. Crystals of the highest quality are obtained, as required for X-ray diffraction and electrical conductivity studies, if the crystal growth is accomplished using low applied currents and is halted at ~50% conversion. Properties are recorded in Table I.

Anal. Calcd. for $(TMTSF)_2[ClO_4]$: C, 24.13; H, 2.43. Found: C, 24.12; H, 2.41.

C. TETRABUTYLAMMONIUM HEXAFLUOROARSENATE AND BIS(4,4′,5,5′-TETRAMETHYL-2,2′-BI-1,3-DISELENOLYLIDENE) RADICAL ION (1+) HEXAFLUOROARSENATE

$$[CH_3(CH_2)_3]_4N[I] + K[AsF_6] \xrightarrow{H_2O} [CH_3(CH_2)_3]_4N[AsF_6] + KI$$

$$[CH_3(CH_2)_3]_4N[AsF_6] + 2\ TMTSF \xrightarrow{4.0\ \mu A} (TMTSF)_2[AsF_6]$$

Submitted by STEVE COMPTON,* HAU H. WANG,† and JACK M. WILLIAMS†
Checked by MARSHA M. LEE‡

1. Tetrabutylammonium Hexafluoroarsenate

Procedure

Initially, 3.97 g of potassium hexafluoroarsenate (17.4 mmol) in 36 mL of distilled water is added slowly to 140 mL of an aqueous solution that contains 6.42 g of tetrabutylammonium iodide [Alfa Inorganics] (17.4 mmol) at 60°. A pale yellow precipitate of tetrabutylammonium hexafluoroarsenate forms immediately. The mixture is stirred for 10 min and then allowed to cool to room temperature. The precipitate is removed by filtration and is washed with five 10-mL portions of distilled water. The crude product is recrystallized twice from hot methanol. The yield is 5.77 g of white needles of $[CH_3(CH_2)_3]_4N[AsF_6]$ (13.4 mmol, 80%) (mp 245–248°).

Anal. Calcd. for $[CH_3(CH_2)_3]_4N[AsF_6]$: C, 44.55; H, 8.41; N, 3.25; F. 26.42. Found: C, 44.42; H, 8.41; N, 3.01; F. 26.21.

2. Bis(4,4′,5,5′-tetramethyl-2,2′-bi-1,3-diselenolylidene) Radical Ion (1+) Hexafluoroarsenate

Procedure

For details of the electrolysis, see Section B. The solutions needed for the cell consist of 0.647 g (1.5 mmol) of $[CH_3(CH_2)_3]_4N[AsF_6]$ in 10 mL of dry TCE

*Research participant sponsored by the Argonne Division of Educational Programs from Dartmouth College, Hanover, NH.
†Chemistry Division, Argonne National Laboratory, Argonne, IL 60439.
‡Department of Chemistry, The Johns Hopkins University, Baltimore, MD 21218 (electrolytic synthesis only).

and 0.027 g (0.06 mmol) of TMTSF in 5 mL of dry TCE. After 3–5 days at 4.0 μA, the crystals are harvested when the anode solution loses its pink color. The yield is 20.82 mg (64.0%) of metallic black needles of bis(4,4′,5,5′-tetramethyl-2,2′-bi-1,3-diselenolylidene) hexafluoroarsenate $(TMTSF)_2[AsF_6]$. Properties are given in Table I.

Anal. Calcd. for $(TMTSF)_2[AsF_6]$: C, 22.14; H, 2.23. Found: C, 22.22; H, 2.46.

D. TETRABUTYLAMMONIUM TETRAFLUOROBORATE AND BIS(4,4′,5,5′-TETRAMETHYL-2,2′-BI-1,3-DISELENOLYLIDENE) RADICAL ION (1+) TETRAFLUOROBORATE

$$[CH_3(CH_2)_3]_4NOH + H[BF_4] \rightarrow [CH_3(CH_2)_3]_4N[BF_4] + H_2O$$

$$[CH_3(CH_2)_3]_4N[BF_4] + 2\ TMTSF \xrightarrow{5\ \mu A} (TMTSF)_2[BF_4]$$

Submitted by ANN E. REHAN,* ROBERT A. BARKHAU,* and JACK M. WILLIAMS†
Checked by MARSHA M. LEE‡

1. Tetrabutylammonium Tetrafluoroborate

Procedure

■ **Caution.** *Hydrogen tetrafluoroborate reacts with glass. Polyethylene equipment and a well-ventilated hood must be used.*

Initially, 1.67 mL of hydrogen tetrafluoroborate (48 wt % in H_2O) [Aldrich] in 10 mL of distilled water is added dropwise (plastic syringe) with stirring (Teflon stir bar) to 22.78 mL of tetrabutylammonium hydroxide (40 wt % in H_2O) [Aldrich] in a polyethylene beaker. A white precipitate, tetrabutylammonium tetrafluoroborate, forms immediately. The mixture is allowed to stir for 15 min and then cooled to 0° for 1 hr. The product is removed by filtering (polyethylene funnel and filter flask) and is dried in a desiccator (plastic) over $CaSO_4$ for 12 hr. The filtrate is concentrated under vacuum to give additional product. The

*Research participants sponsored by the Argonne Division of Educational Programs: Ann E. Rehan, Carroll College, Helena, MT, and Robert A. Barkhau, Carthage College, Kenosha, WI.

†Chemistry Division, Argonne National Laboratory, Argonne, IL 60439.

‡Department of Chemistry, The Johns Hopkins University, Baltimore, MD 21218 (electrolytic synthesis only).

products are combined, recrystallized from hot ethyl acetate, and washed with two 10-mL aliquots of ice-cold distilled water to give 2.02 g (67.3%) of white crystals of tetrabutylammonium tetrafluoroborate.

Anal. Calcd. for $[CH_3(CH_2)_3]_4N[BF_4]$: C, 58.36; H, 11.02; N, 4.25, Found: C, 58.20; H, 11.00; N, 4.20.

2. Bis(4,4′,5,5′-tetramethyl-2,2′-bi-1,3-diselenolylidene) Radical Ion (1+) Tetrafluoroborate

Procedure

For details of the electrolysis, see Section B. The solutions needed for the cell consist of 0.494 g (1.5 mmol) of $[CH_3(CH_2)_3]_4N[BF_4]$ in 10 mL of dry TCE and 0.027 g (0.06 mmol) of TMTSF in 5 mL of dry TCE. The crystals are grown for approximately six days (5 μA), after which time they are harvested to give 22.15 mg (75.0%) of metallic black needles of bis(4,4′,5,5′-tetramethyl-2,2′-bi-1,3-diselenolylidene) tetrafluoroborate, $(TMTSF)_2[BF_4]$. Properties are given in Table I.

Anal. Calcd. for $(TMTSF)_2[BF_4]$: C, 24.44; H, 2.86. Found: C, 24.81; H, 2.74.

E. TETRABUTYLAMMONIUM HEXAFLUOROPHOSPHATE AND BIS(4,4′,5,5′-TETRAMETHYL-2,2′-BI-1,3-DISELENOLYLIDENE) RADICAL ION (1+) HEXAFLUOROPHOSPHATE

$$[CH_3(CH_2)_3]_4NHSO_4 + H[PF_6] \xrightarrow{H_2O} [CH_3(CH_2)_3]_4N[PF_6] + H_2SO_4$$

$$[CH_3(CH_2)_3]_4N[PF_6] + 2TMTSF \xrightarrow{5\ \mu A} (TMTSF)_2[PF_6]$$

Submitted by DENNIS A. STEPHENS* and JACK M. WILLIAMS†
Checked by MARSHA M. LEE‡

1. Tetrabutylammonium Hexafluorophosphate

Procedure

■ **Caution.** *Hydrogen hexafluorophosphate reacts with glass to produce H_2SiF_6. Polyethylene equipment and a well-ventilated hood must be used.*
In a polyethylene beaker, 18.0 g (53.0 mmol) of tetrabutylammonium hydrogen sulfate [Aldrich] is dissolved in 150 mL of distilled water by stirring with a Teflon stirring bar. Next, a solution of 10 mL (41.1 mmol) of hydrogen hexafluorophosphate [Alpha Products] (60 wt % in H_2O) in 25 mL of distilled water (polyethylene beaker) is added to the $[CH_3(CH_2)_3]_4NHSO_4$ solution. The mixture is stirred for 5 min and is then placed in an ice bath for 45 min. The resulting precipitate is removed by filtering and rinsed with 10-mL aliquots of ice-cold distilled water until the pH of the final rinse is 7. The white crystals are dried in a vacuum desiccator over $CaSO_4$ for 2–3 hr and recrystallized twice from hot ethyl acetate. Before final drying under vacuum, the crystals are rinsed with five 10-mL aliquots of distilled water. This process yields 15.0 g (94.3%) of white crystals of tetrabutylammonium hexafluorophosphate.

Anal. Calcd. for $[CH_3(CH_2)_3]_4N[PF_6]$: C, 49.60; H, 9.36; N, 3.62; P, 8.00; F, 29.42. Found: C, 49.78; H, 8.92; N, 3.60; P, 7.92; F, 29.51.

*Research participant sponsored by the Argonne Division of Educational programs from St. Xavier College, Chicago, IL.
†Chemistry Division, Argonne National Laboratory, Argonne, IL 60439.
‡Department of Chemistry, The Johns Hopkins University, Baltimore, MD 21218 (electrolytic synthesis only).

2. Bis(4,4′,5,5′-Tetramethyl-2,2′-bi-1,3-diselenolylidene) Radical Ion (1+) Hexafluorophosphate

Procedure

For details of the electrolysis, see Section B. The solutions needed for the cell consist of 0.581 g (1.5 mmol) of $[CH_3(CH_2)_3]_4N[PF_6]$ in 10 mL of dry TCE and 0.027 g (0.06 mmol) of TMTSF in 5 mL of dry TCE. After electrolysis for 3–5 days at 5 μA, the crystals are harvested when the TMTSF solution begins to lose its pink color. This procedure yields 25 mg (80.0%) of metallic black crystals of bis(4,4′,5,5′-tetramethyl-2,2′-bi-1,3-diselenolylidene) hexafluorophosphate, $(TMTSF)_2[PF_6]$.

Anal. Calcd. for $(TMTSF)_2[PF_6]$: C, 23.07; H, 2.32. Found: C, 23.64: H, 2.44.

All of the synthetic metals discussed herein are isostructural and triclinic (space group $P\bar{1}$). The identifying crystal lattice parameters are given in Table I.

Properties (Summary)

The unusual electrical properties of the 2:1 TMTSF salts are summarized and discussed elsewhere.[3,5,8] For additional characterization purposes beyond chemical analyses, the crystallographic lattice constants for the isostructural (triclinic, space group $P\bar{1}$) TMTSF derivatives discussed here are given in Table I.

TABLE I. Crystallographic Lattice Constants for TMTSF Derivatives.

	Crystal Lattice Constants (Å)			Angles (deg)			
Compound	*a*	*b*	*c*	α	β	γ	V_{cell}, 298 K (Å^3)
$(TMTSF)_2[ClO_4]$	7.266	7.678	13.275	84.58	86.73	70.43	694.3
$(TMTSF)_2[AsF_6]$	7.277	7.711	13.651	83.16	86.00	71.27	719.9
$(TMTSF)_2[BF_4]$	7.255	7.647	13.218	82.23	87.15	70.36	688.1
$(TMTSF)_2[PF_6]$	7.297	7.711	13.522	83.39	86.27	71.01	714.3

References

1. K. Bechgaard, D. O. Cowan, and A. N. Bloch, *J. Chem. Soc. Chem. Commun.*, **(1974),** 937.
2. K. Bechgaard, D. O. Cowan, A. N. Bloch, and L. Henriksen, *J. Org. Chem.*, **40,** 746 (1975).
3. K. Bechgaard, *Mol. Cryst. Liq. Cryst.*, **79,** 1 (1982).
4. F. Wudl, E. Aharon-Shalom, and S. H. Bertz, J. Org. Chem., **46,** 4612 (1981).
5. L.-Y. Chiang, D. O. Cowan, T. O. Poehler, and A. N. Bloch, *Mol. Cryst. Liq. Cryst.*, **86,** 27 (1982).

6. D. F. Shriver, *Manipulation of Air-Sensitive Compounds*, McGraw-Hill, New York, 1969.
7. M. M. Lee, J. P. Stokes, F. M. Wiygul, T. J. Kistenmacher, D. O. Cowan, T. O. Poehler, A. N. Bloch, W. W. Fuller, and D. U. Gubser, *Mol. Cryst. Liq. Cryst.*, **79,** 145 (1982).
8. J. M. Williams, *Prog. Inorg. Chem.*, **33,** 183 (1985).

43. METHYLMERCURY(II) NITRATE AND METHYLMERCURY(II) TRIFLUOROACETATE

Submitted by ROBERT D. BACH* and HARSHA B. VARDHAN*
Checked by DONALD W. GOEBEL, JR.,* and JOHN P. OLIVER*

Methylmercury(II) nitrate and methylmercury(II) trifluoroacetate have been utilized in the study of the interaction of [MeHg(II)] species with biological substrates such as nucleosides.[1,2] We now report improved isolation procedures and reproducible syntheses[2,3] for these compounds that give excellent yields.

■ **Caution.** *All methylmercury compounds are extremely toxic. Contact of either the solid or concentrated solution of [MeHg]NO_3 or [MeHg]$OCOCF_3$ with the skin causes painful blisters. The entire procedure should be carried out in a well-ventilated hood, and all samples should be kept tightly sealed. All wastes (such as precipitates, filter papers, and wash solutions) should be collected in sealed containers and disposed of properly. The glassware used is detoxified by rinsing with alkaline Na[BH_4](~0.25 M) followed by liberal rinsing with water. Proper toxic chemical handling procedures (such as the use of latex gloves and laboratory coat) must be followed.*

A. PREPARATION OF METHYLMERCURY(II) IODIDE

$$CH_3I + Mg + HgI_2 \xrightarrow{Et_2O} CH_3HgI + MgI_2$$

Both [MeHg]NO_3 and [MeHg]$OCOCF_3$ are prepared from methylmercury(II) iodide.

Procedure

Iodomethylmagnesium is prepared by the addition of a solution of MeI (125 mL, 2.0 moles) in dry diethyl ether (250 mL) to a well-stirred suspension of

*Department of Chemistry, Wayne State University, Detroit, MI 48202.

Mg (51.0 g, 2.1 g-atom) in dry diethyl ether (750 mL). The reaction is carried out in a 2-L three-necked round-bottomed flask equipped with a mechanical stirrer, a reflux condenser, and an argon inlet. The flask is flame-dried and then cooled in a slow stream of argon prior to the addition of reactants. The entire synthesis of methylmercury(II) iodide is carried out under a positive pressure of argon (or nitrogen). The rate of addition of MeI is adjusted to keep the reaction mixture at a slow and steady reflux. After the addition is completed (~6 hr), the reaction mixture is stirred for an additional hour. The second step of the synthesis is carried out in a 5-L three-necked round-bottomed flask equipped with a mechanical stirrer, a reflux condenser, and an argon inlet. The flask is flame-dried and cooled in a slow stream of argon. The Grignard reagent is filtered through a plug of glass wool (under argon) into a pressure-equalizing addition funnel. The solution is added slowly to a vigorously stirred slurry of red HgI_2 (900 g, 1.98 moles) in dry diethyl ether (2.5 L). A positive pressure of argon is maintained during the reaction. After the addition of the Grignard reagent is complete (~7 hr), the reaction mixture is heated gently at reflux for 0.5 hr (until the red color is completely discharged). An off-white precipitate is formed. The reaction mixture is cooled to 0° and quenched by very slow addition of ice cold water (200 mL). The supernatant solution is decanted from the precipitate, and the diethyl ether is removed under vacuum to obtain a slurry. The precipitate and the slurry are combined and washed with a solution of glacial acetic acid (100 mL) in water (1 L). The pale yellow crystalline solid is filtered, washed with water (1.5 L), and dried. Yield: 518 g (76%). The MeHgI obtained is used without further purification.

Properties

Methylmercury(II) iodide is a pale yellow crystalline solid, mp 148–149° (lit 144°).[4] ^1H NMR ($CDCl_3$) δ 1.24 ppm, $J_{^{199}Hg-^1H}$ = 185 Hz. IR (KBr): 3000 (w), 2920 (w), 2780 (w), 2300 (w), 770 (s), and 520 (m) cm^{-1}.

B. PREPARATION OF METHYLMERCURY(II) NITRATE

$$MeHgI + AgNO_3 \rightarrow [MeHg]NO_3 + AgI$$

Procedure

To a solution of 8.5 g (50 mmol) of $AgNO_3$ in 100 mL of absolute ethanol in a 500-mL flask is added 17.5 g (51 mmol) of MeHgI, and the suspension is stirred for 4 hr. Water (100 mL) is added, and the precipitated AgI is filtered and washed with water (50 mL). The combined filtrate and washings are carefully concentrated to dryness under reduced pressure with gentle heating, using a rotary evaporator equipped with a Dry Ice/acetone trap. The flask containing the solution should not be overheated, to avoid loss of material during concentration. The resulting crude off-white solid is transferred to an all-glass sublimator and slowly sublimed (48 hr) at 50°/0.15 torr, yielding 13.1 g (94%) of fine white crystals.

Properties

Methylmercury(II) nitrate is a hygroscopic white crystalline solid, mp 59–60°. Its IR spectrum (KBr) exhibits a strong absorption at 1385 cm^{-1}; ^{1}H NMR ($CDCl_3$) δ 1.35 ppm, $J_{^{199}Hg—^{1}H} = 236$ Hz.

C. PREPARATION OF METHYLMERCURY(II) TRIFLUOROACETATE

$$MeHgI + AgOCOCF_3 \rightarrow [MeHg]OCOCF_3 + AgI$$

Procedure

To a solution of 11.2 g (51 mmol) of $AgOCOCF_3$ [Aldrich] in 100 mL of absolute ethanol in a 500 mL-flask is added 17.13 g (50 mmol) of MeHgI, and the suspension is stirred for 4 hr. Water (100 mL) is added, and the precipitated AgI is filtered and washed with water (50 mL). The combined filtrate and washings are concentrated to dryness in a rotary evaporator, taking care not to overheat the solution during concentration. The off-white crude product is transferred to a glass sublimator and slowly sublimed at 50°/0.15 torr to give 13.31 g (81%) of fine white crystals.

Properties

Methylmercury(II) trifluoroacetate is a hygroscopic white crystalline solid, mp 81–82°C (lit 81–82°).[5] ^{1}H NMR ($CDCl_3$) δ 1.33 ppm, $J_{^{199}Hg—^{1}H} = 226$ Hz. IR (KBr): 1675 (s), 1570 (w), 1435 (w), 1200 (m), 835 (m), 800 (m), and 720 (m) cm^{-1}. Crude methylmercury(II) acetate prepared similarly from MeHgI and $AgOCOCH_3$ can not be purified satisfactorily by direct sublimation. It is best purified by initial recrystallization from absolute ethanol. It may then be sublimed if desired.

References

1. P. L. Prizant, M. J. Olivier, P. Rivest, and A. L. Beauchamp, *Can. J. Chem.*, **59,** 1311 (1981).
2. A. J. Canty and R. S. Tobias, *Inorg. Chem.*, **18,** 413 (1979).
3. J. Kuyper, *Inorg. Chem.*, **17,** 1458 (1978).
4. J. L. Maynard, *J. Am. Chem. Soc.*, **54,** 2108 (1932).
5. V. N. Kalinin, I. A. Federov, K. G. Gasanov, E. I. Fedin, and L. I. Zakharin, *Izv. Akad. Nauk SSSR, Ser. Khim.*, **10,** 2402 (1970); *Chem. Abstr.*, **75,** 13216y (1971).

Chapter Three

TRANSITION METAL ORGANOMETALLIC COMPOUNDS

44. DICARBONYLBIS(η^5-CYCLOPENTADIENYL) COMPLEXES OF TITANIUM, ZIRCONIUM, AND HAFNIUM

Submitted by DAVID J. SIKORA,* KEVIN J. MORIARTY,* and MARVIN D. RAUSCH*
Checked by A. RAY BULLS,† JOHN E. BERCAW,†, VIKRAM D. PATEL,‡ and ARTHUR J. CARTY‡

Dicarbonylbis(η^5-cyclopentadienyl)titanium, $Ti(CO)_2(\eta^5\text{-}C_5H_5)_2$, was first synthesized in 1959. It was the first carbonyl complex of a Group 4 metal.[1,2] Since the original synthesis, several preparations of $Ti(CO)_2(\eta^5\text{-}C_5H_5)_2$ have been reported.[3–9] The corresponding zirconium and hafnium analogs of $Ti(CO)_2(\eta^5\text{-}C_5H_5)_2$ were not described until 1976. The preparation of $Zr(CO)_2(\eta^5\text{-}C_5H_5)_2$ was reported independently by three different research groups, each employing different methods.[7,8,10] One of these groups also described the only known procedure for the synthesis of $Hf(CO)_2(\eta^5\text{-}C_5H_5)_2$.[7]

Because of our interest in the structure and reactivity of $M(CO)_2(\eta^5\text{-}C_5H_5)_2$ (M = Zr, Hf),[11,12] we sought a more convenient method of preparation for these

*Department of Chemistry, University of Massachusetts, Amherst, MA 01003.
†Department of Chemistry, California Institute of Technology, Pasadena, CA 91125.
‡Department of Chemistry, University of Waterloo, Waterloo, Ontario, Canada N2L 3G1.

compounds, since existing syntheses were of low yield and/or required severe reaction conditions.[7,8,10] We have subsequently developed facile routes to the syntheses of $M(CO)_2(\eta^5\text{-}C_5H_5)_2$ (M = Ti, Zr, Hf) utilizing the reductive carbonylation of $MCl_2(\eta^5\text{-}C_5H_5)_2$. The synthesis of $Ti(CO)_2(\eta^5\text{-}C_5H_5)_2$ described here is a modification of the convenient procedure reported originally by Demerseman et al.[6] Our procedure involves magnesium metal activated in situ by mercury(II) chloride. The advantages of this method are as follows: (1) no high-pressure apparatus is required, since the reaction occurs readily at 1 atm CO pressure[8,10]; (2) the reaction does not involve large amounts of elemental mercury or sodium amalgam[7]; (3) each synthesis is a one-step procedure that avoids the preparation of intermediates[8]; (4) the yields are reproducible and give ample amounts of product for further studies[7]; and (5) the starting metallocene dichlorides are commercially available.

Due to the current interest in the structure and chemistry of the corresponding pentamethylcyclopentadienyl analogs $M(CO)_2(\eta^5\text{-}C_5Me_5)_2$ (M = Ti, Zr, Hf),[13–15] we have extended our procedure to the syntheses of $Ti(CO)_2(\eta^5\text{-}C_5Me_5)_2$,[16,17] $Zr(CO)_2(\eta^5\text{-}C_5Me_5)_2$,[18] and the heretofore unknown $Hf(CO)_2(\eta^5\text{-}C_5Me_5)$.* Dichlorobis($\eta^5$-pentamethylcyclopentadienyl)titanium and -zirconium are readily reduced with activated magnesium powder. The hafnium analog $HfCl_2(\eta^5\text{-}C_5Me_5)_2$, however, is resistant to reduction under these conditions. A more reactive form, Rieke magnesium[20] activated with mercury(II) chloride, is therefore utilized. A main advantage of these procedures for the preparation of $M(CO)_2(\eta^5\text{-}C_5Me_5)_2$ (M = Ti, Zr, Hf) is that $MCl_2(\eta^5\text{-}C_5Me_5)_2$ is used directly in the synthesis without the necessity of preparing reaction intermediates.[16,18] A detailed method for the synthesis of $HfCl_2(\eta^5\text{-}C_5Me_5)_2$ is included. A higher yield preparation of the latter compound has been reported.[21]

All procedures are performed using standard Schlenk tube techniques[22] under an atmosphere of dry oxygen-free argon. All glassware is oven-dried and then flame-dried under vacuum and allowed to cool to room temperature while under argon. The tetrahydrofuran (THF) and 1,2-dimethoxyethane (DME) are predried over potassium hydroxide flakes, further dried over sodium wire, and finally distilled under argon from the sodium ketyl of benzophenone. Hexane is dried over calcium hydride and freshly distilled under argon prior to use. CAMAG neutral grade alumina [Alfa Products] is heated by means of a heat gun on a rotary evaporator operating at 10^{-2} torr (vacuum pump) for 2 hr and then allowed to cool under an argon atmosphere to room temperature. Five percent by weight of degassed water is added to the alumina, the mixture is shaken until thoroughly mixed, and the alumina is stored under argon. Water is purged with argon for 15 min, heated at reflux under argon for 12 hr, and then allowed to cool under argon.

*$Hf(CO)_2(\eta^5\text{-}C_5Me_5)_2$ has recently been reported spectroscopically; see Reference 19.

■ **Caution.** *Because of the known toxicity of carbon monoxide and metal carbonyls, all preparations must be carried out in an efficiently operating fume hood. The residual material contained in the reaction vessel after the initial filtration includes activated magnesium and metallic mercury. The magnesium metal may be decomposed by the careful addition of water (100 mL) (except for Rieke magnesium, which should be decomposed with 2-propanol). Although this reaction vigorously evolves hydrogen, at no time has ignition been observed. The metallic mercury should be disposed of properly.*[22]

A. DICARBONYLBIS(η^5-CYCLOPENTADIENYL)TITANIUM

$$TiCl_2(\eta^5\text{-}C_5H_5)_2 + Mg + 2CO \rightarrow Ti(CO)_2(\eta^5\text{-}C_5H_5)_2 + MgCl_2$$

Procedure

Dichlorobis(η^5-cyclopentadienyl)titanium [Alfa Products] (5.00 g, 20.1 mmol), magnesium turnings* (1.62 g, 66.6 mmol) and 100 mL of THF are placed into a 250-mL Schlenk tube and stirred magnetically. Recrystallization of the titanium compound from xylene is recommended. The tube is flushed with carbon monoxide for 5 min. Mercury(II) chloride (3.60 g, 13.3 mmol) is then added while carbon monoxide is allowed to flow slowly over the solution through the side-arm stopcock and out to a mercury overpressure valve. A small amount of heat is generated during the amalgamation of the magnesium. Although this does not seem to have any detrimental effects on the reduction, it is recommended that the Schlenk tube be cooled (water bath) during the addition of mercury(II) chloride and subsequent amalgamation. After 5 min of stirring, the bath may be removed and the reaction run at room temperature. The reaction mixture is stirred in the carbon monoxide atmosphere for 12 hr at room temperature, during which time the color changes from bright red to dark green and finally to dark red. The reaction vessel is then flushed with argon, and the solution is poured into a fritted funnel[22] containing a plug (12 × 4 cm) of 5% deactivated alumina covered with 1.5 cm of sea sand. The reaction mixture is allowed to pass through the plug in order to remove the magnesium chloride formed in the reaction. Hexane is subsequently used to elute the remaining material until the eluate is colorless. The THF/hexane solution is concentrated to dryness under reduced pressure, leaving the crude dark red $Ti(CO)_2(\eta^5\text{-}C_5H_5)_2$. The product is purified by dissolving it in ~100 mL of hexane and passing this solution through another plug (8 × 4 cm) of deactivated alumina. The plug is eluted with fresh hexane until the solution emerging from the fritted funnel is colorless. The solvent is

*Fisher Brand magnesium turnings—for Grignard reaction [Fisher Scientific].

then removed under reduced pressure, leaving 4.1 g (87%) of $Ti(CO)_2(\eta^5\text{-}C_5H_5)_2$. The purity of this material is satisfactory for further reactions. However, the product can be conveniently recrystallized from hexane at $-20°$.

Properties

Dicarbonylbis(η^5-cyclopentadienyl)titanium is a maroon red air-sensitive solid that is soluble in both aliphatic and aromatic solvents. The ^1H NMR spectrum (C_6D_6) exhibits a singlet at δ 4.62 ppm (external TMS). The IR spectrum shows two metal carbonyl stretching vibrations at 1977 and 1899 cm^{-1} in hexane, or at 1965 and 1883 cm^{-1} in THF.

B. DICARBONYLBIS(η^5-CYCLOPENTADIENYL)ZIRCONIUM

$$ZrCl_2(\eta^5\text{-}C_5H_5)_2 + Mg + 2CO \rightarrow Zr(CO)_2(\eta^5\text{-}C_5H_5)_2 + MgCl_2$$

Procedure

Dichlorobis(η^5-cyclopentadienyl)zirconium [Alfa Products] (2.00 g, 6.84 mmol) together with magnesium turnings [Fisher] (0.83 g, 34.2 mmol) and 50 mL of THF are placed in a 100-mL Schlenk tube and magnetically stirred. On dissolution of the $ZrCl_2(\eta^5\text{-}C_5H_5)_2$, mercury(II) chloride (1.85 g, 6.81 mmol) is added to the mixture, at which time carbon monoxide is allowed to flow slowly over the solution through the side-arm stopcock and out to a mercury overpressure valve. The solution is stirred in the carbon monoxide atmosphere for 24 hr at room temperature, during which time the solution changes from colorless to dark green and finally to dark red. The reaction vessel is flushed with argon, and the solution is poured into a fritted funnel[22] containing a plug (10 × 3 cm) of 5% deactivated alumina covered with 1.5 cm of sea sand. The reaction mixture is allowed to pass through the plug in order to remove the magnesium chloride formed in the reaction. The plug is then eluted with hexane until the eluate is colorless. The dark red reaction solution appears green when passing through the plug and dark reddish-green on exiting, depending on how the solution is viewed. The THF/hexane solution is then concentrated to dryness, leaving a dark solid. The $Zr(CO)_2(\eta^5\text{-}C_5H_5)_2$ is purified by dissolving it in ~75 mL of hexane and passing the solution through another plug (5 × 3 cm) of 5% deactivated alumina. The plug is eluted with fresh hexane until the solution emerging from the fritted funnel is colorless. The hexane is then concentrated under vacuum until black needle-like crystals begin to form. At this point the solution is cooled to $-20°$, resulting in further crystal formation. The hexane is decanted from the crystals into another Schlenk tube and further concentrated and cooled. The resulting black crystal crops are dried under vacuum and combined, yielding 1.00 g (53%) of $Zr(CO)_2(\eta^5\text{-}C_5H_5)_2$.*

Properties

Dicarbonylbis(η^5-cyclopentadienyl)zirconium is a black air-sensitive solid that upon dissolution in aliphatic or aromatic solvents yields dark reddish-green solutions. The ^1H NMR spectrum (C_6D_6) exhibits a singlet at δ 4.95 ppm (external TMS). The IR spectrum displays two metal carbonyl stretching vibrations at 1975 and 1885 cm^{-1} in hexane, or at 1967 and 1872 cm^{-1} in THF.

C. DICARBONYLBIS(η^5-CYCLOPENTADIENYL)HAFNIUM

$$HfCl_2(\eta^5\text{-}C_5H_5)_2 + Mg + 2CO \rightarrow Hf(CO)_2(\eta^5\text{-}C_5H_5)_2 + MgCl_2$$

Procedure

Dichlorobis(η^5-cyclopentadienyl)hafnium [Alfa Products] (2.00 g, 5.27 mmol), magnesium powder [RMC-50/100-UM (~50–100 mesh), Reade Manufacturing] (0.50 g, 20.6 mmol), and 50 mL of THF are placed into a 100-mL Schlenk tube and stirred magnetically. The use of magnesium powder for this preparation as opposed to magnesium turnings is critical to the success of the reaction. The use of magnesium powder for the preparation of $Zr(CO)_2(\eta^5\text{-}C_5H_5)_2$, however, was found to result in lower yields of product. When the solution is stirred, a deep narrow vortex is desirable. This can be accomplished with a small stirring bar. On dissolution of the $HfCl_2(\eta^5\text{-}C_5H_5)_2$, mercury(II) chloride (1.00 g, 3.68 mmol) is added to the mixture, at which time carbon monoxide is allowed to flow slowly over the solution through the side-arm stopcock and out to a mercury overpressure valve. The solution is stirred in the carbon monoxide atmosphere for 24 hr, during which time the color changes from colorless to dark green and finally to dark red. The reaction vessel is flushed with argon, and the solution is poured into a fritted funnel[22] containing a plug (10 $\times$ 3 cm) of 5% deactivated alumina covered with 1.5 cm of sea sand. The reaction mixture is allowed to pass through the plug in order to remove the magnesium chloride formed in the reaction. Hexane is then used to elute the remaining material until the eluate is colorless. The THF/hexane solution is concentrated to dryness, leaving a purple solid. The $Hf(CO)_2(\eta^5\text{-}C_5H_5)_2$ is purified by dissolving it in ~75 mL of hexane and passing this solution through another plug (5 $\times$ 3 cm) of 5% deactivated

*The checkers report that the same percentage yield is obtained when the amounts of the starting materials are doubled.

alumina. This plug is eluted with fresh hexane until the solution emerging from the fritted funnel is colorless. The hexane is concentrated under vacuum until purple needle-like crystals begin to form. The solution is then cooled to $-20°$, resulting in further crystal formation. The hexane is decanted from the crystals into another Schlenk tube and further concentrated and cooled. The resulting crystal crops are dried under vacuum and combined, yielding 0.58 g (30%) of $Hf(CO)_2(\eta^5\text{-}C_5H_5)_2$.

Properties

Dicarbonylbis(η^5-cyclopentadienyl)hafnium is a purple air-sensitive solid that is soluble in both aliphatic and aromatic solvents. The ^1H NMR spectrum (C_6D_6) exhibits a singlet at δ 4.81 ppm (external TMS). The IR spectrum displays two metal carbonyl stretching vibrations at 1969 and 1878 cm^{-1} in hexane, or at 1960 and 1861 cm^{-1} in THF.

D. DICARBONYLBIS(η^5-PENTAMETHYLCYCLOPENTADIENYL)TITANIUM

$$TiCl_2(\eta^5\text{-}C_5Me_5)_2 + Mg + 2CO \rightarrow Ti(CO)_2(\eta^5\text{-}C_5Me_5)_2 + MgCl_2$$

Procedure

Dichlorobis(η^5-pentamethylcyclopentadienyl)titanium[16] [Strem Chemicals] (2.00 g, 5.14 mmol), magnesium powder [RMC-50/100-UM(~50-100 mesh), Reade] (0.62 g, 25.5 mmol), and 50 mL of THF are placed in a 100-mL Schlenk tube and stirred magnetically. On dissolution of the $TiCl_2(\eta^5\text{-}C_5Me_5)_2$, mercury(II) chloride (1.39 g, 5.12 mmol) is added to the mixture, at which time carbon monoxide is allowed to flow slowly over the solution through the side-arm stopcock and out to a mercury overpressure valve. The solution is stirred under a carbon monoxide atmosphere for 24 hr at room temperature, during which time the color changes from red to green and finally to red. The THF is removed under reduced pressure, leaving a red residue. The residue is extracted with hexane until the extracts are colorless. The hexane solution is poured into a fritted funnel[22] containing a plug (5×3 cm) of 5% deactivated alumina covered with 1.5 cm of sea sand. The plug is eluted with hexane until the eluate is colorless. The resulting red solution is then concentrated under reduced pressure to approximately one-fourth of its original volume and cooled to $-20°$. Upon crystal formation, the hexane is decanted into another Schlenk tube and further concentrated and cooled. The resulting crystal crops are dried under vacuum and combined, yielding 1.25 g (65%) of $Ti(CO)_2(\eta^5\text{-}C_5Me_5)_2$ as lustrous brick-red needles.

Properties

Dicarbonylbis(η^5-pentamethylcyclopentadienyl)titanium is moderately stable in air, but it is best handled and stored under an inert atmosphere. It is extremely soluble in both aliphatic and aromatic solvents. The ^1H NMR spectrum (C_6D_6) exhibits a singlet at δ 1.67 ppm (external TMS). The IR spectrum displays two metal carbonyl stretching vibrations at 1940 and 1858 cm^{-1} (hexane). The mass spectrum shows a parent ion at *m/e* 374.

E. DICARBONYLBIS-(η^5-PENTAMETHYLCYCLOPENTADIENYL)ZIRCONIUM

$$ZrCl_2(\eta^5\text{-}C_5Me_5)_2 + Mg + 2CO \rightarrow Zr(CO)_2(\eta^5\text{-}C_5Me_5)_2 + MgCl_2$$

Procedure

Dichlorobis(η^5-pentamethylcyclopentadienyl)zirconium[23] [Strem Chemicals] (0.80 g, 1.85 mmol), magnesium powder [RMC-50/100-UM(~50–100 mesh), Reade] (0.22 g, 9.05 mmol), and 20 mL of THF are placed in a 50-mL Schlenk tube and stirred magnetically such that a deep narrow vortex is formed. On dissolution of the $ZrCl_2(\eta^5\text{-}C_5Me_5)_2$, mercury(II) chloride (0.50 g, 1.84 mmol) is added to the mixture, at which time carbon monoxide is allowed to flow slowly over the solution through the side-arm stopcock and out to a mercury overpressure valve. The solution is stirred under a carbon monoxide atmosphere for 24 hr at room temperature, during which time the color changes from pale yellow to dark reddish green. The THF is removed under reduced pressure, leaving a dark-colored solid. This solid is extracted with hexane until the extracts are colorless. The hexane solution is poured into a fritted funnel[22] containing a plug (5 $\times$ 3 cm) of 5% deactivated alumina covered with 1.5 cm of sea sand. The plug is eluted with hexane until the eluate is colorless. The resulting dark reddish-green solution is then concentrated under reduced pressure to approximately one-fourth of its original volume and cooled to $-20°$. Upon crystal formation, the hexane is decanted into another Schlenk tube and further concentrated and cooled. The resulting crystal crops are dried under vacuum and combined, yielding 0.62 g (80%) of $Zr(CO)_2(\eta^5\text{-}C_5Me_5)_2$ as lustrous black needles.

Properties

Dicarbonylbis(η^5-pentamethylcyclopentadienyl)zirconium, unlike its titanium analog, is highly air-sensitive. It is easily soluble in both aliphatic and aromatic solvents. The ^1H NMR spectrum (C_6D_6) exhibits a singlet at δ 1.73 ppm (external TMS), and the IR spectrum displays two metal carbonyl stretching vibrations at 1945 and 1852 cm^{-1} (hexane). The mass spectrum shows a parent ion at *m/e* 416.

F. DICHLOROBIS-(η^5-PENTAMETHYLCYCLOPENTADIENYL)HAFNIUM

$$2C_5Me_5Li + HfCl_4 \rightarrow HfCl_2(\eta^5\text{-}C_5Me_5)_2 + 2LiCl$$

Procedure

(Pentamethylcyclopentadienyl)lithium[24] (6.00 g, 42.2 mmol) and 125 mL of 1,2-dimethoxyethane (DME) are placed in a 250-mL three-necked round-bottomed flask that is fitted with a gas inlet and a condenser attached to a mercury over-pressure valve. The slurry is cooled to $-78°$, and freshly sublimed hafnium tetrachloride [Research Organic*] (6.73 g, 21.0 mmol) is added. The reaction mixture is allowed to warm to room temperature and is then heated at reflux for 4 days. The DME is removed under reduced pressure, leaving a yellow solid. The use of argon as a protective atmosphere is discontinued at this point. The residue is then dissolved in 70 mL of chloroform, and 28 mL of 6 *N* hydrochloric acid is added. The organic layer is washed with 50 mL of water and separated, and the aqueous layer is washed with 50 mL of chloroform and separated. The organic layers are then combined and dried over anhydrous sodium sulfate. After filtration, the yellow solution is concentrated to ~5 mL, yielding pale yellow crystals of $HfCl_2(\eta^5\text{-}C_5Me_5)_2$ (2.10 g, 19%), which are subsequently collected on a Büchner funnel, washed with pentane, and dried.

Properties

Dichlorobis(η^5-pentamethylcyclopentadienyl)hafnium is a yellow air-stable solid that is soluble in both aromatic and chlorinated solvents. The ^1H NMR spectrum ($CDCl_3$) exhibits a singlet at δ 2.05 ppm.

*The material had a listed purity of 99.9% (Spectro grade) but was further purified by zone refinement at 280–290°/ 10^{-3} torr.

G. DICARBONYLBIS-(η^5-PENTAMETHYLCYCLOPENTADIENYL)HAFNIUM

$$2K + MgCl_2 \rightarrow Mg + 2KCl$$

$$HfCl_2(\eta^5\text{-}C_5Me_5)_2 + Mg + 2CO \rightarrow Hf(CO)_2(\eta^5\text{-}C_5Me_5)_2 + MgCl_2$$

Procedure

A 100-mL three-necked round-bottomed flask is fitted with a reflux condenser and gas inlet and outlet tubes. Approximately 60 mL of THF is added to the flask together with freshly cut potassium metal (1.11 g, 28.4 mmol), anhydrous $MgCl_2$ (1.90 g, 20.0 mmol), and KI (0.50 g, 3.0 mmol).

■ **Caution.** *Potassium metal reacts explosively on contact with water and may also form potentially dangerous superoxides. Consult reference 25 for proper handling procedure.*

The reaction mixture is stirred and slowly heated to reflux, at which point the potassium metal melts and the reaction commences. After refluxing for 3 hr, a very finely divided dark gray suspension of magnesium metal can be observed.[20] The reaction mixture is then cooled to room temperature and purged with carbon monoxide for 5 min. Dichlorobis(η^5-pentamethylcyclopentadienyl)hafnium (1.50 g, 2.89 mmol) is added to the magnesium slurry, and the mixture is allowed to stir under a carbon monoxide atmosphere for 12 hr, with a slow, continuous flow of carbon monoxide through the flask. Mercury(II) chloride (0.50 g, 1.84 mmol) is added, and the mixture is allowed to stir further under a stream of carbon monoxide for an additional 12 hr. The dark reaction mixture is then filtered through a fritted funnel[22] in order to separate it from the magnesium. This filtration may be slow, due to the magnesium particles clogging the pores of the frit. The use of deoxygenated Celite helps alleviate this problem. (Celite was deoxygenated in a manner analogous to the deoxygenation of alumina; however, it was not deactivated with water.) The frit is then washed with fresh THF, and the dark red filtrate is concentrated to dryness under reduced pressure, leaving a solid residue. The residue is extracted with hexane until the extracts are colorless. The hexane solution is poured into a fritted funnel[22] containing a plug (5 × 3 cm) of 5% deactivated alumina covered with 1.5 cm of sea sand. The plug is eluted with hexane until the eluate is colorless. The resulting purple solution is concentrated to approximately one-fourth of its original volume and cooled to −20°. Upon crystal formation, the hexane is decanted into another Schlenk tube and is further concentrated and cooled. The resulting crystal crops are dried under vacuum and combined, yielding 0.36 g (25%) of $Hf(CO)_2(\eta^5\text{-}C_5Me_5)_2$ as lustrous purple needles.

Anal. Calcd. for $C_{22}H_{30}HfO_2$: C, 52.32; H, 5.99. Found: C, 52.19; H, 5.95.

■ **Caution.** *The residual material contained in the reaction flask and fritted funnel includes highly activated magnesium metal and metallic mercury. The residual material must not come into contact with water, since the hydrogen that is generated would ignite. The magnesium should be decomposed by reaction with 2-propanol. The metallic mercury should be disposed of properly.*[22]

Properties

Dicarbonylbis(η^5-pentamethylcyclopentadienyl)hafnium is a purple air-sensitive solid that is very soluble in both aliphatic and aromatic solvents. The ^{1}H NMR spectrum (C_6D_6) exhibits a singlet at δ 1.74 ppm (external TMS) and the IR spectrum displays two metal carbonyl stretching vibrations at 1940 and 1844 cm^{-1} (hexane). The mass spectrum shows a parent ion at *m/e* 506.

References

1. J. G. Murray, *J. Am. Chem. Soc.*, **81,** 752 (1959).
2. J. G. Murray, *J. Am. Chem. Soc.*, **83,** 1287 (1961).
3. P. C. Wailes, R. S. P. Coutts, and H. Weigold, *The Organometallic Chemistry of Titanium, Zirconium, and Hafnium,* Academic Press, New York, 1974.
4. M. D. Rausch and H. Alt, *J. Am. Chem. Soc.*, **96,** 5936 (1974).
5. B. Demerseman, G. Bouquet, and M. Bigorgne, *J. Organomet. Chem.*, **93,** 199 (1975).
6. B. Demerseman, G. Bouquet, and M. Bigorgne, *J. Organomet. Chem.*, **101,** C24 (1975).
7. J. T. Thomas and K. T. Brown, *J. Organomet. Chem.*, **111,** 297 (1976).
8. G. Fachinetti, G. Fochi, and C. Floriani, *J. Chem. Soc., Chem. Commun.*, **1976,** 230.
9. B. Demerseman, G. Bouquet, and M. Bigorgne, *J. Organomet. Chem.*, **145,** 41 (1978).
10. B. Demerseman, G. Bouquet, and M. Bigorgne, *J. Organomet. Chem.*, **107,** C19 (1976).
11. D. J. Sikora, M. D. Rausch, R. D. Rogers, and J. L. Atwood, *J. Am. Chem. Soc.*, **101,** 5079 (1979).
12. D. J. Sikora and M. D. Rausch, *J. Organomet. Chem.*, **276,** 21 (1984).
13. P. T. Wolczanski and J. E. Bercaw, *Acc. Chem. Res.*, **13,** 121 (1980).
14. D. J. Sikora, M. D. Rausch, R. D. Rogers, and J. L. Atwood, *J. Am. Chem. Soc.*, **103,** 982 (1981).
15. D. J. Sikora, M. D. Rausch, R. D. Rogers, and J. L. Atwood, *J. Am. Chem. Soc.*, **103,** 1265 (1981).
16. J. E. Bercaw, R. H. Marvich, L. G. Bell, and H. H. Brintzinger, *J. Am. Chem. Soc.*, **94,** 1219 (1972).
17. B. Demerseman, G. Bouquet, and M. Bigorgne, *J. Organomet. Chem.*, **132,** 223 (1977).
18. J. M. Manriquez, D. R. McAlister, R. D. Sanner, and J. E. Bercaw, *J. Am. Chem. Soc.*, **98,** 6733 (1976).
19. J. A. Marsella, C. J. Curtis, J. E. Bercaw, and K. G. Caulton, *J. Am. Chem. Soc.*, **102,** 7244 (1980).
20. R. E. Rieke and S. E. Bales, *J. Am. Chem. Soc.*, **96,** 1775 (1974).
21. D. M. Roddick, M. D. Fryzuk, P. F. Seidler, G. L. Hillhouse, and J. E. Bercaw, *Organometallics,* **4,** 97, 1694 (1985).
22. D. F. Shriver, *The Manipulation of Air-Sensitive Compounds,* McGraw-Hill, New York, 1978.

23. J. M. Manriquez, D. R. McAlister, E. Rosenberg, A. M. Shiller, K. L. Williamson, S. I. Chan, and J. E. Bercaw, *J. Am. Chem. Soc.*, **100,** 3078 (1978).
24. D. W. Macomber and M. D. Rausch, *J. Am. Chem. Soc.*, **105,** 5325 (1983).
25. L. F. Fieser and M. Fieser, *Reagents for Organic Synthesis*, Wiley, New York, 1967, p. 950.

45. SODIUM CARBONYL FERRATES, $Na_2[Fe(CO)_4]$, $Na_2[Fe_2(CO)_8]$, AND $Na_2[Fe_3(CO)_{11}]$. BIS[μ-NITRIDO-BIS(TRIPHENYLPHOSPHORUS)(1+)] UNDECACARBONYLTRIFERRATE(2−), $[(Ph_3P)_2N]_2[Fe_3(CO)_{11}]$

$$Fe(CO)_5 + 2NaC_{10}H_8 \xrightarrow{\text{THF}} Na_2[Fe(CO)_4] + 2C_{10}H_{18} + CO$$

$$2Fe(CO)_5 + 2NaC_{10}H_8 \xrightarrow{\text{THF}} Na_2[Fe_2(CO)_8] + 2C_{10}H_8 + 2CO$$

$$Fe_3(CO)_{12} + 2NaC_{10}H_8 \xrightarrow{\text{THF}} Na_2[Fe_3(CO)_{11}] + C_{10}H_8 + CO$$

$$Na_2[Fe_3(CO)_{11}] + 2[(Ph_3P)_2N]Cl \xrightarrow{\text{MeOH}} [(Ph_3P)_2N]_2[Fe_3(CO)_{11}] + 2NaCl$$

Submitted by HENRY STRONG,* PAUL J. KRUSIC,† and JOSEPH SAN FILIPPO, JR.*
Checked by SCOTT KEENAN‡ and RICHARD G. FINKE‡

Carbonylferrates have been the subject of many studies. The well-defined mono-, di-, and trinuclear species $[Fe(CO)_4]^{2-}$, $[Fe_2(CO)_8]^{2-}$, and $[Fe_3(CO)_{11}]^{2-}$ have been obtained by a variety of methods[1-3] in varying yields and degrees of convenience. The procedures described here provide uniform, convenient, high-purity, high-yield syntheses of the sodium salts of these three important reagents. In addition, the preparation of the bis[μ-nitrido-bis(triphenylphosphorus)(1+)] salt of $[Fe_3(CO)_{11}]^{2-}$ by metathesis of $[(Ph_3P)_2N]Cl$ with $Na_2[Fe_3(CO)_{11}]$ is presented.

Procedure

■ **Caution.** *The toxic nature of the reagents and products requires that these reactions be performed in a well-ventilated fume hood.*

*Department of Chemistry, Rutgers University, New Brunswick, NJ 08903.
†Central Research and Development Department, E. I. du Pont de Nemours and Company, Wilmington, DE 19898.
‡Department of Chemistry, University of Oregon, Eugene, OR 97403.

■ **Caution.** *The use of $Li[AlH_4]$ in purifying THF is dangerous. It should not be attempted until it is ascertained that the THF is peroxide-free and also not grossly wet.*[4]

Peroxide-free tetrahydrofuran (THF) is distilled under nitrogen from lithium tetrahydroaluminate. Pentane is distilled under nitrogen from P_4O_{10}. Iron pentacarbonyl [Pressure Chemical] is freshly distilled prior to each use.[5] Commercial triiron dodecacarbonyl [Alfa Products] is obtained wet with methanol, which is removed by subjecting the sample to vacuum (0.1 torr) overnight. Commercial naphthalene (recrystallized quality) is used without further purification. All manipulations are carried out in a nitrogen-flushed dry box or in standard Schlenk apparatus under a nitrogen atmosphere.

Disodium tetracarbonylferrate(2-), $Na_2[Fe(CO)_4]$, is prepared in a 1-L three-necked round-bottomed flask, one arm of which is modified to permit the contents of the flask to be filtered under an inert atmosphere.[5] The flask is equipped with a Teflon-coated magnetic stirrer bar and a 200-mL addition funnel. All remaining inlets are sealed with rubber serum stoppers [Ace Scientific], and the vessel is flame-dried.* Under a flush of nitrogen the flask is charged with a weighed quantity (3.45 g, 75.0 mmol) of sodium dispersion (20 microns, 50% by weight) in paraffin [Alfa Products]. It is then placed in an ice bath, and a solution of naphthalene (9.90 g, 77.0 mmol) in tetrahydrofuran (500 mL) is added via a stainless steel cannula. The contents of the flask are stirred for 2 hr at 0° and the resulting deep-green solution of (naphthalene)sodium is chilled at $\lesssim -70°$ in a Dry Ice/acetone bath. A solution of freshly distilled iron pentacarbonyl (7.02 g, 36.0 mmol) in tetrahydrofuran (100 mL) is added slowly over a 30-min period, attended by vigorous stirring. Failure to use freshly distilled $Fe(CO)_5$ leads to diminished yields and purity. The deep green color is gradually replaced by a persistent beige color. At this point, addition is discontinued, and the resulting mixture is stirred an additional hour before being permitted to warm to ambient temperature.

Pentane§ (200 mL) is added by cannula to the reaction mixture, which is then stirred for an additional 30 min before the flask is tilted and its contents filtered under a positive pressure of nitrogen through a coarse-frit glass disk filter. The collected snow-white precipitate of $Na_2[Fe(CO)_4]$ is rinsed with two 100-mL portions of pentane,* and the flask is transferred to the dry box, where the contents are dried under vacuum (0.1 torr) for 4 hr to give 7.39 g [96%† based on $Fe(CO)_5$] of disodium tetracarbonylferrate(2-).‡ Approximate elapsed time

*Checkers dried the flask in an oven at 150° and flushed it immediately with dry nitrogen.

§Checkers used hexane.

†Checkers obtained a yield of 83%. They report difficulty in dissolving all of the sodium in THF.

‡$Na_2[Fe(CO)_4]\cdot 1.5$ dioxane is available commercially [Aldrich].[2a]

for total synthesis is 12 hr. (■ **Caution.** *Solid $Na_2[Fe(CO)_4]$ is an exceedingly pyrophoric material.*)

Disodium octacarbonyldiferrate(2-), $Na_2[Fe_2(CO)_8]$, is prepared by a procedure similar to that described above. Thus, a solution of iron pentacarbonyl (14.04 g, 72.0 mmol) in tetrahydrofuran (200 mL) is added over a 30-min period with stirring to the previously described solution of (naphthalene)sodium. Following a workup procedure equivalent to that described above, the orange precipitate that is obtained is rinsed with three 200-mL portions of pentane. Upon drying under vacuum the orange solid yields 13.6 g (99%* based on iron pentacarbonyl) of bright yellow $Na_2[Fe_2(CO)_8]$. The addition and removal of THF causes a reversible color change.[2b] Elapsed time for the total synthesis is ~3 hr.

■ **Caution.** *Dry $Na_2[Fe_2(CO)_8]$ is a pyrophoric substance.*

Disodium undecacarbonyltriferrate(2-), $Na_2[Fe_3(CO)_{11}]$, is prepared by a modification of the above procedure. Thus, in a nitrogen flushed dry box, a 250-mL single-necked, round-bottomed flask equipped with a Teflon-coated stirring bar is charged with 1.84 g (44 mmol) of sodium dispersion and capped with a rubber septum stopper. The flask is removed from the dry box and cooled in an ice bath; a solution of naphthalene (6.00 g, 47.0 mmol) in tetrahydrofuran (150 mL) is introduced by cannula, and the resulting mixture is stirred for 2 hr.

A modified (see above) 1-L three-necked flask equipped with addition funnel and Teflon-coated stirrer bar is charged with 10.07 g (20.0 mmol) of $Fe_3(CO)_{12}$, capped with a rubber septum, and flushed with nitrogen before adding THF (125 mL). The flask is placed in a Dry Ice/acetone bath, and the solution of sodium naphthalide is transferred through a cannula into the 200-mL addition funnel. This solution is added slowly over a period of 1 hr to the chilled, well-stirred solution of $Fe_3(CO)_{12}$ in THF. This order of addition is essential; reversal of the indicated order leads to substantial contamination of the product by unidentified side products. The resulting mixture is stirred for an additional 2 hr before it is permitted to warm to ambient temperature. The flask is then transferred to the dry box, and the contents are concentrated under vacuum to dryness. The remaining dark red-brown solid is rinsed with three 200-mL portions of pentane and dried under vacuum once again. The isolated yield of $Na_2[Fe_3(CO)_{11}]$ is 10.2 g [98%† based on $Fe_3(CO)_{12}$]. Approximate elapsed time for total synthesis is 4 hr.

■ **Caution.** *Dry $Na_2[Fe_3(CO)_{11}]$ is a pyrophoric substance.*

Bis[μ-nitrido-bis(triphenylphosphorus)(1+)] undecacarbonyltriferrate(2−) is obtained by treating a solution of $Na_2[Fe_3(CO)_{11}]$ (1.2 g, 2.3 mmol) in 25 mL of anhydrous methanol, which is distilled from $Mg(OCH_3)_2$ and is contained in a 250-mL single-necked round-bottomed flask with a solution of $[(Ph_3P)_2N]Cl$[6]

*Checkers obtained a yield of 88% after having initial difficulties in dissolving sodium.
†Checkers obtained 57% yield.

[Aldrich] (3.0 g, 5.2 mmol) in methanol (25 mL). The dark red-brown solid that precipitates is collected by suction filtration on a medium-porosity frit under an inert atmosphere. Recrystallization from dichloromethane as previously described[3] yields 2.9 g (81%)† of crystalline, dark red-brown crystals of $[(Ph_3P)_2N]_2[Fe_3(CO)_{11}]$.

Properties

Disodium tetracarbonylferrate(2−) is a snow-white solid that is extremely sensitive to oxygen. It has a reported[2] solubility of 7×10^{-3} *M* in THF and can be stored for moderate periods of time in an inert atmosphere, at room temperature, if kept in the dark. The IR spectrum, recorded in *N,N*-dimethylformamide (DMF), exhibits a stretching frequency at 1730 cm^{-1}, consistent with previous literature reports.[7] The structure of the $[Fe(CO)_4]^{2-}$ anion has been established by X-ray,[8] and the utility of this reagent has been discussed.[9]

Disodium octacarbonyldiferrate(2−) has been reported previously.[2b] This extremely air-sensitive solid is largely insoluble in most organic solvents, with only marginal solubility in THF. Its IR spectrum, recorded in DMF, shows the following CO stretching vibrations: 1835 (w), 1860 (s), and 1910 (m) cm^{-1}, consistent with previously reported values.[2b] A single-crystal X-ray structure determination of the $[Fe_2(CO)_8]^{2-}$ anion has been carried out.[10]

Disodium undecacarbonyltriferrate(2−) is a well-known substance that has also been characterized structurally.[11] The IR spectrum of this material, recorded in DMF, shows CO stretching bands at 1940 (s), 1915 (m), and 1880 (w) cm^{-1}, consistent with values observed previously.[7,11]

References

1. R. B. King and F. G. A. Stone, *Inorg. Synth.,* **7,** 197 (1963).
2. (a) J. P. Collman, R. G. Finke, J. N. Cawse, and J. I. Brauman, *J. Am. Chem. Soc.,* **99,** 2515 (1977); (b) J. P. Collman, R. G. Finke, P. L. Matlock, R. Wahren, R. G. Komoto, and J. I. Brauman, ibid., **100,** 1119 (1978).
3. H. A. Hodali and D. F. Shriver, *Inorg. Synth.,* **20,** 222 (1980).
4. Anon., *Inorg. Synth.,* **12,** 317 (1970).
5. D. F. Shriver, *Manipulation of Air-Sensitive Compounds,* McGraw-Hill, New York, 1969.
6. J. K. Ruff and W. J. Schlientz, *Inorg. Synth.,* **15,** 84 (1974).
7. W. F. Edgell, M. T. Yang, B. J. Bulkin, R. Bayer, and N. Koizumi, *J. Am. Chem. Soc.,* **87,** 3080 (1965).
8. R. G. Teller, R. G. Finke, J. P. Collman, H. B. Chin, and R. Bau, *J. Am. Chem. Soc.,* **99,** 1104 (1977); J. P. Collman, R. G. Finke, P. L. Matlock, and J. I. Brauman, ibid., **98,** 4685 (1976).
9. J. P. Collman, *Accounts Chem. Res.,* **8,** 342 (1975). A preparative scale synthesis of the

†Checkers obtained 71% yield.

solvated complex $Na_2Fe(CO)_4 \cdot 1.5$ dioxane was previously reported; see *Org. Synth.*, **59,** 102 (1980).
10. H. B. Chin, M. B. Smith, R. D. Wilson, and R. Bau, *J. Am. Chem. Soc.*, **96,** 5285 (1974).
11. F. Y-K. Lo, G. Longoni, P. Chini, L. D. Lower, and L. F. Dahl, *J. Am. Chem. Soc.*, **102,** 7691 (1980).

46. TRICARBONYL (η^5-CYCLOPENTADIENYL)IRON(1+) TRIFLUOROMETHANESULFONATE, $[CpFe(CO)_3](CF_3SO_3)$

$$Na[CpFe(CO)_2] + ClC(O)OMe \rightarrow CpFe(CO)_2[C(O)OMe] + NaCl$$

$$CpFe(CO)_2[C(O)OMe] + HOSO_2CF_3 \rightarrow [CpFe(CO)_3](CF_3SO_3) + MeOH$$

Submitted by MONO M. SINGH* and ROBERT J. ANGELICI*
Checked by COLIN P. HORWITZ†

Earlier methods[1–3] of preparing $[CpFe(CO)_3]^+$ (Cp = η^5-C_5H_5) from $CpFe(CO)_2X$ (X = Cl, Br, I) used high CO pressures and temperatures and long reaction times. In a modified procedure,[4] atmospheric CO pressure was used, but the yield of the cation was low. The complex can also be prepared, in low yields, from the dicarbonyl cations[5,6] $[CpFe(CO)_2(\text{acetone})]^+$ and $[CpFe(CO)_2(OH_2)]^+$. Recently, $[CpFe(CO)_3][BF_4]$ was prepared by reaction of $[CpFe(CO)_2]^-$ with CO_2 at $-80°$ followed by $H[BF_4]$.[7] The $[CpFe(CO)_3][PF_6]$ salt has been prepared by oxidation of $Cp_2Fe_2(CO)_4$ with $NO[PF_6]$ in the presence of CO.[8,9] Herein, we describe a reliable method of preparing $[CpFe(CO)_3]^+$ from $Cp_2Fe_2(CO)_4$ that takes advantage of the known reaction of alkoxycarbonyl complexes, $CpFe(CO)_2[C(O)OR]$, with an acid to generate the cation $[CpFe(CO)_3]^+$.[10–12]

Procedure

The starting material $Cp_2Fe_2(CO)_4$ may be prepared from $Fe(CO)_5$ and cyclopentadiene dimer[13] or purchased commercially [Strem Chemicals]. Before use it may be purified by recrystallization from CH_2Cl_2/hexane at $-20°$. Unless otherwise mentioned, all procedures are carried out under N_2. A yellow-brown solution of $Na[CpFe(CO)_2]$ in 250 mL of dry tetrahydrofuran (THF) is prepared

*Department of Chemistry and Ames Laboratory, U.S.D.O.E., Iowa State University, Ames, IA 50011.
†Department of Chemistry, Northwestern University, Evanston, IL 60201.

by reducing 10.0 g (28.3 mmol) of $Cp_2Fe_2(CO)_4$ with Na/Hg amalgam for 2 hr.[14] After removing the excess sodium amalgam from the reaction vessel,[14] a solution of 5 mL (65 mmol) of methyl chloroformate (ethyl chloroformate may also be used) in 30 mL of dry THF is added in 3–5-mL portions to the rapidly stirring $[CpFe(CO)_2]^-$ mixture. After being stirred for an additional 45 min, the contents of the flask are decanted into another round-bottomed flask. From this step forward, the procedure may be carried out in air, but this should be done as rapidly as possible until the CF_3SO_3H (triflic acid) is added (see below). The solvent is removed from the mixture on a rotary evaporator, the residue is extracted with 200 mL of benzene* added in several portions, and the extracts are filtered through Celite filter aid. The flask and the precipitates are washed with another 50 mL of benzene.*

■ **Caution.** *Benzene is a suspected carcinogen and should be used only in a well-ventilated fume hood, and gloves should be worn at all times.*

The extracts and the washings are collected in a 500-mL conical flask. While the mixture is stirred vigorously with a magnetic stirring bar, 5–6 mL of trifluoromethanesulfonic acid, CF_3SO_3H (triflic acid), is added.

■ **Caution.** *Triflic acid is very corrosive and fumes in air. A well-ventilated fume hood must be used. Rubber gloves should be worn.*

The mixture is allowed to stand for at least 1 hr at 0° (or better, overnight in a refrigerator) and is then filtered. The precipitate is washed with several portions of diethyl ether until the washings are colorless. The yellow microcrystalline compound is dried in air and purified further by dissolving in a minimum volume of warm, dry acetone and adding diethyl ether to the solution dropwise until the point of crystallization is reached. Cooling this mixture to $-20°$ gives bright yellow crystals of $[CpFe(CO)_3](CF_3SO_3)$, which are collected, washed with diethyl ether, and dried in air. A second crop of crystals may be obtained after reducing the volume of the filtrate, adding more diethyl ether, and cooling at $-20°$. Total yield: 7.5–11 g (38–55%).

Anal. Calcd. for $C_9H_5F_3FeO_6S$: C, 30.53; H, 1.42. Found: C, 30.84; H, 1.38.

Properties

The salt $[CpFe(CO)_3](CF_3SO_3)$ is stable in air, but it changes from yellow to brown slowly if stored in a desiccator over a long period of time. It is soluble in acetone and acetonitrile but only slightly soluble in dichloromethane. When treated with triphenylphosphine in acetone, $[CpFe(CO)_3]^+$ forms $[CpFe(CO)_2(PPh_3)]^+$.[15] The cation reacts with halide ions[16] to form the neutral complexes $CpFe(CO)_2X$ (X = Cl, Br, and I). Treatment of $[CpFe(CO)_3]^+$ with aliphatic

*The checker has suggested that toluene works as well as benzene.

amines and alkoxides leads to the formation of carbamoyl and alkoxycarbonyl derivatives, respectively.[10,17] Spectral data for $[CpFe(CO)_3](CF_3SO_3)$ are: IR (in CH_3CN), ν_{CO} 2123 (vs), 2078 (vs) cm^{-1}; ^1H and ^{13}C NMR (in CD_3CN), $^1H_{Cp}$ δ 5.80, $^{13}C_{CO}$ δ 203.16, $^{13}C_{Cp}$ δ 91.10 ppm.

References

1. A. Davison, M. L. H. Green, and G. Wilkinson, *J. Chem. Soc.*, **1961,** 3172.
2. E. O. Fisher and K. Fichtel, *Chem. Ber.*, **94,** 1200 (1961).
3. R. B. King, *Inorg. Chem.*, **1,** 964 (1962).
4. A. E. Kruse and R. J. Angelici, *J. Organomet. Chem.*, **24,** 231 (1970).
5. E. C. Johnson, T. J. Meyer, and N. Winterton, *Inorg. Chem.*, **10,** 1673 (1971).
6. B. D. Dombek and R. J. Angelici, *Inorg. Chim. Acta*, **7,** 345 (1973).
7. T. Bodnar, E. Coman, K. Menard, and A. Cutler, *Inorg. Chem.*, **21,** 1275 (1982).
8. R. H. Reimann and E. Singleton, *J. Organomet. Chem.*, **32,** C44 (1971).
9. M. M. Singh and R. J. Angelici, *Inorg. Chem.*, **23,** 2691 (1984).
10. L. Busetto and R. J. Angelici, *Inorg. Chim. Acta*, **2,** 391 (1968).
11. R. B. King, M. B. Bisnette, and A. Fronzaglia, *J. Organomet. Chem.*, **5,** 341 (1966).
12. M. H. Quick and R. J. Angelici, *J. Organomet. Chem.*, **160,** 231 (1978).
13. R. B. King and F. G. A. Stone, *Inorg. Synth.*, **7,** 110 (1963).
14. B. D. Dombek and R. J. Angelici, *Inorg. Synth.*, **17,** 100 (1977).
15. B. V. Johnson, P. J. Ouseph, J. S. Hsieh, A. L. Steinmetz, and J. E. Shade, *Inorg. Chem.*, **18,** 1796 (1979).
16. R. K. Kochhar and R. Pettit, *J. Organomet. Chem.*, **6,** 272 (1966).
17. R. J. Angelici, *Acc. Chem. Res.*, **5,** 335 (1972).

47. DICARBONYL(η^5-CYCLOPENTADIENYL)(2-METHYL-1-PROPENYL-κ*C*[1])IRON AND DICARBONYL(η^5-CYCLOPENTADIENYL)(η^2-2-METHYL-1-PROPENE)IRON(1+) TETRAFLUOROBORATE

Submitted by MYRON ROSENBLUM,* WARREN P. GIERING,† and SARI-BETH SAMUELS‡
Checked by PAUL J. FAGAN§

Cationic olefin complexes of dicarbonyl(η^5-cyclopentadienyl) iron have been of wide interest in syntheses for a number of years. These complexes, generally isolated as their tetrafluoroborate or hexafluorophosphate salts, have been prepared by the reaction of $Fe(\eta^5\text{-}C_5H_5)(CO)_2Br$ with simple olefins in the presence

*Department of Chemistry, Brandeis University, Waltham, MA 02254.
†Department of Chemistry, Boston University, Boston, MA 02215.
‡Union Carbide Corp., Bound Brook, NJ 08805.
§Central Research and Development Department, E. I. du Pont de Nemours & Co., Wilmington, DE 19898.

of Lewis acid catalysts,[1] by protonation of allyl ligands in Fe(η^5-C_5H_5)$(CO)_2$[(allyl)κC^1] complexes,[2] or by treatment of these with cationic electrophiles,[3] by hydride abstraction from Fe(η^5-C_5H_5)$(CO)_2$(alkyl) complexes,[4] through reaction of epoxides with Fe(η^5-C_5H_5)$(CO)_2$ anion followed by protonation,[5] or by thermally induced ligand exchange between [Fe(η^5-C_5H_5)$(CO)_2$(η^2-2-methyl-1-propene)][BF_4][5–7] or [Fe(η^5-C_5H_5)$(CO)_2$(tetrahydrofuran)][BF_4][8] and excess olefin.

The latter two methods are often the most convenient. Dicarbonyl(η^5-cyclopentadienyl)(η^2-2-methyl-1-propene)iron(1+) tetrafluoroborate is a readily synthesized crystalline solid that can be stored indefinitely at −20°. When solutions of the salt are heated in 1,2-dichloroethane (65–70°, 5–10 min) or in dichloromethane (40°, 3–4 hr) in the presence of 2–3 *M* equivalents of an olefin, ligand exchange occurs, yielding the derived [Fe(η^5-C_5H_5)$(CO)_2$(olefin)][BF_4] complex.[3,5] The exchange reaction is limited to the preparation of those complexes that are thermodynamically more stable than the 2-methyl-1-propene complex itself under the conditions of the exchange reaction. These generally include terminal, alkyl-substituted olefins, 1,2-dialkyl-substituted olefins, and cycloalkenes. Heteroatoms such as O, N, and S present in the olefin may interfere with formation of the olefin complex through competitive complexation.

The procedure given here can be completed easily within a day. Although specific for the preparation of the 2-methyl-1-propene complex it can be adapted readily as an alternative method for the preparation of [Fe(η^5-C_5H_5)$(CO)_2$(η^2-olefin)][BF_4] complexes through metallation of an allyl halide or tosylate, followed by protonation of the monohapto-allyliron complex.

A. DICARBONYL(η^5-CYCLOPENTADIENYL)(2-METHYL-1-PROPENYL-κC^1)IRON

$$[\mathrm{Fe}(\eta^5\text{-}\mathrm{C_5H_5})(\mathrm{CO})_2]_2 + 2\mathrm{Na(Hg)} \xrightarrow{\mathrm{THF}} 2\mathrm{Na}[\mathrm{Fe}(\mathrm{CO})_2(\eta^5\text{-}\mathrm{C_5H_5})]^* + 2\mathrm{Hg}$$

$$\mathrm{Na}[\mathrm{Fe}(\mathrm{CO})_2(\eta^5\text{-}\mathrm{C_5H_5})] + \mathrm{C_4H_7Cl} \xrightarrow{\mathrm{THF}} \mathrm{Fe}(\eta^5\text{-}\mathrm{C_5H_5})(\mathrm{CO})_2(2\text{-}\mathrm{MeC_3H_4}\text{-}\kappa C^1) + \mathrm{NaCl}$$

Procedure

■ **Caution.** *Care should be exercised in the preparation of the mercury amalgam because the initial reaction is highly exothermic. This and all subsequent operations should be carried out in a well-ventilated fume hood.*

*For smaller scale preparations, K[Fe$(CO)_2$(η^5-C_5H_5)], available from Alfa Division of Ventron Corporation or preparable by reduction of [Fe(η^5-C_5H_5)$(CO)_2$]$_2$ with potassium metal,[9] may be used.

A 500-mL three-necked flask with a stopcock at the bottom is fitted with a nitrogen inlet and a motor-driven mechanical stirrer with a Teflon paddle. The flask is flushed thoroughly with nitrogen while being flame-dried, and then 30 mL of mercury is introduced. A pan may be placed under the flask in case of breakage. The mercury is stirred vigorously as 4.5 g (0.196 mole) of sodium metal, cut into small pieces, is slowly added to it. The flask is capped with a rubber septum. (■ **Caution.** *The amalgamation of sodium is highly exothermic. Small pieces of sodium must be added to mercury behind a shield.*) After the resulting hot amalgam has cooled to room temperature, 200 mL of tetrahydrofuran (THF), which is predried over KOH and then freshly distilled under a nitrogen atmosphere from sodium benzophenone ketyl, is added. In general, transfer of large volumes of dry solvent or of solutions is best made by a 10 gauge cannula [Hamilton] inserted through rubber septa capping both delivery and receiver vessels. Transfer is made by positive nitrogen pressure applied through a hypodermic needle, while a second needle in the receiver is used as a vent. Eighteen grams (0.051 mole) of dicarbonyl(η^5-cyclopentadienyl)iron dimer[10] [Alfa or Strem] is added at once, and vigorous stirring is continued at room temperature for 30–45 min. The progress of the reaction can be monitored by following the changes in the carbonyl region of the IR spectrum of the solution, employing carefully dried sodium chloride liquid sample cells filled by syringe under nitrogen. Carbonyl absorption bands of the dimer $[Fe(CO)_2(\eta^5\text{-}C_5H_5)]_2$ at 1995, 1950, and 1780 cm^{-1} are replaced by those of the salt, which exhibits strong absorption bands at 1877 and 1806 cm^{-1} due to the tight ion pair as well as weaker absorptions at 1862, 1786, and 1770 due to solvent-separated and carbonyl-bridged ion pairs.[11] The amalgam is drained, and the amber-red solution of sodium dicarbonyl (η^5-cyclopentadienyl)ferrate(1 −) is ready for use without further purification.

The solution is cooled in an ice bath and is stirred rapidly as 9.65 g (0.107 mole) of 1-chloro-2-methyl-1-propene (isobutenyl chloride) [Aldrich] is added over a period of 5 min. The reaction can be followed by observing the changes in the IR spectrum of the solution. The carbonyl absorption bands characteristic of the anion are replaced by those typical of the product at 1998 and 1950 cm^{-1}. Upon completion of the addition of 1-chloro-2-methyl-1-propene, stirring at 0° is continued for 1 hr to ensure completion of the reaction. The resulting solution of dicarbonyl (η^5-cyclopentadienyl)(2-methyl-1-propenyl-κ*C*^{1})iron may be used directly without purification. Alternatively, the product can be isolated and purified by removing solvent under reduced pressure, followed by chromatography of the residue on 300 g of neutral activity III alumina. The column is made up in anhydrous diethyl ether, and after dissolving the crude product in petroleum ether, elution under N_2 is carried out with this solvent. The product may be further purified by short-path distillation at pressures less than 10^{-4} mm (pot temperature less than 40°). It is then sufficiently pure to be stored at −20° for prolonged periods without decomposition. The yield of dark amber oil is typically 19–20 g (80–90%).

Anal. Calcd. for $C_{11}H_{12}FeO_2$: C, 56.93; H, 5.21. Found: C, 57.19; H, 5.35.

The IR spectrum of the complex in dichloromethane solution is characterized by two strong bands in the carbonyl region, at 2003 and 1945 cm^{-1}. A ^1H NMR spectrum of the complex in chloroform-*d* exhibits the following resonances; δ 4.63 (s, 5, Cp), δ 4.47 (m, 2, CH_2=), δ 2.11 (s, 2, CH_2), and δ 1.77 (s, 3, CH_3).

B. DICARBONYL(η^5-CYCLOPENTADIENYL)(η^2-2-METHYL-1-PROPENE)IRON(1+) TETRAFLUOROBORATE

$$Fe(\eta^5\text{-}C_5H_5)(CO)_2(2\text{-}MeC_3H_4\text{-}\kappa C^1) + H[BF_4] \rightarrow [Fe(\eta^5\text{-}C_5H_5)(CO)_2(\eta^2\text{-}C_4H_8)][BF_4]$$

The above product is placed in a dry 1-L single-necked round-bottomed flask with a side arm. The flask is flushed with nitrogen and fitted with a magnetic stirring bar and a rubber septum. Anhydrous diethyl ether (300 mL) is degassed by purging for several minutes with a stream of dry nitrogen, using a gas dispersion tube, and is then transferred to the 1-L flask by cannula. The solution is cooled to 0° in an ice bath. Then 17 mL of 48% aqueous hydrogen tetrafluoroborate(1−) [Ozark-Mahoning] (0.12 mole) is added slowly by syringe while the solution is stirred vigorously. Manual shaking may be necessary at the end to ensure mixing of the reactants. A yellow-orange precipitate forms immediately. The septum is removed, and the mixture is transferred by a 2.5-mm cannula to a Schlenk tube[12] fitted with a coarse-porosity sintered glass filter. The product is washed with anhydrous diethyl ether until the washings are colorless and is then dried by passing a stream of dry nitrogen through the Schlenk tube.

The crude product may be purified as follows. The Schlenk tube receiver is replaced by a 500-mL round-bottomed flask with a magnetic stirring bar. The Schlenk tube outlet stopcock is closed, and the crude salt is taken up in 30 mL of dichloromethane that was previously dried over 4A molecular sieves and then deoxygenated by nitrogen purge. The stopcock is then opened, and the resulting cherry red solution is vacuum-filtered into the round-bottomed flask. The process is repeated several times with smaller portions of dichloromethane until the washings are colorless. The Schlenk tube is then replaced by a rubber septum, and the dichloromethane solution is cooled in an ice bath and stirred vigorously as 250 mL of anhydrous diethyl ether is added over a period of 5 min. The resulting golden yellow solid is transferred as before to a Schlenk tube with a medium-porosity filter. The filter cake is washed several times with small portions of diethyl ether, dried under nitrogen in the Schlenk tube, and finally dried under vacuum. The yield of yellow crystalline dicarbonyl(η^5-cyclopentadienyl)(η^2-2-methyl-1-propene)iron(1+) tetrafluoroborate is 25–28 g (78–88%).

Anal. Calcd. for $C_{11}H_{13}BF_4FeO_2$: C, 41.30; H, 4.10. Found: C, 41.19; H, 3.78. IR (CH_3NO_2) ν_{CO} 2030, 2070 cm^{-1}. NMR (CD_3NO_2): δ 5.64 (s, 5, Cp), 3.91 (s, 2, CH_2=), 1.96 (s, 6, CH_3).

The product may be stored indefinitely under nitrogen at −20° without decomposition. It is soluble in dichloromethane, acetone, and nitromethane but insoluble in hydrocarbons and in diethyl ether.

References

1. E. O. Fischer and K. Fichtel, *Chem. Ber.*, **94,** 1200 (1961); **95,** 2063 (1962); E. O. Fischer and E. Moser, *Inorg. Synth.*, **12,** 38 (1970).
2. M. L. H. Green and P. L. I. Nagy, *J. Chem. Soc.*, **1963,** 189.
3. A. Cutler, D. Ehntholt, P. Lennon, K. Nicholas, D. F. Marten, M. Madhavarao, S. Raghu, A. Rosan, and M. Rosenblum, *J. Am. Chem. Soc.*, **97,** 3149 (1975).
4. M. L. H. Green and P. L. I. Nagy, *J. Organometal. Chem.*, **1,** 58 (1963).
5. W. P. Giering, M. Rosenblum, and J. Tancrede, *J. Am. Chem. Soc.*, **94,** 7170 (1972); A. Cutler, D. Ehntholt, W. P. Giering, P. Lennon, S. Raghu, A. Rosan, M. Rosenblum, J. Tancrede, and D. Wells, ibid., **98,** 3495 (1976).
6. W. P. Giering and M. Rosenblum, *J. Chem. Soc., Chem. Commun.*, **1971,** 441.
7. B. Foxman, D. Marten, A. Rosan, S. Raghu, and M. Rosenblum, *J. Am. Chem., Soc.*, **99,** 2160 (1977).
8. D. L. Reger and C. Coleman, *J. Organometal. Chem.*, **131,** 153 (1977).
9. J. S. Plotkin and S. G. Shore, *Inorg. Chem.*, **20,** 284 (1981).
10. J. J. Eisch and R. B. King (eds.), *Organometallic Syntheses,* Vol. 1, Academic Press, New York, p. 114.
11. M. Nitay and M. Rosenblum, *J. Organometal. Chem.*, **186,** C23 (1977).
12. D. F. Shriver, *The Manipulation of Air-Sensitive Compounds,* McGraw-Hill, New York, 1969, p. 147.

48. TETRAETHYLAMMONIUM μ-CARBONYL-1κ*C*:2κ*C*-DECACARBONYL-1κ^3*C*,2κ^3*C*,3κ^4*C*-μ-HYDRIDO-1κ:2κ-*triangulo*-TRIRUTHENATE(1−)

$$2Ru_3(CO)_{12} + 2Na[BH_4] \xrightarrow{THF} 2Na[HRu_3(CO)_{11}] + 2CO + B_2H_6$$

$$Na[HRu_3(CO)_{11}] + Et_4NBr \xrightarrow{MeOH} Et_4N[HRu_3(CO)_{11}] + NaBr$$

Submitted by GEORG SÜSS-FINK*
Checked by NANCY JONES,† HERBERT D. KAESZ,† and JOSEPH W. KOLIS‡

There are several literature reports[1–7] on the formation of the trinuclear cluster anion $[HRu_3(CO)_{11}]^-$, all of which involve reactions of $Ru_3(CO)_{12}$ with basic reagents. Of synthetic value are the reactions with $Na[BH_4]$,[2] $(Et_4N)[BH_4]$,[3] and KH.[4] The standard preparation, sodium tetrahydroborate reduction of $Ru_3(CO)_{12}$,[2] affords $Na[HRu_3(CO)_{11}]$ in tetrahydrofuran (THF) solution almost quantitatively within 30 min. The anionic product can be isolated as the tetraethylammonium salt by crystallization from methanol, giving yields up to 85%. Alternatively, the isolation as the bis(triphenylphosphoranylidene)ammonium salt is also possible.[2] The cluster anion $[HRu_3(CO)_{11}]^-$ acts as a homogeneous catalyst in hydroformylation,[8,9] hydrogenation,[9] hydrosilylation,[9,10] and silacarbonylation,[9,11] in the water-gas shift reaction,[4,12,13] and in syn-gas reactions.[14,15]

Procedure

The reaction can be conducted with standard Schlenk techniques.[16] All manipulations must be carried out under purified nitrogen, and all solvents must be distilled over drying agents and saturated with nitrogen prior to use.[16] In a 250-mL Schlenk tube, 640 mg (1 mmol) of $Ru_3(CO)_{12}$[17] [Alfa] and 160 mg (4 mmol) of $Na[BH_4]$ are dissolved in 100 mL of tetrahydrofuran. The solution, which turns dark red, is stirred at 25° for 30 min. The reaction mixture is filtered through a 2-cm layer of filter pulp, using a 250-mL Schlenk frit, and the filtrate is evaporated to dryness. The residue is dissolved in methanol (30 mL), and after addition of Et_4NBr (250 mg, 1.2 mmol) in methanol (10 mL), the solution is concentrated to 10 mL. The product is allowed to crystallize at room temperature for 6 hr; then the solution is cooled to −78°. After 15 hr, the crystalline

*Laboratorium für Anorganische Chemie der Universität Bayreuth, Postfach 3008, D-8580 Bayreuth, F.R.G.
†Department of Chemistry, University of California, Los Angeles, CA 90024.
‡Department of Chemistry, Northwestern University, Evanston, IL 60201.

precipitate of analytically pure $(Et_4N)[HRu_3(CO)_{11}]$ is isolated, washed three times with methanol (2 mL) at $-78°$, and dried under vacuum (10^{-5} torr). A typical yield is 600 mg (81%).

Anal. Calcd. for $C_{19}H_{21}NO_{11}Ru_3$: C, 30.73; H, 2.86; N, 1.89. Found: C, 31.00; H, 3.00; N, 1.98.

Properties

The title compound, $(Et_4N)[HRu_3(CO)_{11}]$, is a deep red, almost black, crystalline material. The crystals are only slightly air-sensitive and decompose rather indistinctly over the range 325–330°. They dissolve in polar solvents such as THF, *N,N*-dimethylformamide, CH_2Cl_2, CH_3CN, or CH_3OH. The blood-red solutions that result are much more sensitive to oxygen than the solid. The IR spectrum of $(Et_4N)[HRu_3(CO)_{11}]$ (in CH_2Cl_2 solution) displays characteristic absorptions at 2074 (w), 2017 (vs), 1988 (s), 1955 (m), and 1696 (w) cm^{-1}

References

1. J. Knight and M. J. Mays, *J. Chem. Soc., Dalton Trans.*, **1972,** 1022.
2. B. F. G. Johnson, J. Lewis, P. R. Raithby, and G. Süss-Fink, *J. Chem. Soc., Dalton Trans.*, **1979,** 1356.
3. D. H. Gibson, F. U. Ahmed, and K. R. Phillips, *J. Organometal. Chem.*, **218,** 325 (1981).
4. J. C. Bricker, C. C. Nagel, and S. G. Shore, *J. Am. Chem. Soc.*, **104,** 1444 (1982).
5. J. B. Keister, *J. Chem. Soc., Chem. Commun.*, **1979,** 214.
6. C. Ungermann, V. Landis, S. A. Moya, H. Cohen, H. Walter, R. G. Pearson, R. G. Rinker, and P. C. Ford, *J. Am. Chem. Soc.*, **101,** 5922 (1979).
7. C. R. Eady, P. F. Jackson, B. F. G. Johnson, J. Lewis, M. C. Malatesta, M. McPartlin, and W. J. H. Nelson, *J. Chem. Soc., Dalton Trans.*, **1980,** 383.
8. G. Süss-Fink, *J. Organometal. Chem.*, **193,** C20 (1980).
9. G. Süss-Fink and J. Reiner, *J. Mol. Catal.*, **16,** 231 (1982).
10. G. Süss-Fink and J. Reiner, *J. Organometal. Chem.*, **221,** C36 (1981).
11. G. Süss-Fink, *Angew. Chem.*, **94,** 72 (1982); *Angew. Chem. Int. Ed. Engl.*, **21,** 73 (1982); *Angew. Chem. Suppl.*, **1982,** 71.
12. R. M. Laine, *Ann. N. Y. Acad. Sci.*, **333,** 124 (1980).
13. P. C. Ford, *Acc. Chem. Res.*, **14,** 31 (1981).
14. B. D. Dombek, *J. Am. Chem. Soc.*, **103,** 6508 (1981).
15. J. F. Knifton, *J. Am. Chem. Soc.*, **103,** 3959 (1981).
16. S. Herzog and J. Dehnert, *Z. Chem.*, **4,** 1 (1964).
17. M. I. Bruce, J. G. Matisons, R. C. Wallis, B. W. Skeleton, and A. H. White, *J. Chem. Soc., Dalton Trans.*, *1983*, 2365.

49. BROMO(η^5-CYCLOPENTADIENYL)-[1,2-ETHANEDIYLBIS(DIPHENYLPHOSPHINE)]IRON AND BROMO(η^5-CYCLOPENTADIENYL)-[1,2-ETHANEDIYLBIS(DIPHENYLPHOSPHINE)]-BIS(TETRAHYDROFURAN)IRONMAGNESIUM(*Fe-Mg*)

Submitted by S. G. DAVIES,* H. FELKIN,* and O. WATTS*
Checked by S. J. SIMPSON†

Fe(OC)(CO)Br + $\langle$PPh$_2$ / PPh$_2$ $\xrightarrow{h\nu}$ Fe(Ph$_2$P)(PPh$_2$)Br $\xrightarrow[2THF]{Mg}$ Fe(Ph$_2$P)(PPh$_2$)MBr·2thf

thf = tetrahydrofuran

The preparation of bromo(η^5-cyclopentadienyl)[1,2-ethanediylbis(diphenylphosphine)]iron described here (up to 63% yield) is an improvement of the original literature reaction (41% yield).[1] This compound is also formed when cyclopentadienylthallium(I) is reacted with $Fe(Ph_2PCH_2CH_2PPh_2)Br_2$.[2] It readily undergoes one-electron oxidation[3] and reacts with a variety of nucleophiles, with replacement of bromide.[4]

The use of activated magnesium powder[5] for the preparation of bromo(η^5-cyclopentadienyl)[1,2-ethanediylbis(diphenylphosphine)]bis(tetrahydrofuran) ironmagnesium(*Fe-Mg*) leads to more reproducible, and consistently higher, yields than magnesium turnings. The crystal structure and reactions of $Fe[MgBr(thf)_2](\eta^5\text{-}C_5H_5)(Ph_2PCH_2CH_2PPh_2)$ (thf = tetrahydrofuran) have been described elsewhere.[4,6] This iron-magnesium complex is one of the few examples of an inorgano-Grignard that contains a transition metal–magnesium bond[7] rather than having the magnesium bonded to a CO ligand.[8] The iron-magnesium Grignard reagent is a valuable source of electron-rich iron complexes due to its nucleophilicity.

*Institut de Chimie des Substances Naturelles, C.N.R.S., 91190 Gif-sur-Yvette, France.
†Department of Chemistry, University of Salford, Salford, England M5 4WT.

A. BROMO(η^5-CYCLOPENTADIENYL)[1,2-ETHANEDIYLBIS(DIPHENYLPHOSPHINE)]IRON

Procedure

All manipulations are carried out under a nitrogen (or argon where stated) atmosphere using conventional Schlenk tube techniques.[9] All the solvents used are deoxygenated (saturated with nitrogen). To achieve this, the solvent is boiled briefly at room temperature under reduced pressure and then shaken briefly with nitrogen at atmospheric pressure. This procedure is repeated twice (three times in all).

The compounds $Fe(\eta^5\text{-}C_5H_5)(CO)_2Br$[10] (1.54 g, 6.0 mmol) and 1,2-ethanediylbis(diphenylphosphine) (2.46 g, 6.2 mmol) are dissolved in anhydrous benzene (200 mL distilled from CaH_2) and transferred into a standard quartz, water-cooled, deoxygenated photochemical vessel fitted with a magnetic stirrer. The concentration of $Fe(\eta^5\text{-}C_5H_5)(CO)_2Br$ must be $\leq 0.03\ M$.

■ **Caution.** *Benzene is a suspected carcinogen. It should be handled in a well-ventilated fume hood, and gloves should be worn at all times.*

During irradiation of the stirred red solution with a medium-pressure Hanovia 450-W UV lamp (no. 679A) for 18 hr, argon is introduced through a frit at the bottom of the reaction vessel. Its purpose is to remove the carbon monoxide that forms in the photolysis. The resultant violet solution is filtered through Celite (3 cm). This filtration removes a pale green precipitate (~20% yield) consisting of a mixture of $[CpFe(diphos)(CO)_2]Br$ and [CpFe(diphos)CO]Br. The former compound can be converted into the latter by treatment with Me_3NO. The filtrate is evaporated under reduced pressure (0.1 torr) to yield a violet gum. This gum is dissolved in dichloromethane (20 mL), and pentane (100 mL) is added. Black crystals form slowly (about 12 hr at room temperature) and are separated by filtration. The supernatant is evaporated to a violet gum and crystallized from the same solvent mixture. Both crops of crystals are washed quickly with cold ($< -10°$) acetone (2 × 5 mL) in order to remove a small amount of an unidentified fluffy off-white precipitate. They are recrystallized from dichloromethane (15 mL) by addition of pentane (130 mL) and cooling to −30° for 14 hr. Filtration and drying under vacuum gives 2.15–2.26 g (60–63%) of $Fe(\eta^5\text{-}C_5H_5)(Ph_2PCH_2CH_2PPh_2)Br$; mp 208–211°.

Anal. Calcd. for $[C_{31}H_{29}BrFeP_2]$: C, 62.1; H, 4.9; P, 10.3; Br, 13.3. Found: C, 61.9; H, 4.9; P, 10.3; Br, 13.5.

Properties

Bromo(η^5-cyclopentadienyl)[1,2-ethanediylbis(diphenylphosphine)]iron is an air-sensitive black crystalline solid that is soluble in many organic solvents (benzene, dichloromethane, diethyl ether, tetrahydrofuran, carbon disulfide, and chloroform). The ^1H NMR spectrum (60 MHz, CS_2) shows a triplet at δ 3.92 (J = 2 Hz, 5H, C_5H_5), multiplets at δ 7–8 (20H, aryl-H), and a multiplet at δ 2.1–2.55 (4H).

B. BROMO(η^5-CYCLOPENTANEDIENYL)[1,2-ETHANEDIYLBIS(DIPHENYLPHOSPHINE)]-BIS(TETRAHYDROFURAN)IRONMAGNESIUM(*Fe-Mg*)

Procedure

All reactions and manipulations are carried out under a nitrogen atmosphere (dry and oxygen-free) using conventional Schlenk tube techniques.[9] Tetrahydrofuran (THF) is distilled under nitrogen from sodium-benzophenone ketyl. The glassware is flamed under vacuum, and the filter paper and magnesium turnings are oven-dried (100°). Filtration is carried out using a dry polyethylene or steel tube fitted with a double layer of dry filter paper.

1,2-Dibromoethane (0.38 mL, 4.6 mmol) is added to a magnetically stirred suspension of magnesium turnings (140 mg, 5.8 mg-atom) in anhydrous THF (50 mL). Ethylene is given off. (■ **Caution.** *This reaction can be quite vigorous.*) The mixture is then heated under reflux for 20 min. The hot $MgBr_2$ solution is filtered onto good quality potassium metal[5] (328 mg, 8.4 mg-atom) and is refluxed for 2 hr. The solvent is removed from the resulting fine black magnesium powder by filtration, as above. (■ **Caution.** *Magnesium powder is pyrophoric in air.*) The powder must not be washed with THF, as a small amount of $MgBr_2$ is essential for the following reaction. To the magnesium powder is added $Fe(\eta^5\text{-}C_5H_5)(Ph_2PCH_2CH_2PPh_2)Br$ (500 mg, 0.8 mmol) in anhydrous THF (15 mL). The mixture is stirred magnetically for 16 hr at 20°. During this time the color of the solution changes from violet to deep red. Filtration, which can be slow, gives a solution of $Fe[MgBr(thf)_2](\eta^5\text{-}C_5H_5)(Ph_2PCH_2CH_2PPh_2)$ suitable for most purposes.[4,6] On cooling (−30°), red crystals of $Fe[MgBr(thf)_2](\eta^5\text{-}C_5H_5)(Ph_2PCH_2CH_2PPh_2)$ which contain two THF molecules of crystallization are obtained (280–420 mg, 40–60%).

Properties

Bromo(η^5-cyclopentadienyl)[1,2-ethanediylbis(diphenylphosphine)]bis(tetrahydrofuran)ironmagnesium(*Fe-Mg*) is extremely sensitive to oxygen and moisture. It is soluble in THF and benzene. The ^1H NMR spectrum of the THF adduct (60 MHz, C_6D_6) shows δ 1.3 (16H, multiplet), δ 3.42 (12H, multiplet, —CH_2—O—,THF), δ 4.23 (5H, broad, C_5H_5), and δ 7.0–8.0 (20H, multiplet, aryl-H). Reaction of the solution of $Fe[MgBr(thf)_2](\eta^5-C_5H_5)(Ph_2PCH_2CH_2PPh_2)$ with EtOH at $-78°$ followed by evaporation gave $Fe(\eta^5-C_5H_5)(Ph_2PCH_2CH_2PPh_2)H$ (381 mg, 88%) after chromatography (neutral alumina, pentane-diethyl ether 1:1).

References

1. R. B. King, L. W. Houk, and K. H. Pannell, *Inorg. Chem.*, **8,** 1042 (1969).
2. R. D. Adams, A. Davison, and J. P. Selegue, *J. Am. Chem. Soc.*, **101,** 7232 (1979).
3. P. M. Treichel, K. P. Wagner, and H. J. Mueh, *J. Organometal. Chem.*, **86,** C13 (1975).
4. H. Felkin, P. J. Knowles, and B. Meunier, *J. Organometal. Chem.*, **146,** 151 (1978).
5. R. D. Rieke, *Acc. Chem. Res.*, **10,** 301 (1977).
6. H. Felkin, P. J. Knowles, B. Meunier, A. Mitschler, L. Ricard, and R. Weiss, *J. Chem. Soc., Chem. Commun.*, **1974,** 44; H. Felkin, B. Meunier, C. Pascard, and T. Prangé, *J. Organometal. Chem.*, **135,** 361 (1977). H. Felkin and B. Meunier, *Nouveau J. Chim.*, **1,** 281 (1977).
7. S. G. Davies and M. L. H. Green, *J. Chem. Soc., Dalton Trans.*, **1978,** 1510.
8. G. B. McVicker, *Inorg. Synth.*, **16,** 56 (1975).
9. D. F. Shriver, *The Manipulation of Air-Sensitive Compounds*, McGraw-Hill, New York, 1969.
10. B. F. Hallam and P. L. Pauson, *J. Chem. Soc.*, **1956,** 3030.

50. (η^4-1,5-CYCLOOCTADIENE)(PYRIDINE)-(TRICYCLOHEXYLPHOSPHINE)IRIDIUM(I) HEXAFLUOROPHOSPHATE

Submitted by ROBERT H. CRABTREE* and SHEILA M. MOREHOUSE†
Checked by JENNIFER M. QUIRK‡

Homogeneous hydrogenation catalysts that selectively reduce unhindered C═C bonds are well known.[1] Some catalysts[2] also reduce tetrasubstituted C═C bonds if there are activating substituents (—NHAc,—CO_2Me) present. Only one catalyst[3] has been reported to reduce hindered olefins whether or not activating groups are present: $[Ir(cod)(py)(tcyp)][PF_6]$ (cod = 1,5-cyclooctadiene, py = pyridine, tcyp = tricyclohexylphosphine). The catalyst is also highly selective, preferentially reducing less hindered C═C groups before more hindered groups,[4] and, in certain cases, causing hydrogen addition to one face only of a chiral molecule.[4,5] Certain groups, especially —OH, on one face of a substrate can direct the attack on a nearby C═C group of the catalyst from that face by prior

*Department of Chemistry, Yale University, 225 Prospect Street, New Haven, CT 06520.
†Department of Chemistry, Manhattanville College, Purchase, NY 10577.
‡Union Carbide, Tarrytown, NY 10591.

coordination to iridium. Selectivities of 1000:1 have been observed in such cases.[5] Hydroxyl groups also appear to direct attack on nearby C═C groups rather than distant ones. Neither Pd/C nor classical homogeneous catalysts such as $RhCl(PPh_3)_3$ behave in this way. The catalyst is also insensitive to oxidizing functionality, such as carbon-halogen bonds or O_2, but is poisoned by —CO_2H, —CN, —NH_2, and to some extent —OH, but not by —CO_2R, —NHAc, —$OSiR_3$, keto, or cyclopropyl groups.

An important feature of preparation A is the use of the highly ionizing solvent Me_2CO/H_2O. No other solvent mixture tried gives such good results. In particular, other solvents lead to contamination of the product with [IrCl(cod)(py)].[6]

A. (η^4-1,5-CYCLOOCTADIENE)BIS(PYRIDINE)IRIDIUM(I) HEXAFLUOROPHOSPHATE

$$[IrCl(C_8H_{12})]_2 + 4C_5H_5N + 2NH_4[PF_6] \rightarrow 2\,[Ir(C_8H_{12})(C_5H_5N)_2][PF_6] + 2NH_4Cl$$

Procedure

Twenty milliliters of deoxygenated acetone/water (1:1) and 0.7 mL of pyridine are placed in a 100-mL Schlenk tube equipped for magnetic stirring under nitrogen. The reagent-grade solvents and reagents in this and the subsequent step do not need to be specially pure or dry. To this solution, 0.42 g (0.62 mmol) of di-μ-chloro-bis(η^4-1,5-cyclooctadiene)diiridium(I)[7] [Strem Chemical] and 0.34 g (1.9 mmol) of $K[PF_6]$ are added. The mixture is allowed to stir under nitrogen for approximately 3 hr at room temperature (20°), or until the red solid $[Ir(cod)Cl]_2$ has dissolved and the color of the mixture appears distinctly yellow. The mixture should not be heated. Care is needed to exclude air.[8]

The more volatile acetone is then largely removed by evaporating the solution gently at room temperature on a vacuum line for 10 min or until the volume of the mixture falls to about 10 mL. The yellow solid $[Ir(cod)py_2][PF_6]$ precipitates during this process and can be isolated by filtration through a frit under nitrogen. The solid is washed with three portions of 5 mL of degassed water, and dried under vacuum. Yield: 0.73 g (97%).

Anal. Calcd. for $C_{18}H_{22}N_2PF_6Ir$: C, 35.83; H, 3.68; N, 4.64. Found: C, 35.89; H, 3.76; N, 4.52.

Properties

The complex is slightly air-sensitive even in the solid state and should be kept under nitrogen or used immediately for the following preparation. The product

[Ir(cod)py$_2$][PF$_6$] can be recrystallized from dichloromethane/diethyl ether and is identified by its ^{1}H NMR spectrum. The cyclooctadiene vinylic protons appear at δ 3.83, with the complex aromatic proton resonances occuring at δ 7.1–7.7 and δ 8.6–8.8 (CDCl$_3$, 25°, shifts in ppm relative to internal TMS).

B. (η^4-1,5-CYCLOOCTADIENE)(PYRIDINE)-(TRICYCLOHEXYLPHOSPHINE)IRIDIUM(I) HEXAFLUOROPHOSPHATE

$$[Ir(C_8H_{12})(C_6H_5N)_2][PF_6] + P(C_6H_{11})_3 \rightarrow [Ir(C_8H_{12})\{P(C_6H_{11})_3\}(C_6H_5N)][PF_6] + C_6H_5N$$

Procedure

In a 100-mL Schlenk tube equipped with a stirring bar, 0.30 g (0.5 mmol) of (η^4-1,5-cyclooctadiene)bis(pyridine)iridium(I) hexafluorophosphate is dissolved in 20 mL of deoxygenated methanol under nitrogen.[8] To this solution is added 0.17 g (0.6 mmol) of tricyclohexylphosphine (tcyp) [Strem Chemical]. The orange solution is stirred for 15 min at room temperature, during which time the product begins to crystallize from the solution. The mixture must not be heated. The methanol is removed under reduced pressure at room temperature until about 5 mL remains. Diethyl ether (10 mL) is added, and the reaction mixture is cooled in an ice bath for 30 min. The orange crystalline product is isolated by filtration, washed three times with 5-mL portions of diethyl ether, and dried under vacuum. Yield: 0.39 g (97%).

Anal. Calcd. for $C_{31}H_{50}NP_2F_6Ir$: C, 46.27; H, 6.26; N, 1.74. Found C, 45.72; H, 6.23; N, 1.63.

Properties

This complex is air-stable both in the solid state and in solution. It is soluble in CH_2Cl_2, $CHCl_3$, and Me_2CO but insoluble in alcohols, water, benzene, diethyl ether, and hexane. The compound [Ir(cod)(tcyp)(py)][PF$_6$] can be recrystallized from dichloromethane/diethyl ether and identified by its ^{1}H NMR spectrum. The cyclooctadiene vinyl protons appear at δ 4.02 in the NMR. Other resonances are observed for cyclohexylphosphine (δ 0.8–2.5 complex), and for the aryl groups(δ 7.6–7.9 and δ 8.7–8.8 complex) (CDCl$_3$, 25°). The details for the use of the complex in hydrogenation are described elsewhere.[3–5]

Acknowledgment

We thank Johnson Matthey Co. for a loan of iridium, and the Petroleum Research Fund, administered by the American Chemical Society, for funding.

References

1. J. A. Osborn, F. H. Jardine, J. F. Young, and G. Wilkinson, *J. Chem. Soc., Ser. A,* **1966,** 1711.
2. H. Kagan and T-P. Dang, *J. Am. Chem. Soc.,* **94,** 6429 (1972).
3. R. H. Crabtree, H. Felkin, and G. E. Morris, *J. Organometal. Chem.,* **141,** 205 (1977).
4. J. W. Suggs, S. D. Cox, R. H. Crabtree, and J. M. Quirk, *Tetrahedron Lett.,* **22,** 303 (1981).
5. (a) R. H. Crabtree and M. W. Davis, *Organometallics,* **2,** 681 (1983); (b) G. Stork and D. E. Kahne, *J. Am. Chem. Soc.,* **105,** 1072 (1983).
6. R. H. Crabtree and G. E. Morris, *J. Organometal. Chem.,* **135,** 395 (1977).
7. R. H. Crabtree, J. M. Quirk, H. Felkin, and T. Fillebeen-Kahn, *Synth. React. Inorg. Met-Org. Chem.,* **12,** 407 (1982).
8. D. F. Shriver, *The Manipulation of Air-Sensitive Compounds,* McGraw-Hill, New York, 1969.

51. TETRACARBONYL(η^2-METHYL ACRYLATE)RUTHENIUM

$$Ru_3(CO)_{12} + 3CH_2{=}CH{-}CO_2CH_3 \xrightarrow[t \leq 15^\circ]{h\nu(\lambda \geq 370\ nm)} 3Ru(\eta^2\text{-}CH_2{=}CHCO_2CH_3)(CO)_4$$

Submitted by F.-W. GREVELS,* J. G. A. REUVERS,* and J. TAKATS†
Checked by B. F. G. JOHNSON‡

Although various synthetic routes have become available for the syntheses of tetracarbonyl(η^2-olefin) complexes of iron,[1] a general high-yield procedure for the preparation of the analogous ruthenium compounds has long been lacking. Photolysis of $Ru_3(CO)_{12}$ in the presence of excess olefin has been reported to yield $Ru(\eta^2\text{-olefin})(CO)_4$ complexes of ethylene,[2] 1-pentene,[3] ethyl acrylate,[4] and diethyl fumarate.[4] However, in no case were analytically pure materials obtained, decomposition often occurring while the excess olefin was being removed from the reaction mixture. A simple method for the photochemical preparation

*Max-Planck-Institut für Strahlenchemie, Stiftstrasse 34-36, D-4330 Mülheim a.d. Ruhr, FRG.
†Department of Chemistry, University of Alberta, Edmonton, Alberta, Canada T6G 2G2.
‡University Chemical Laboratory, Lensfield Road, Cambridge, CB2 1EW, U.K.

of tetracarbonyl(η^2-methyl acrylate)ruthenium is described here. A similar procedure affords the corresponding (η^2-dimethyl fumarate) and (η^2-dimethyl maleate) complexes in nearly quantitative yield.[5]

Procedure

■ **Caution.** *Ruthenium carbonyl complexes must be handled as toxic compounds in a well-ventilated fume hood. In particular, any kind of bodily contamination, orally or via skin contact, must be strictly avoided. Gloves should be worn.*

All operations are carried out under an atmosphere of argon. Dodecacarbonyltriruthenium can be purchased [Strem Chemicals] or it can be prepared from $RuCl_3 \cdot 3H_2O$.[6,7] Methyl acrylate (synthetic grade, 99%) [Aldrich] is used as received. Hexane (95%) [Aldrich] is distilled under argon before use.

Dodecacarbonyltriruthenium (3.20 g, 5.0 mmol), methyl acrylate (8.6 g, 100 mmol), and hexane (250 mL) are placed in a 300-mL photochemical reaction vessel (Fig. 1). The light source, a high-pressure mercury lamp Philips HPK 125 W, is located inside the reactor and is surrounded by a GWV cutoff filter

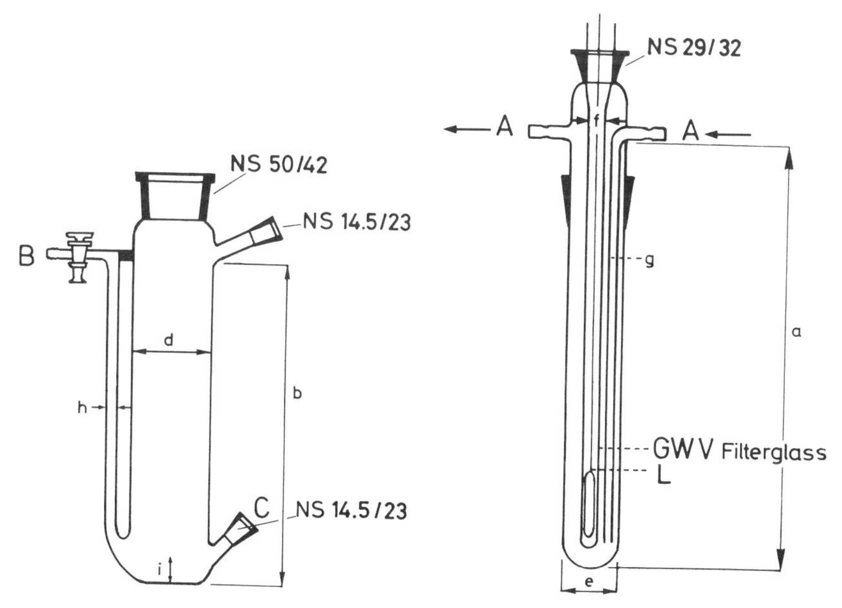

Fig. 1. Irradiation apparatus. **A**, *water cooling;* **B**, *Ar inlet;* **C**, *septum (rubber) for withdrawing IR samples;* **L** = *high-pressure mercury lamp (Philips HPK, 125-W), used in connection with a Philips VG1/HP 125-W power supply converter unit. Dimensions (in mm):* **a** = *400,* **b** = *240,* **d** = *70,* **e** = *44,* **f** = *28,* **g** = *6,* **h** = *10,* **i** = *10.*

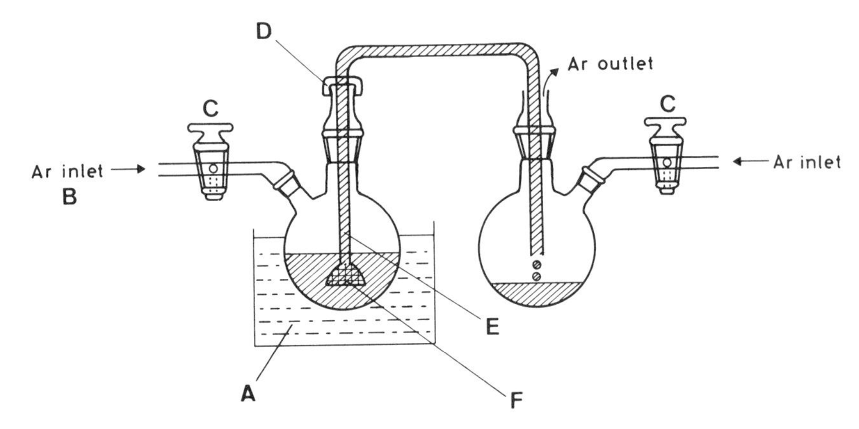

Fig. 2. Inverse filtration at low temperature. **A**, *cooling bath, Dry Ice/acetone;* **B**, *argon pressure maintained at 10–20 torr above atmospheric pressure;* **C**, *three-way stopcock;* **D**, *rubber cap;* **E**, *polyethylene tube, 3 mm, widened to 10 mm at one end;* **F**, *filter wad, cotton or glass wool.*

tube,* $\lambda \geq 370$ nm. Argon is bubbled through the solution, via inlet *B*, before the light source is turned on. As the reaction proceeds, upon irradiation at 10–15°, the solid $Ru_3(CO)_{12}$ gradually dissolves. Irradiation is continued until the orange-yellow color of $Ru_3(CO)_{12}$ has disappeared (5–8 hr). The reaction can also be monitored conveniently by means of IR spectroscopy, which shows the exclusive formation of Ru(η^2-methyl acrylate)$(CO)_4$ (as well as the disappearance of the ν_{CO} bands at 2061 (vs), 2031 (s), 2017 (w), and 2011 (m) cm^{-1} due to $Ru_3(CO)_{12}$) with ν_{CO} bands at 2121 (w), 2049.5 (s), 2035 (s), and 2008.5 (s) cm^{-1}

The solution is filtered if necessary, cooled to $-78°$, and allowed to remain at this temperature for several days. The complex precipitates as colorless crystals. The supernatant solution is removed by inverse filtration (Fig. 2) and the crystals are dried under vacuum at $-30°$. Yield: 3.50 g of Ru(η^2-methyl acrylate)$(CO)_4$ (78%).† A second crop can be obtained from the mother liquor by concentrating it to one-fourth of its original volume and cooling to $-78°$ for several days. As described above, the colorless crystals are isolated and dried under vacuum at $-30°$ (0.81 g; 18%). Ru(η^2-methyl acrylate)$(CO)_4$ may be

*Glaswerk Wertheim, Ernst-Abbe-Strasse 1, D-6980 Wertheim/Main, FRG. The Max-Planck-Institut für Strahlenchemie will pass on tube material at cost.

†The checkers found that isolation of the product is difficult largely because of its instability above $-30°$. However, since its use is that of a precursor to other materials, it can be used in situ without having been isolated.

recrystallized from hexane (precooled, $\leq -30°$) to which 0.5% methyl acrylate is added.

Anal. Calcd. for $C_8H_6O_6Ru$: C, 32.11; H, 2.02. Found: C, 32.16; H, 1.88.

Properties

Tetracarbonyl(η^2-methyl acrylate)ruthenium is obtained as a colorless solid that is stable indefinitely at temperatures below $-30°$ under argon. The complex is soluble in organic solvents but decomposes unless free methyl acrylate is added to the solvent. It appears that an equilibrium is established (Eq. 1), involving the complex, methyl acrylate, and the species $[Ru(CO)_4]$. Excess free methyl acrylate shifts the equilibrium to the left, thereby preventing decomposition of the complex and facilitating the workup of the reaction mixture. This moderate stability of Ru(η^2-methyl acrylate)$(CO)_4$ at room temperature establishes it as a useful source of the moiety $[Ru(CO)_4]$ under mild conditions. For example, addition of a cooled ($-30°$) hexane solution of a suitable ligand L to a similar solution of (**I**), followed by the slow warming up of the reaction mixture to ambient temperature, leads to the mononuclear complex $LRu(CO)_4$ (Eq. 1, L = dimethyl fumarate, fumaronitrile, maleic anhydride, trimethyl phosphite, triphenylphosphine).[5,8] In addition to these ligand exchange reactions, compound (**I**) reacts at ambient temperature (20°) with methyl sorbate and diethyl 2,4-hexadienedioate, respectively, to yield Ru(η^4-diene)$(CO)_3$ complexes, or with diethyl 2,4-hexadienedioate at slightly lower temperature, to a novel triruthenium cluster: undecacarbonyl μ^3-(1-η^1:2,3-η^2:4-η^1-diethyl) (2,4-hexadienedioate) triruthenium.[8] Dimethyl acetylenedicarboxylate is trimerized by (**I**) to hexamethyl benzenehexacarboxylate, $C_6(CO_2CH_3)_6$, at $T \leq 25°$.[8]

The use of a cutoff filter with $\lambda \geq 370$ nm is recommended in order to prevent secondary photoreactions, such as the substitution of carbon monoxide for an olefin ligand to give Ru(η^2-methyl acrylate)$_2(CO)_3$.[9] The colorless Ru(η^2-methyl acrylate)$(CO)_4$ (**I**) is transparent at $\lambda \geq 370$ nm, that is, in the region of the long-wavelength absorption maximum of the starting material $Ru_3(CO)_{12}$ at 390 nm. The absorption curve of (**I**) increases almost monotonically from about 370 nm to a maximum at 268 nm ($\epsilon \approx 7000$, in hexane that contains 0.5% methyl acrylate; the same solution is used in the reference cell). The IR spectrum of **I** exhibits four bands in the metal carbonyl region at 2121 (w), 2049.5 (s), 2035 (s), and 2008.5 (s) cm^{-1} and an ester carbonyl band at 1715 cm^{-1}. This is consistent with a trigonal-bipyramidal geometry in which the olefin occupies an equatorial position (C_{2v} local symmetry). ^{1}H NMR data: δ 1.74 (dd, 3 Hz, 8.1 Hz), H^1; 2.47 (dd, 3 Hz, 11.1 Hz), H^2; 2.81 (dd, 8.1 Hz, 11.1 Hz), H^3; 3.29 s, H^4; in toluene-d_8 at $-40°$. ^{13}C NMR data: δ 35.5 [d, 159 Hz, C(1)]; 23.9 [t, 161 Hz, C(2)]; 51.1 [q, 148 Hz, C(4)]; 176.3 [s, C(3)]; and 193.6, 194.8, 195.5,

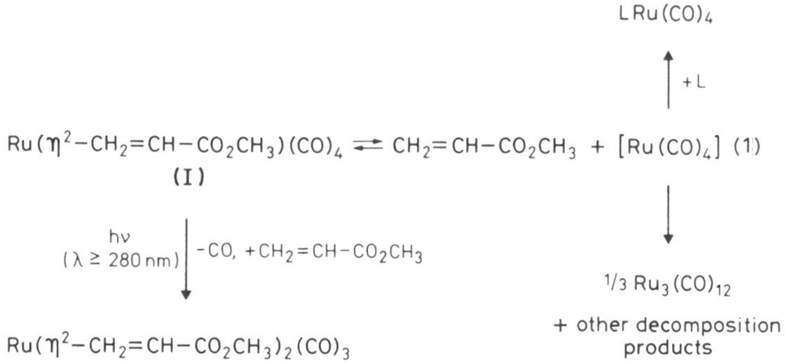

$$Ru(\eta^2\text{-}CH_2{=}CH{-}CO_2CH_3)(CO)_4 \rightleftharpoons CH_2{=}CH{-}CO_2CH_3 + [Ru(CO)_4] \quad (1)$$

(I)

$[Ru(CO)_4] \xrightarrow{+L} LRu(CO)_4$

$[Ru(CO)_4] \rightarrow 1/3\ Ru_3(CO)_{12}$ + other decomposition products

$$\text{(I)} \xrightarrow[h\nu\ (\lambda \geq 280\ nm)]{-CO,\ +CH_2=CH-CO_2CH_3} Ru(\eta^2\text{-}CH_2{=}CH{-}CO_2CH_3)_2(CO)_3$$

and 197.6 (CO); in toluene-d_8 at $-50°$ (Bruker WH 270; 270 and 67.89 MHz, respectively) (Fig. 3).

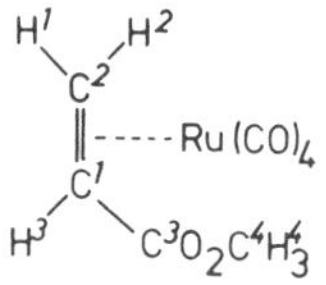

Fig. 3. Tetracarbonyl (η^2-methyl acrylate)ruthenium.

The experimental procedure described here for the preparation of Ru(η^2-methyl acrylate)$(CO)_4$ is applicable to a variety of other olefins as manifested by the syntheses of Ru(η^2-olefin)$(CO)_4$ complexes of, for example, dimethyl fumarate, dimethyl maleate, allyl acrylate, methyl vinyl ketone (3-butene-2-one), and acrylonitrile.[5]

Acknowledgment

This work was supported by the Alexander von Humboldt Foundation through a stipend (to J. T.). We thank Mr. K. Schneider for technical assistance.

References

1. R. B. King, in *The Organic Chemistry of Iron*, Vol. 1, E. A. Koerner von Gustorf, F.-W. Grevels, and I. Fischler (eds.), Academic Press, New York, 1978, p. 397.
2. B. F. G. Johnson, J. Lewis, and M. V. Twigg, *J. Organomet. Chem.*, **67,** C75 (1974).
3. R. G. Austin, R. S. Paonessa, P. J. Giordano, and M. S. Wrighton, *Adv. Chem. Ser.*, **168,** 189 (1978).
4. L. Kruczynski, J. L. Martin, and J. Takats, *J. Organomet. Chem.*, **80,** C9 (1974).
5. F.-W. Grevels, J. G. A. Reuvers, and J. Takats, *J. Am. Chem. Soc.*, **103,** 4069 (1981).
6. A. Mantovani and S. Cenini, *Inorg. Synth.*, **17,** 47 (1976).
7. C. R. Eady, P. F. Jackson, B. F. G. Johnson, J. Lewis, M. C. Malatesta, M. McPartlin, and W. J. H. Nelson, *J. Chem. Soc., Dalton Trans.*, **1980,** 383.
8. F.-W. Grevels and J. G. A. Reuvers, unpublished results; reported in part at the 10th International Conference on Organometallic Chemistry, Toronto, Canada, August 9–15, 1981, Contribution No. 1A-06.
9. F.-W. Grevels, J. G. A. Reuvers, and J. Takats, *Angew. Chem.*, **93,** 475 (1981); *Angew. Chem. Int. Ed. Engl.*, **20,** 452 (1981).

Chapter Four

TRANSITION METAL COMPOUNDS AND COMPLEXES

52. TITANIUM(II) CHLORIDE

$$Me_3SiSiMe_3 + TiCl_4 \rightarrow 2Me_3SiCl + TiCl_2$$

Submitted by SURAJ P. NARULA* and HEMANT K. SHARMA*
Checked by OM DUTT GUPTA†

Titanium(II) chloride has been prepared by the thermal decomposition of titanium(III) chloride[1,2] and also by the reduction of titanium(IV) chloride with metals.[3–5] This compound can be obtained in relatively high purity by the direct reaction of titanium(IV) chloride and hexamethyldisilane.[6] The procedure described below is superior to the previously reported methods because simpler equipment is used and large quantities can be processed with a resultant saving in time.

Procedure

A 50-mL three-necked round-bottomed flask is equipped with a reflux condenser having a calcium chloride moisture-guard tube, a nitrogen inlet, and a dropping

*Department of Chemistry, Panjab University, Chandigarh-160014, India.
†Department of Chemistry, University of Idaho, Moscow, ID 83843.

funnel. The flask is flame dried, and dry, oxygen-free nitrogen gas is admitted for 1 hr to expel the last traces of moisture.‡ Titanium tetrachloride, 2.68 g (14 mmol), is added to the flask, and the flow of nitrogen gas is slowed. Hexamethyldisilane [Aldrich], 2.15 g (14 mmol), is added dropwise and with continuous shaking. The reaction mixture becomes yellow, and a solid compound appears. Since the reaction is exothermic, the reaction flask is kept in an ice bath during the course of the addition. The contents of the flask are brought to room temperature and then heated at reflux at 120° for 4 hr in an oil bath. A dark reddish-brown solid is formed. The supernatant liquid is removed, and the solid is washed repeatedly with dry carbon tetrachloride using a hypodermic syringe under a dry nitrogen atmosphere. The reddish-brown solid is dried under vacuum and weighed. Yield: 1.12 g (70%).

Anal. Calcd. for $TiCl_2$: Ti, 40.33; Cl, 59.66. Found: Ti, 40.72; Cl, 59.25.

Properties

Titanium(II) chloride is a dark reddish-brown solid. It is a strong reducing agent, and it deliquesces in air. It is sensitive to oxygen, and it is decomposed by water. Titanium(II) chloride is insoluble in diethyl ether, chloroform, carbon tetrachloride, dichloromethane, benzene, and hexane and soluble in absolute ethanol.

References

1. W. C. Schumb and R. F. Sundström, *J. Am. Chem. Soc.*, **55,** 596 (1933).
2. D. G. Clifton and G. E. McWood, *J. Phys. Chem.*, **60,** 311 (1956).
3. O. Ruff and F. Neumann, *Z. Anorg. Allgem. Chem.*, **128,** 81 (1923).
4. W. Klemm and L. Grimm, *Z. Anorg. Allgem. Chem.*, **249,** 198 (1942).
5. P. Ehrlich, H. J. Hein, and H. Kuhnl, *Z. Anorg. Allgem. Chem.*, **292,** 139 (1957).
6. R. C. Paul, A. Arneja, and S. P. Narula, *Inorg. Nucl. Chem. Lett.*, **5,** 1013 (1969).

‡The checker reports that oven-dried glass apparatus is adequate for this reaction.

53. TRIS(2,2,6,6-TETRAMETHYL-3,5-HEPTANEDIONATO)CHROMIUM(III)

$$6(CH_3)_3C\underset{\underset{O}{\|}}{C}CH_2\underset{\underset{O}{\|}}{C}C(CH_3)_3 + 2CrCl_3\cdot 6H_2O + 3NH_2\overset{\overset{O}{\|}}{C}NH_2 \rightarrow$$

$$2(C_{11}H_{19}O_2)_3Cr + 6NH_4Cl + 3CO_2 + 9H_2O$$

Submitted by DALE STILLE* and J. R. DOYLE*
Checked by JAMES E. FINHOLT† and GRETCHEN E. MCGUIRE†

This procedure is a modification of the method reported previously for the preparation of tris(2,2,6,6-tetramethyl-3,5-heptanedionato)chromium(III)[1] and is based on the technique developed for the preparation of tris(2,4-pentanedionato)chromium(III).[2]

Tris(2,4-pentanedionato)chromium(III) and tris(2,2,6,6-tetramethyl-3,5-heptanedionato)chromium(III) have been utilized as so-called shiftless spin-lattice relaxation reagents.[3] These reagents are also of value in suppressing unfavorable nuclear Overhauser effects, in making ^{13}C NMR data quantitative, and as NMR spin labels to assign spectral lines in ^{13}C spectra.

Procedure

A solution of 3.0 g (0.011 mole) of chromium(III) chloride hexahydrate ($CrCl_3\cdot 6H_2O$) in a mixture of 25 mL of water and 65 mL of absolute ethanol is prepared in a 250-mL round-bottomed flask. To this solution is added 20.0 g (0.33 mole) of urea, 5.0 g (0.027 mole) of 2,2,6,6-tetramethyl-3,5-heptanedione (dipivaloylmethane) [Aldrich], and a magnetic stirring bar. The flask is fitted with a reflux condenser, and the mixture is heated to reflux (~85°) with stirring for 24 hr. During this time the solution changes from deep green to a very dark purple and a dark-colored solid separates from the reaction mixture. The solution is cooled to room temperature and diluted with 100 mL of water. The product is separated by filtering the mixture and is then washed with three 50-mL portions of water. The precipitate, 5.4 g, is dried overnight in air at room

*Department of Chemistry, University of Iowa, Iowa City, IA 52242.
†Department of Chemistry, Carleton College, Northfield, MN 55057.

temperature and then sublimed under vacuum (180°/0.1 torr).* Care should be exercised during the sublimation to prevent contamination of the dark purple product by a light green, slightly less volatile, contaminant. If the sublimation temperature is maintained near 180°, the light green solid will not sublime at the pressure specified. The yield of the sublimed product is about 4.5 g. The product can be purified further by dissolving the pulverized sublimate in 150 mL of boiling ethanol (slow dissolution), filtering, and evaporating the filtrate to 75 mL. The product separates upon cooling to room temperature and can be recovered by suction filtration. The yield of recrystallized product is 3.6 g. A second crop of crystals can be obtained by evaporation of the filtrate to 20 ml and cooling. The yield of the second crop is 0.5 g. The combined yield is 4.1 g (76%).†

Anal. Calcd. for $C_{33}CrH_{57}O_6$: C, 65.86; H, 9.55. Found: C, 66.39, H, 9.98.

Properties

The tris(2,2,6,6-tetramethyl-3,5-heptanedionato)chromium(III) crystals are usually purple platelets, but a ruby red needle-shaped polymorph may be formed on the slow evaporation of a 95% ethanol solution of the product. Bands in the IR spectrum are 2965 (s), 1593 (s), 1540 (s), 1505 (s), 1450 (m), 1387 (s), 1356 (s), 1300 (vw), 1250 (m), 1227 (m), 1180 (m), 1150 (m), 1025 (w), 965 (w), 875 (m), 793 (m), 760 (w), 740 (w), 640 (m), and 450–505 (broad m) cm^{-1}. The product is stable in air and readily soluble in both polar and nonpolar solvents, except water.

References

1. G. S. Hammond, D. C. Nonhebel, and C. S. Wu, *Inorg. Chem.*, **2,** 73 (1963).
2. W. C. Fernelius and J. E. Blanch, *Inorg. Synth.*, **5,** 130 (1957).
3. G. C. Levy, U. Edlund, and J. G. Hexem, *J. Mag. Res.*, **19,** 259 (1975).

*Checkers report that since the crude material is a light, fluffy powder, the sublimation procedure was improved by putting a wad of glass wool over the crude material and below the cold section of the sublimation apparatus. Sublimation of 5 g requires about 1.5 hr.

†Checkers report an overall yield of 65%. *Anal.* Cr: Calcd.: 8.54. Found: 8.61.

54. *cis*-BIS(1,2-ETHANEDIAMINE)DIFLUOROCHROMIUM(III) IODIDE

Submitted by DANNY T. FAGAN,* JOANN S. FRIGERIO,* and JOE W. VAUGHN*
Checked by ANI HYSLOP† and JAMES E. FINHOLT†

The complex *cis*-$[Cr(en)_2F_2]I$ is a useful starting material for the preparation of compounds of the type *cis*-$[Cr(en)_2FX]I$ ($X = Cl^-$, Br^-, NCS^-, H_2O, and NH_3).[1-3] The initial preparation[4] of *cis*-$[Cr(en)_2F_2]I$ involved the reaction of anhydrous 1,2-ethanediamine with $CrF_2 \cdot xH_2O$ that was suspended in a dry diethyl ether solution of hydrogen fluoride. Workup of the crude product from hydroiodic acid gave the desired compound. The present method, which is similar to that reported previously,[2] utilizes the direct reaction of chromium(III) fluoride-water (1/3.5) with excess dry 1,2-ethanediamine to yield *cis*-$[Cr(en)_2F_2][Cr(en)F_4] \cdot xH_2O$, from which the desired cation can be easily isolated. This method avoids the difficulty of preparing easily oxidized Cr(II) compounds and working in an atmosphere of nitrogen.

A. *cis*-BIS(1,2-ETHANEDIAMINE)DIFLUOROCHROMIUM(III) (1,2-ETHANEDIAMINE)TETRAFLUOROCHROMATE(III)

$$2CrF_3 \cdot 3.5H_2O + 3en \rightarrow [Cr(en)_2F_2][Cr(en)F_4] \cdot xH_2O + (7-x)H_2O$$

Procedure

Powdered chromium(III) fluoride-water (1/3.5) [Alfa Products], 43.0 g (0.250 mole), is added to 52.5 g (0.875 mole) of previously dried 1,2-ethanediamine contained in a 500-mL polyethylene beaker. The 1,2-ethanediamine is dried by distillation from solid sodium hydroxide pellets. The fraction that boils from 116–118° is collected.

■ **Caution.** *This and all subsequent steps until after the crude product has been air-dried must be carried out in a hood, and protective gloves must be worn.*

The reaction mixture is heated on a steam cone with frequent stirring until the reaction starts. It is best if a sturdy metal spatula is used to stir the reaction

*Michael Faraday Laboratories, Department of Chemistry, Northern Illinois University, DeKalb, IL 60115.
†Department of Chemistry, Carleton College, Northfield, MN 55057.

mixture, since the reaction produces a dense, sticky product. The time required for the reaction to start depends upon the amounts of $CrF_3 \cdot 3.5H_2O$ and 1,2-ethanediamine used as well as the purity of the $CrF_3 \cdot 3.5H_2O$. The reaction usually starts about 10–30 min after the heating begins. This point is identified by a rapid color change to purple and the evolution of considerable heat. Once the reaction has started, stirring is continued until the reaction mixture turns to a pasty purple solid. The solid is heated for an additional 15 min, after which it is placed in a large evaporating dish on a steam cone and heated for 3–4 hr. The solid is pulverized as it dries, to speed the removal of the excess amine. It is then stirred with 200 mL of acetone at room temperature for 15 min before being collected by filtration, washed with 40 mL of fresh acetone, air-dried, and ground to a powder. The stirring with acetone is repeated until the dry crude material no longer smells of unreacted diamine. The number of acetone extractions can be reduced by drying the crude product in an oven at 125° for several hours. (■ **Caution.** *This should be carried out in a hood.*) Yield: 50–51 g (96–98%). Although the crude material is suitable for further syntheses, recrystallization is necessary to obtain a pure product.

A 5.0-g sample of crude substance is added slowly to 35–40 mL of water at room temperature with constant stirring. The solution is filtered, using a 9.0-cm Büchner funnel and medium-porosity filter paper, and the stirred filtrate is slowly diluted with 40 mL of 95% ethanol to precipitate the product as a purple paste. The precipitate is collected, washed once with 20 mL of acetone, and then washed three times with 10-mL portions of diethyl ether. It is air-dried, and finally it is dried for 12 hr at 125°. Yield: 2.5 g (50%).*

Anal. Calcd. for *cis*-$[Cr(en)_2F_2][Cr(en)F_4] \cdot 0.5H_2O$: Cr, 25.55, C, 17.69; H, 6.14; N, 20.63; F, 28.01. Found: Cr, 25.49; C, 17.65; H, 5.96; N, 20.92; F, 27.18.

B. (±)-*cis*-BIS(1,2-ETHANEDIAMINE)DIFLUOROCHROMIUM(III) IODIDE

$$[Cr(en)_2F_2][Cr(en)F_4] + HI \rightarrow [Cr(en)_2F_2]I + H[Cr(en)F_4]$$

Procedure

Thirty grams (0.072 mole) of crude $[Cr(en)_2F_2][Cr(en)F_4]$ is added in approximately 5-g portions with constant stirring to 100–110 mL of water at room temperature. The purple solution is suction-filtered by using a 12.0-cm Büchner funnel and coarse filter paper. The stirred purple filtrate is acidified with 40 mL

*Checkers obtained a yield of 75% for recrystallized material.

of 47% hydriodic acid and is then slowly diluted with 450 mL of 95% ethanol followed by 300 mL of diethyl ether. The pink solid is collected by filtration, washed with three 25-mL portions of 95% ethanol, followed by 50 mL of acetone, and air-dried. Yield: 18.5 g (69%). Although the compound obtained at this point is suitable for further use in syntheses, recrystallization is necessary to obtain the pure material. A filtered solution of 9.2 g (0.027 mole) of crude $[Cr(en)_2F_2]I$ in 80 mL of water is diluted with 11 mL of 47% hydroiodic acid. This stirred solution is diluted with 190 mL of 95% ethanol and is cooled to room temperature in an ice bath. The pink solid is collected by filtration, washed three times with 10-mL portions of 95% ethanol, followed by 25 mL of acetone, and air-dried. The yield is 4.8–5.2 g (52–57%).* The compound is dried at 90–95° for 3 hr prior to analysis.

Anal. Calcd. for *cis*-$[Cr(en)_2F_2]I$: Cr, 15.43; C, 14.24; H, 4.75; N, 16.62; I, 37.68; F, 11.28. Found: Cr, 15.28; C, 14.20; H, 4.66; N, 16.69; I, 37.41; F, 11.23.

Properties

The electronic spectrum of a *cis*-$[Cr(en)_2F_2][Cr(en)F_4]$ in aqueous solution is characterized by λ_{max} 525 nm, ϵ 97.6 M^{-1} cm^{-1}; λ_{min} 439, ϵ 27.5; and λ_{max} 382, ϵ 55.2. Since the double salt undergoes rather rapid hydrolysis in aqueous solution, the spectrum should be determined at 10° using a freshly prepared solution. The electronic spectrum of *cis*-$[Cr(en)_2F_2]I$ in aqueous solution is characterized by λ_{max} 515, ϵ 75.5; λ_{min} 430, ϵ 9.8; and λ_{max} 375, ϵ 39.3[5] The geometry of the *cis*-$[Cr(en)_2F_2]^+$ cation has been confirmed by the resolution of the racemic mixture into its Λ and Δ forms.

References

1. J. W. Vaughn and A. M. Yeoman, *Syn. React. Inorg. Metal-Org. Chem.*, **7,** 165 (1977).
2. J. W. Vaughn and A. M. Yeoman, *Inorg. Chem.*, **15,** 2320 (1976).
3. J. W. Vaughn, *Inorg. Chem.*, **22,** 844 (1983).
4. K. R. A. Fehrmann and C. S. Garner, *J. Am. Chem. Soc.*, **82,** 6294 (1960).
5. J. W. Vaughn and B. J. Krainc, *Inorg. Chem.*, **4,** 1077 (1965).

*Checkers obtained a yield of 78% for recrystallized material.

55. ORGANIC INTERCALATED IONIC FERROMAGNETS OF CHROMIUM(II): BIS(ALKYLAMMONIUM) TETRACHLOROCHROMATE(II) COMPOUNDS

$$Cr_{metal} + 4HCl_{gas} \xrightarrow[\text{reflux}]{\text{EtOH}} [CrCl_4]^{2-} + 2H^+ + H_2$$

$$[CrCl_4]^{2-} + 2C_nH_{2n+1}NH_3Cl \xrightarrow{\text{EtOH}} (C_nH_{2n+1}NH_3)_2[CrCl_4] + 2Cl^- \qquad (n = 1,2, \ldots 8)$$

Submitted by CARLO BELLITTO*
Checked by GARY L. GARD†

The ferromagnetic properties of layer perovskites $(RNH_3)_2[CrX_4]$, where R is an alkyl or aryl group and X = Cl or Br, are of considerable interest because only a few samples of ferromagnetic insulators are known.[1] In addition, they are transparent in the visible region, and therefore they are interesting not only for academic reasons[2] but also because of their potential technological applications in optical modulation devices.[3]

Two different methods of preparation of these compounds are reported in the literature.[4,5] The first involves aqueous procedures and subsequent dehydration. Therefore, only a microcrystalline product is obtained. Here the second one is described.[5] A further improvement in obtaining high quality crystals suitable for magneto-optical experiments is reported elsewhere.[6]

■ **Caution.** *All reactions involving noxious reagents (methanamine, ethanamine, etc.) or corrosive substances such as gaseous hydrogen chloride and must be carried out in a well-ventilated fume hood.*

Procedure

Alkylammonium chlorides are prepared from the corresponding amines and HCl in EtOH. The first two of the series, CH_3NH_3Cl and $C_2H_5NH_3Cl$, are commercial reagents [Pfalz and Bauer]. They are all recrystallized (EtOH/HCl) prior to use and dried under vacuum over P_4O_{10}. All solvents are deoxygenated by purging with dry high-purity nitrogen for at least 20 min before use. All the reactions

*Istituto Teoria, Struttura Elettronica e Comportamento Spettrochimico dei Composti di Coordinazione del C.N.R., Area della Ricerca di Roma, 00016 Monterotondo Staz., Italy.
†Department of Chemistry, Portland State University, Portland, OR 97207.

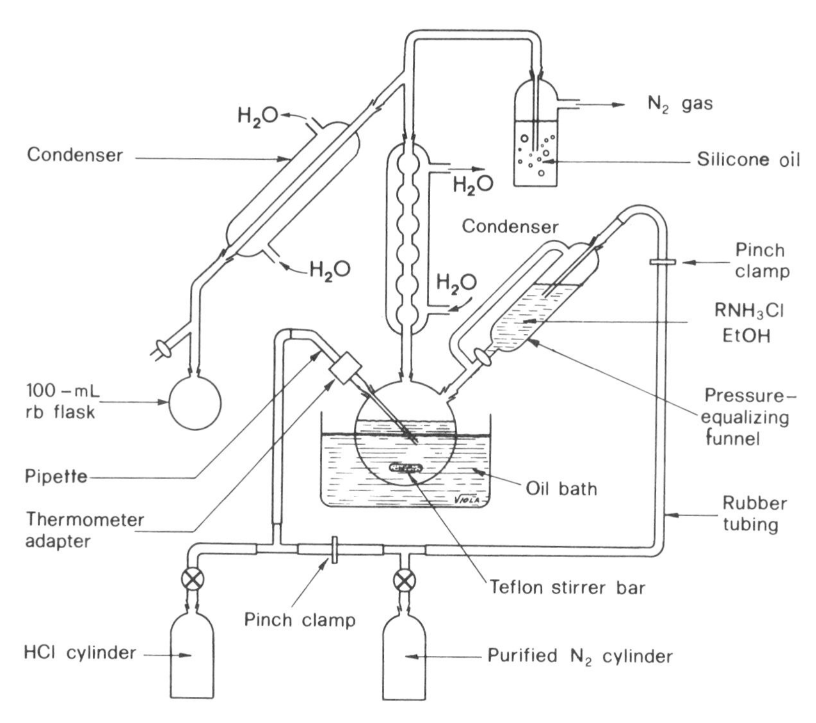

Fig. 1. Apparatus for syntheses of bis(alkylammonium)tetrachlorochromate(II) compounds.

are carried out under purified N_2, using the apparatus shown in Fig. 1. Oxygen and water must be rigorously excluded. A solution of 3.40 g (~50 mmol) of CH_3NH_3Cl in 170 mL of hot absolute ethanol is prepared by simple addition of the amine hydrochloride to solvent at 30–40°. This solution is then transferred to a 300-mL pressure-equalizing funnel under a stream of N_2 gas.

Finely divided electrolytic chromium metal [Alfa Products] (1.306 g, 25.1 mmol) is placed in a three-necked 500-mL flask containing a Teflon-coated stirring bar with 50 mL of absolute ethanol, and the N_2 gas is bubbled through the solvent for 20 min. The chromium metal is dissolved completely by passing anhydrous HCl gas under reflux through the alcohol until a deep-blue solution is obtained. The addition of HCl is then stopped while N_2 gas is bubbled through the solution.

■ **Caution.** *The N_2 exhaust containing excess of HCl must be bubbled through a water trap in a well-ventilated fume hood.*

The CH_3NH_3Cl/EtOH or other amine hydrochloride solution, degassed previously, is added dropwise to the chromium (II) solution. Immediately a yellow-green compound separates. The complex is then filtered and dried under vacuum with a Schlenk filtration apparatus. The yield averages 3.5 g (54%).

Anal. Calcd. for $C_2H_{12}N_2CrCl_4$: C, 9.30; H, 4.65; N, 10.85; Cr, 20.15; Cl, 55.00. Found: C, 9.70; H, 4.70; N, 10.90; Cr, 20.80; Cl, 54.10. Good crystals are prepared by cooling a hot saturated ethanol solution of the title compound.

The other members of the series are prepared in a similar manner. Increasing the number of carbon atoms in the cation increases the solubility of the corresponding compound in ethanol. If the solid product does not separate when the alkylammonium chloride is added to the blue solution of chromium(II), the excess solvent should be removed from the flask by distillation. At the point where the crystals begin to appear, the oil bath is lowered and the solution (deep-blue color) is allowed to cool slowly. Beautiful shiny platelet crystals separate. An alternative procedure is to start with a smaller volume of solvent and saturated solutions of alkylammonium chlorides (e.g., in the case of $(C_2H_5NH_3)_2[CrCl_4]$, 40 mL of EtOH).

Properties

The bis(alkylammonium)tetrachlorochromate(II) compounds are yellow-green platelets. These compounds are very air-sensitive and absorb water to give hydrated Cr(II) complexes. The visible absorption spectrum is peculiar because it shows a pair of sharp, well-resolved bands at 15,800 and 18,760 cm^{-1} and a charge-transfer edge starting at 24,000 cm^{-1}. The possible presence of chromium(III) as a contaminant in the compound can be seen in a UV/vis spectrum. Chromium(III) in ethanol has a broad band at 21,500 cm^{-1}, whereas chromium(II) has no absorption band at this wavelength. Solid samples can be checked for Cr(III) content by dissolving small quantities in deoxygenated spectroscopic grade ethanol in a glove bag, sealing in the solution in a 1-cm spectrophotometer cell and recording the spectrum.

References

1. P. Day, *Acc. Chem. Res.*, **12,** 236 (1979).
2. L. J. DeJongh and A. R. Miedema, *Adv. Phys.*, **23,** 1 (1974).
3. R. Wolfe, A. J. Kurtzig, and R. C. LeCraw, *J. Appl. Phys.*, **41,** 1218, (1970).
4. C. F. Larkworthy and A. Yavari, *Inorg. Chim. Acta*, **20,** L9 (1976); *J. Chem. Soc., Dalton Trans.*, **1978,** 1236.
5. C. Bellitto and P. Day, *J. Chem. Soc., Chem. Commun.*, **1976,** 870. *J. Chem. Soc., Dalton Trans.*, **1978,** 1207.
6. C. Bellitto and P. Day, *J. Cryst. Growth,* **58,** 641 (1982).

56. YELLOW MOLYBDENUM(VI) OXIDE DIHYDRATE

$$Na_2MoO_4 \cdot 2H_2O(aq) + 2HClO_4(aq) \rightarrow MoO_3 \cdot 2H_2O(s) + 2NaClO_4(aq) + H_2O$$

Submitted by J. B. B. HEYNS* and J. J. CRUYWAGEN*
Checked by SCOTT A. KINKEAD†

The known methods for preparing yellow $MoO_3 \cdot 2H_2O$ involve precipitation from nitric acid medium[1,2] or from hydrochloric acid medium.[3] We have found the yield by the former method, as modified by Freedman, to be much lower and the crystallization time much longer than that indicated. Moreover, the product obtained has been shown to be impure.[4] Although a pure, well-crystallized product was obtained from hydrochloric acid,[4–6] the procedure is tedious and the crystallization time also much longer than that originally stated.[5,6]

The facile method described here has been developed empirically by varying the concentrations (and concentration ratio) of molybdate and perchloric acid in a series of solutions. Conditions were thus established that are favorable for the precipitation of $MoO_3 \cdot 2H_2O$ in pure form within a reasonably short time, while inhibiting the formation of any of the other oxides (or mixtures of oxides) of molybdenum.

Procedure

Precipitation is effected from a solution that is initially 0.3 *M* in Na_2MoO_4 and 3.0 *M* in $HClO_4$. Typically, a reaction mixture is prepared by slowly adding, with constant stirring, 25 mL of a 0.60 *M* aqueous solution of $Na_2MoO_4 \cdot 2H_2O$ to 25 mL of a 6.0 *M* aqueous solution of $HClO_4$ contained in a 200-mL polyethylene beaker. Too rapid mixing causes precipitation of a white oxide. However, this redissolves slowly upon stirring. The final clear solution is left to stand in a polyethylene beaker (rather than glass, to avoid possible contamination with silicates) at ambient temperature for 4 weeks. Crystallization begins spontaneously within 2–3 weeks and is practically complete after 4 weeks. Seeding with $MoO_3 \cdot 2H_2O$ reduces the time for the onset and completion of precipitation by about 1 week. It will take longer for crystallization to begin if a container with a very smooth surface is used. A used polyethylene beaker with some fine

*Department of Chemistry, University of Stellenbosch, Stellenbosch 7600, Republic of South Africa.
†Department of Chemistry, University of Idaho, Moscow, ID 83843.

scratch marks is recommended. The precipitate is collected in a glass filtering crucible (fine porosity), thoroughly washed with water, and air-dried at room temperature in a dust-free environment. When using 15–75 mmol of $Na_2MoO_4 \cdot 2H_2O$, yields of $MoO_3 \cdot 2H_2O$ range from 80 to 84% of the theoretical.

Anal. Calcd. for $MoO_3 \cdot 2H_2O$: Mo, 53.3; H_2O, 20.0. Found: Mo, 53.1; H_2O, 19.9.

Properties

The product obtained by this procedure is a microcrystalline, bright yellow powder. The oxide easily loses one molecule of water of crystallization, and for this reason it cannot be stored over a desiccant. It can in fact be converted quantitatively to the yellow monohydrate, $MoO_3 \cdot H_2O$, by drying at 100° for 2 hr[2] (to date, the only route to this monohydrate). The substance is readily characterized by its X-ray powder diffraction pattern. Some data are given below. A complete listing is given in the literature.[7]

d (Å)	I/I_1	*d* (Å)	I/I_1	*d*(Å)	I/I_1
6.90	100	3.31	45	2.305	12
3.77	30	3.24	45	1.968	12
3.67	30	2.650	15	1.953	14
3.45	35	2.618	10	1.838	10

References

1. A. Rosenheim, *Z. Anorg. Allgem. Chem.*, **50**, 320 (1906).
2. M. L. Freedman, *J. Inorg. Nucl. Chem.*, **81**, 3834 (1959).
3. G. Carpéni, *Bull. Soc. Chim. Fr.*, **1947**, 484.
4. I. Lindqvist, *Acta Chem. Scand.*, **4**, 650 (1950).
5. J. R. Günter, *J. Solid State Chem.*, **5**, 354 (1972).
6. J. M. Adams and J. R. Fowler, *J. Chem. Soc., Dalton Trans.*, **1976**, 201.
7. W. F. McClune (ed.), *Powder Diffraction File*, JCPDS International Center for Diffraction Data, Swarthmore, Pennsylvania, 1983, entry 16-497.

57. TRICHLOROTRIS(TETRAHYDROFURAN)-MOLYBDENUM(III)

$$4[MoCl_4(C_4H_8O)_2] + 4C_4H_8O + Sn \rightarrow 4[MoCl_3(C_4H_8O)_3] + SnCl_4$$

Submitted by JONATHAN R. DILWORTH* and JON ZUBIETA†
Checked by T. ADRIAN GEORGE‡ and JOHN SMITH‡

Complex compounds of molybdenum in oxidation states lower than 4 and free from oxo or carbonyl ligands, can be prepared readily from the convenient starting material $[MoCl_3(thf)_3]$ (thf = tetrahydrofuran). The complex $[MoCl_3(thf)_3]$ provides an excellent starting material for the syntheses both of molybdenum(III) complexes[1] and of novel mononuclear phosphine and dinitrogen complexes of molybdenum(II) and molybdenum(0).[2] However, the reported synthesis[3] using zinc as reductant for $[MoCl_4(thf)_2]$ is difficult to control, and the product can be contaminated with intensely colored by-products. The use of tin powder as reducing agent is much more convenient, producing the complex in high yield and generally free of colored contaminants.

Procedure

All reactions are carried out under nitrogen using dried solvents in conventional Schlenk apparatus. The complex $[MoCl_4(thf)_2]$ is prepared by the published method.[3]

Tetrachlorobis(tetrahydrofuran)molybdenum(IV), 5.0 g, is suspended in 60 mL of tetrahydrofuran and stirred with 10 g of coarse tin powder, 20 mesh, at room temperature for 20 min. The solution is filtered, and any $[MoCl_3(thf)_3]$ product on the sinter is freed from tin by washing through with ~20 mL of dry dichloromethane. The solution is evaporated at 10^{-2} torr to ~30 mL, and the complex is removed by filtration as a pale orange crystalline material. Yield: 3.4 g (62%). The complex is stored under dry argon in a freezer and in the dark. Care should be taken, since the product is extremely moisture-sensitive.

Anal. Calcd. for $C_{12}H_{24}Cl_3O_3Mo$: C, 34.4; H, 5.73. Found: C, 34.1; H, 5.79.

Properties

The complex $[MoCl_3(thf)_3]$ is crystallized as pale orange needles from dichloromethane/tetrahydrofuran solution. The IR spectrum of the pure complex is free

*ARC Unit of Nitrogen Fixation, University of Sussex, Brighton BNI 9RQ, U.K.
†Department of Chemistry, State University of New York at Albany, Albany, NY 12222.
‡Department of Chemistry, University of Nebraska, Lincoln, NE 68588.

of intense bands in the 900–1000 cm^{-1} region, which is characteristic of molybdenum oxo species. The compound $[MoCl_3(thf)_3]$ reacts readily with certain 1,1-dithio acids to yield the tris(dithioacid)molybdenum(III) monomers in high yield (~70%).[1] Direct reaction with tertiary phosphines in tetrahydrofuran yields complexes of the type $[MoCl_3(PR_3)_x(thf)_{3-x}]$. Reduction of $[MoCl_3(thf)_3]$ by sodium amalgam or metallic magnesium in the presence of an excess of the appropriate organophosphine in tetrahydrofuran yields complexes of the type $[Mo(PRR'_2)_6]$ or $[Mo(PRR'_2)_4]$, depending on the nature of the organo group. Under molecular nitrogen, reaction of $[MoCl_3(thf)_3]$ with 1,2-ethanediylbis(diphenylphosphine) (diphos) yields *trans*-$[Mo(N_2)_2(diphos)_2]$ in high yield.[2]

References

1. J. R. Dilworth and J. A. Zubieta, *J. Chem. Soc., Dalton Trans.*, **1983,** 397.
2. J. Chatt and A. G. Wedd, *J. Organomet. Chem.*, **27,** C15 (1971).
3. J. R. Dilworth and R. L. Richards, *Inorg. Synth.*, **20,** 121 (1980).

58. PHENYLIMIDO COMPLEXES OF TUNGSTEN AND RHENIUM

Submitted by A. J. NIELSON*
Checked by R. E. McCARLEY,† S. L. LAUGHLIN,† and C. D. CARLSON†

The chemistry of organoimido compounds is currently of interest in studies of transition-metal multiple bonds. As a result of the robust M≡NR function, it has been possible to prepare imido complexes of a diversity of metals.[1] For synthetic work with earlier transition metal complexes in high valence states, organoimido compounds are less likely to polymerize than species containing the terminal oxo group. An added advantage is that any polymeric compounds formed are likely to be soluble in organic solvents.

Studies of the chemistry of phenylimido rhenium complexes[2] have normally employed $Re(NPh)Cl_3(PPh_3)_2$ as the starting material because of its easy preparation from aniline and the readily available compound $ReOCl_3(PPh_3)_2$.[3] Tungsten imido complexes cannot be prepared similarly. The lack of good preparative methods leading to suitable starting materials has limited studies with this metal.

Detailed below are procedures for preparing phenylimido tetrachloro complexes of tungsten and rhenium(VI) that are useful materials for further syntheses.[4]

*Department of Chemistry, University of Auckland, Private Bag, Auckland, New Zealand.
†Department of Chemistry, Iowa State University, Ames, IA 50011.

Alkylimido compounds can be prepared similarly, but workup may prove difficult in some cases. The reaction fails for *tert*-butylimido compounds. The reactivity of the α carbon in alkylimido rhenium complexes can render them less useful for general syntheses.[5] Preparations are given for phenylimido complexes of tungsten(V) and -(IV) and a high yield of $Re(NPh)Cl_3(PPh_3)_2$. The phosphines of this complex may be replaced through phosphine interchange reactions.[3]

Starting Materials and General Procedure

The compound $WOCl_4$ is prepared by heating WO_3 in sulfinyl chloride at reflux,[6] and $ReOCl_4$ by heating rhenium(V) chloride in a stream of dry oxygen.[7] Isocyanatobenzene [Aldrich] is used without purification. Benzene and toluene are distilled from sodium wire and tetrahydrofuran (THF) from sodium benzophenone under dry nitrogen. Trimethylphosphine [Strem Chemicals] is prepared by the reaction of MeMgI with $P(OPh)_3$ in diethyl ether.[8] The product is stored in a Schlenk flask and transferred with a syringe. All manipulations are carried out under moisture- and oxygen-free nitrogen by using normal techniques for air-sensitive compounds.[9] When solutions are transferred between flasks, a stainless steel transfer tube is used. Each flask is fitted with a gas inlet tap and a serum cap through which the transfer tube passes. The nitrogen supply to the receiving vessel is turned off, and a vent needle is placed through the septum. With the transfer tube placed below the level of liquid, a positive nitrogen pressure is used to force the solution into the receiving flask.

■ **Caution.** *Benzene is a suspected carcinogen. It should be used only in an efficient hood. Gloves should be worn.*

A. TETRACHLORO(PHENYLIMIDO) COMPLEXES OF TUNGSTEN(VI) AND RHENIUM(VI)

$$[MOCl_4]_2 + 2PhNCO \xrightarrow{\text{benzene}} [MCl_4(PhN)]_2 + 2CO_2 \qquad M = W, Re$$

Procedure

■ **Caution.** *Benzene is a suspected carcinogen. It should be used in a well-ventilated hood. Gloves should be worn. Isocyanates are poisonous.*

1. *Tetrachloro(phenylimido)rhenium(VI).* Tetrachlorooxorhenium[7] (10 g, 29 mmol) is placed in a 250-mL two-necked round-bottomed flask fitted with a gas inlet tap and rubber serum cap. Benzene (50 mL) is added with a syringe, followed by isocyanatobenzene (3.16 mL, 29 mmol). A reflux condenser and nitrogen bubbler are fitted, several boiling chips added, and the mixture is held

under strong reflux for 4 hr (a vigorous effervescence of CO_2 occurs) and then more gently for a further 16 hr. After cooling to room temperature, the solution is filtered under a nitrogen atmosphere, and the remaining solid is washed with several 5-mL portions of benzene. After drying under vacuum, the yield of dark brown tetrachloro(phenylimido)rhenium(VI) is between 10 and 11.2 g (82–92%).

Anal. Calcd. for $C_6H_5Cl_4NRe$: C, 17.2; H, 1.2; Cl, 33.8; N, 3.3. Found; C, 17.8; H, 1.2; Cl, 33.0; N, 3.4.

2. *Tetrachloro(phenylimido)tungsten(VI).* Tetrachlorooxotungsten(VI),[6] (30 g, 87 mmol) and isocyanatobenzene (9.5 mL, 87 mmol) are heated at reflux in 100 mL of benzene in a manner similar to that in part 1. On cooling, the solution is filtered, and the solid is washed with 10-mL portions of benzene until a yellow-green color persists in the washings. After the green solid is dried under vacuum, the yield of analytically pure tetrachloro(phenylimido)tungsten(VI) is 31–33.6 g (85–92%).

Anal. Calcd. for $C_6H_5Cl_4NW$: C, 17.3; H, 1.2; Cl, 34.0; N, 3.4. Found: C, 17.1; H, 1.2; Cl, 34.0; N, 3.38.

Properties

The compounds are air- and moisture-sensitive but can be stored under N_2 for several months without decomposition. They exhibit only slight solubility in benzene, dissolve more appreciably in dichloromethane, and form 1:1 adducts in coordinating solvents such as tetrahydrofuran or propionitrile. Neither compound forms a satisfactory mull in Nujol, and both react with KBr and CsI. They are insufficiently soluble in $CDCl_3$ or CD_2Cl_2 to obtain NMR spectra and cause acetone-d_6 to polymerize. In $CDCl_3$ the ^1H NMR spectrum of the tungsten 1:1 adduct with propionitrile shows resonances at δ 1.46 (t, 3H, CH_3); 2.68 (q, 2H, —CH_2—); 6.62 (m, 1H, *p*-aromatic); 6.93 (m, 2H, *m*-aromatics); 7.36 (m, 2H, *o*-aromatics). ^{13}C NMR spectra: δ (ppm from TMS) 9.35 (CH_3); 11.82 (—CH_2—); 127.23 ($C{\equiv}N$); 131.58 (c-*meta*, C_6H_5); 132.23 (c-*ortho*, C_6H_5); 134.58 (c-*para*, C_6H_5); 148.98 (c-*ipso*, C_6H_5).

B. TRICHLORO(PHENYLIMIDO)BIS(PHOSPHINE)TUNGSTEN(V) AND RHENIUM(V)

$$[MCl_4(PhN)]_2 + 4\ \text{phosphine} \rightarrow 2[MCl_3(PhN)(\text{phosphine})_2]$$

$$M = W,\ \text{phosphine} = PPh_3,\ PMe_2Ph,\ PEt_3,\ PMe_3$$

$$M = Re,\ \text{phosphine},\ PPh_3$$

Procedure

■ **Caution.** *Benzene is a suspected carcinogen. Phosphines are irritants and can be toxic.*

1. *Trichloro(phenylimido)bis(triphenylphosphine)tungsten(V).* Benzene (60 mL) and triphenylphosphine (~ 6–7 g) are placed in a 100-mL round-bottomed flask equipped with an inert gas inlet tap and a rubber serum cap. The solution is added by using a transfer tube to tetrachloro(phenylimido)tungsten(VI) (3 g, 7.1 mmol) in a 250-mL round-bottomed flask, and the mixture is heated at reflux for 16–24 hr, during which time the color turns from yellow to brown. If a yellow gum separates on cooling, further phosphine is added and the solution is heated at reflux for another 5–10 hr. The solution is cooled and filtered. The solvent is removed under vacuum, and the gum is washed several times with petroleum ether. The residue is extracted several times with toluene (5 × 10 mL or until the extracts are no longer yellow-brown). The extracts are combined and reduced in volume to approximately 25 mL. The solution is allowed to stand at −20°, which gives the complex as brown crystals. The crystals are filtered and washed with cold benzene to remove any gummy material. Further crystalline material may be obtained by reducing the volume of the filtrate and repeating the crystallization process. The complex contains one molecule of toluene. Yield: 4.1–4.9 g (71–85%).

Anal. Calcd. for $C_{49}H_{43}Cl_3NP_2W$: C, 59.0; H, 4.3; Cl, 10.7; N, 1.4. Found: C, 59.2; H, 4.3; Cl, 10.9; N, 1.5.

Properties

The brown complexes $WCl_3(PhN)(P)_2$ (P = PPh_3, PMe_2Ph, PEt_3) are air- and moisture-sensitive but may be handled very briefly in air. Under N_2 they are stable indefinitely. They are soluble in most solvents except petroleum ether fractions, hexane, and diethyl ether. In solution the paramagnetic moments (Evans method[10]) are slightly lower than the spin-only value for a d^1 system (1.73 BM). The IR spectra show three bands in the far-IR in the vicinity of 300 cm^{-1}, attributable to $\nu_{(M—Cl)}$ in a *mer* arrangement; for example, $WCl_3(PhN)(PMe_3)_2$ shows bands at 315 (m), 304 (s), and 255 (m) cm^{-1}

2. *Trichloro(phenylimido)bis(trimethylphosphine)tungsten(V).* Trimethylphosphine (4.0 mL, 36 mmol) is added with a syringe to the green solution formed by dissolving tetrachloro(phenylimido)tungsten(VI) (5 g, 12 mmol) in 100 mL of tetrahydrofuran (THF). The mixture is stirred for 15 hr and then filtered, and the solvent is removed to give a brown gum, which is washed several times with petroleum ether to remove excess phosphine. The residue is

extracted with toluene until the solvent is no longer yellow-brown, and the extracts are combined. The volume is reduced to ~25 mL, and the solution is allowed to stand at −20°, whereupon brown crystals of the complex form and are filtered. If a yellow gum also separates during crystallization, it may be removed by washing the crystals quickly with cold THF (−20°). Further crystalline material may be obtained by reducing the volume of the solution and repeating the crystallization procedure. Yield: 4.2–4.9 g (65–77%).

Anal. Calcd. for $C_{12}H_{23}Cl_3NP_2W$: C, 27.0; H, 4.3; N, 2.6. Found: C, 26.8; H, 4.5; N, 2.5.

Properties

The compound $[WCl_3(PhN)(PMe_3)_2]$ is air- and moisture-sensitive, but it can be stored for long periods under N_2. THF solutions are very oxygen-sensitive, turning green to give the complex $WCl_3(PhN)(OPMe_3)_2$, which crystallizes from toluene along with $WCl_3(PhN)(PMe_3)_2$. The former may be removed by washing with cold THF. The complex melts between 184 and 186° and has μ_{eff} of 1.43 BM in chloroform-*d* (Evans method[10]). In the IR spectrum the M—Cl stretching vibrations occur at 315, 300, and 254 cm^{-1}, characteristic of *mer* metal chlorides. The crystal structure shows octahedral geometry with *trans* phosphorus ligands and a lengthened W—Cl bond *trans* to the phenylimido group.[11]

3. *Trichloro(phenylimido)bis(triphenylphosphine)rhenium(V).* Tetrachloro(phenylimido)rhenium(VI) (5 g, 12 mmol) is dissolved in 50 mL of THF, and by using a transfer tube, the green solution is added to a rapidly stirred 50 mL solution of THF containing triphenylphosphine [Aldrich] (9.5 g, 36 mmol). The stirring is continued for up to 5 hr, during which time the complex precipitates as green microcrystals. The complex is removed on a filter and washed several times with THF or benzene to remove any remaining triphenylphosphine. After drying under vacuum, the yield of trichloro(phenylimido)bis(triphenylphosphine)rhenium(V) is 7.6–8.5 g (79–88%).

Anal. Calcd for $C_{42}H_{35}Cl_3NP_2Re$: C, 55.5; H, 3.9; N, 1.5. Found; C, 56.1; H, 4.3; N, 1.6.

Properties

Trichloro(phenylimido)bis(triphenylphosphine)rhenium(V) is an air-stable, green solid that is insoluble in organic solvents; mp 215–218°.[3]

C. PHENYLIMIDO COMPLEXES OF TUNGSTEN(IV)

$$[WCl_4(PhN)]_2 + 6L \xrightarrow{Na/Hg} 2[WCl_2(PhN)L_3]$$

$$L = PMe_2Ph, PMePh_2, PEt_3, PMe_3, Me_3CNC, p\text{-}MePhNC$$

Procedure

Dichloro(phenylimido)tris(trimethylphosphine)tungsten(IV). The experimental conditions outlined are general for the phosphine and isocyanide ligands. Sodium-mercury amalgam is prepared in a 500-mL three-necked flask equipped with a gas inlet tap, mechanical stirrer, and a septum cap, by adding 1 g (43 mmol) of sodium metal in small pieces (approx. 0.1–0.2 cm^3) to 100 g of well-stirred mercury. (■ **Caution.** *The reaction of sodium with mercury is highly exothermic.*) The addition of each piece of sodium is carried out rapidly, and the serum cap is replaced afterwards to prevent loss of components caused by spurting. When cold, the amalgam is washed several times with benzene added by a syringe and is removed by means of a transfer tube placed through the septum. (■ **Caution.** *Benzene is a suspected carcinogen. Good hoods and gloves are mandatory.*) A solution of trimethylphosphine (5.5 mL, 50 mmol) in 100 mL of benzene is then added to the amalgam by means of a transfer tube. In a glove bag, tetrachloro(phenylimido)tungsten(VI) (5 g, 12 mmol) is ground to small particle size and placed in a 250-mL two-necked flask fitted with a gas inlet tap and a septum cap. The flask is removed from the glove bag, 100 mL of benzene is added, and the suspension is transferred rapidly to the flask containing the stirred amalgam, using a transfer tube. The solution is stirred for 5–6 hr, during which time it becomes a deep blue. After filtration, the spent amalgam is washed with 25 mL of benzene and filtered. The extract is added to the bulk solution. The solvent is removed under vacuum to give a gum, which is then washed several times with petroleum ether. The residue is extracted with toluene until the solvent is no longer colored, and the extracts are combined, filtered, and reduced in volume to ~25 mL. Allowing the solution to stand at $-20°$ gives violet-blue crystals of the complex, which are removed by filtering and washed with petroleum ether. Additional complex is obtained by reducing the volume of filtrate and repeating the crystallization. The yield is 5.3–5.9 g (77–86%).

Anal. Calcd. for $C_{15}H_{32}Cl_2NP_3W$: C, 31.4; H, 5.6; Cl, 12.4; N, 2.4. Found: C, 31.6; H, 5.5; Cl, 12.4; N, 2.4.

Properties

The complexes prepared in this manner are blue or brown and are air- and moisture-sensitive. However, they can be stored for several months under N_2. They are diamagnetic and soluble in most organic solvents except petroleum ether and similar hydrocarbons. In the far-IR spectrum, several absorption bands occur in the vicinity of 250–300 cm^{-1}. These are assigned to $\nu_{(M—Cl)}$. The X-ray crystal structure of $WCl_2(PhN)(PMe_3)_3$ shows a *mer* arrangement of phosphine and *cis* chloride ligands, one *trans* to the phenylimido function.[12]

References

1. W. A. Nugent and B. L. Haymore, *Coord. Chem. Rev.*, **31,** 123 (1980).
2. K. W. Chiu, W. Wong, and G. Wilkinson, *J. Chem. Soc., Chem. Commun.*, **1981,** 451; G. La Monica, S. Cenini, and F. Porta, *Inorg. Chim. Acta,* **48**, 91 (1981).
3. N. P. Johnson, C. J. L. Lock, and G. Wilkinson, *Inorg. Synth.*, **9,** 145 (1967).
4. D. C. Bradley, R. J. Errington, M. B. Hursthouse, A. J. Nielson, and R. L. Short, *Polyhedron,* **2,** 843 (1983).
5. J. Chatt, R. J. Dosser, F. King, and G. J. Leigh, *J. Chem. Soc., Dalton Trans.*, **1976,** 2345; J. Chatt, R. J. Dosser, and G. J. Leigh, *J. Chem. Soc., Chem. Commun.*, **1972,** 1243.
6. R. Colton and I. B. Tomkins, *Aust. J. Chem.*, **18,** 447 (1965); A. J. Nielson, *Inorg. Synth.*, **23,** 195 (1985).
7. P. G. Edwards, G. Wilkinson, M. B. Hursthouse, and K. M. A. Malik, *J. Chem. Soc., Dalton Trans.*, **1980,** 2467.
8. W. Wolfsberger and H. Schmidbauer, *Synth. React. Inorg. Met—Org. Chem.*, **4,** 149 (1974).
9. D. F. Shriver, *The Manipulation of Air-Sensitive Compounds,* McGraw-Hill, New York, 1969.
10. D. F. Evans, *J. Chem. Soc.*, **1959,** 203.
11. A. J. Nielson and J. M. Waters, *Aust. J. Chem.*, **36,** 243 (1983).
12. D. C. Bradley, M. B. Hursthouse, K. M. A. Malik, A. J. Nielson, and R. L. Short, *J. Chem. Soc., Dalton Trans.*, **1983,** 2651.

59. LITHIUM INSERTION COMPOUNDS

Submitted by D. W. MURPHY* and S. M. ZAHURAK*
Checked by C. J. CHEN† and M. GREENBLATT†

Many inorganic solids are capable of undergoing insertion reactions with small ions such as H^+, Li^+, and Na^+. The host solid in these reactions undergoes reduction in order to maintain electroneutrality. Inserted ions may be removed by oxidation of the insertion compounds. The structures of the insertion compounds are closely related to those of the respective hosts, with the inserted cation occupying formerly empty sites of the host.

The examples presented here illustrate a variety of reagents for the insertion or removal of lithium ions from inorganic solids. A variety of reagents exhibiting a range of redox potentials allow access to intermediate stoichiometries and control of side reactions. Four reagents are used in these syntheses: butyllithium, a strongly reducing source of lithium; lithium iodide, a mild reducing source; ethanol, a mild oxidant; and iodine, a stronger oxidant. A discussion of the redox levels of these and other reagents may be found elsewhere.[1]

*AT&T Bell Laboratories, Murray Hill, NJ 07974.
†Department of Chemistry, Rutgers University, The State University of New Jersey, New Brunswick, NJ 08903.

A. VANADIUM DISULFIDE

$$2LiVS_2 + I_2 \xrightarrow{CH_3CN} 2\,VS_2 + 2LiI$$

The class of layered transition metal dichalcogenides has been of great interest because of their varied electronic properties and chemical reactions. Most compounds of this class may be prepared by stoichiometric reactions of the elements above 500°. However, the highest vanadium sulfide that can be made in this manner is V_5S_8. An amorphous VS_2 has been prepared by the metathetical reaction of Li_2S and VCl_4.[2] The method presented here allows preparation of polycrystalline VS_2 with the CdI_2 structure.[3]

Procedure

The $LiVS_2$[4] is prepared from an intimate mixture of V_2O_5 (9.094 g, 0.05 mole) and Li_2CO_3 (3.694 g, 0.05 mole). The mixture is placed in a vitreous carbon boat inside a quartz tube in a tube furnace. The tube is connected to a two-way inlet valve for argon (or another inert gas) and hydrogen sulfide. Gas exits through a bubbler filled with oil. An H_2S flow of ~100 mL/min is maintained, and the furnace is heated to 300°, then to 700° at the rate of 100°/hr. (■ **Caution.** *Hydrogen sulfide is a highly poisonous gas. The reaction should be carried out in a well-ventilated fume hood.*) Water and sulfur are deposited on the tube downstream. The reaction is held at 700° for 12 hr. The reaction mixture is cooled under H_2S and then flushed with argon. The boat is removed into a jar filled with argon and placed in a good dry box (a dry atmosphere is sufficient). The mixture is reground and refired in H_2S at 700° for another 16 hr. The $LiVS_2$ prepared in this way is actually $Li_{0.90-0.95}VS_2$. The stoichiometry is adjusted to $Li_{1.0}VS_2$ by treatment with a dilute (0.05 *N*) solution of butyllithium in hexane. See the preparation of Li_2ReO_3 below for details of butyllithium reactions.

A solution of ~0.1 *M* I_2 is prepared from freshly sublimed I_2 and acetonitrile distilled from P_4O_{10}. The solution is standardized by titration with a standard aqueous thiosulfate solution to the disappearance of the I_2 color. Addition of iodide and/or starch gives no color enhancement in acetonitrile.

The solid polycrystalline $LiVS_2$ (4.670 g, 38.3 mmol) is placed in a 300-mL round-bottomed flask under argon. The flask is fitted with a serum cap. A solution of I_2 in CH_3CN (225 mL, 0.091 *M*, 20.5 mmol of I_2) is added to the $LiVS_2$ using a transfer needle through the serum cap. The iodine color rapidly dissipates as solution is added. The heterogeneous reaction mixture is stirred using a magnetic stirrer. After 16 hr the mixture is filtered in air, and the solid product is washed with acetonitrile. The filtrate and washings are titrated with standard aqueous thiosulfate to determine the unreacted excess I_2 (1.35 mmol).

The yield is quantitative.

Combustion anal. Calcd. for VS_2: V, 44.3. Found, 44.5. Atomic absorption for Li gives 180 ppm. X-ray fluorescence shows no iodine.

Properties

The VS_2 produced in this manner is a shiny metallic gray powder. The compound is hexagonal (CdI_2 type) with $a = 3.217$ Å and $c = 5.745$ Å. Sulfur loss occurs in air or inert gas above 300°.

B. LITHIUM DIVANADIUM PENTOXIDE

$$2LiI + 2V_2O_5 \xrightarrow{CH_3CN} 2LiV_2O_5 + I_2$$

Ternary alkali metal vanadium oxide bronzes are well known, including γ-LiV_2O_5.[5] It was recognized that some other composition or structure was formed from the combination of lithium and V_2O_5 at room temperature through electrochemical or butyllithium reactions.[6] It is possible to prepare the low-temperature δ-LiV_2O_5 with butyllithium,[6,7] although irreversible overreduction is difficult to avoid. The use of LiI as reductant avoids any overreduction.[8]

Procedure

The divanadium pentoxide is prepared from reagent grade NH_4VO_3 by heating in air first at 300° for 3 hr and then at 550° for 16 hr. Anhydrous LiI [Alfa Products] is dried under vacuum at 150° prior to use. A solution of ~1.5 *M* LiI in acetonitrile is prepared using acetonitrile freshly distilled from P_4O_{10}. Care is taken to exclude moisture and oxygen in the preparation and storage of this solution. The V_2O_5 powder (5.005 g, 27.5 mmol) is placed in a flask fitted with a serum cap, and an excess of the LiI solution (25.0 mL of 1.44 *M*, 36 mmol) is added via syringe. The supernatant rapidly develops a dark yellow-brown color characteristic of iodine. The reaction mixture is stirred at room temperature for ~24 hr, using a magnetic stirrer. The color of the solid changes from yellow to green to blue-black over several hours. The product is isolated by filtration in air and is washed with acetonitrile. Titration of the filtrate and washings with standard aqueous thiosulfate determines that 13.59 mmol of I_2 is formed in the reaction.

Properties

The LiV_2O_5 formed in this way is dark blue and is stable in air for moderate lengths of time. Long-term storage in a desiccator is satisfactory. The compound is orthorhombic[7,8] with a = 11.272 Å, b = 4.971 Å, and c = 3.389 Å. A reversible first-order structural transformation occurs at 125° to ϵ-LiV_2O_5,[8] which is also orthorhombic with a = 11.335 Å, b = 4.683 Å, and c = 3.589 Å. Above 300° the structure changes irreversibly to that of the thermally stable γ-LiV_2O_5.

C. LITHIUM RHENIUM TRIOXIDES: Li_xReO_3 ($x \leq 0.2$)

$$ReO_3 + 2BuLi \rightarrow Li_2ReO_3 + \text{octane}$$

$$2Li_2ReO_3 + 2EtOH \rightarrow 2LiReO_3 + 2LiOEt + H_2$$

$$ReO_3 + \text{excess LiI} \rightarrow Li_{0.2}ReO_3 + LiI + I_2$$

Rhenium trioxide has one of the simplest extended structures. Octahedral [ReO_6] units share oxygen atoms between units such that Re—O—Re bonds are linear. The symmetry is cubic, and each cell contains one Re and one empty cubeoctahedral cavity. This lattice serves as a starting point for the generation of a number of other structures including perovskites and shear compounds. Since ReO_3 is the simplest of this large family of compounds, an understanding of its behavior with lithium is intrinsic to understanding the class as a whole.

Three phases have been identified in the Li_xReO_3 system.[10] For $x < 0.35$ the structure remains cubic. A line phase at x = 1.0 is rhombohedral, as is a phase at $1.8 \leq x \leq 2.0$.

1. Dilithium Rhenium Trioxide

■ **Caution.** *Concentrated butyllithium (n-BuLi) (~2.0–3.0 M in hexane) is extremely air- and moisture-sensitive and must be handled in an inert atmosphere. Any amount that must be handled in air should be diluted with hexane or other inert solvent before exposure.*

Procedure

Reaction and filtration operations are carried out in a helium-filled glove box. However, the procedures are easily adapted to the use of Schlenk techniques. The hexane used should be distilled from sodium, and the concentrated n-BuLi solution [Alfa Products] standardized by total base titration. If the solution is cloudy as received, it can be filtered in the glove box.

To a flame-dried 100-mL round-bottomed flask is added 4.08 g (17.42 mmol) of ReO_3, along with sufficient hexane (~10.0 mL) to cover the ReO_3. A small excess of an *n*-BuLi solution (12.5 mL, 38.24 mmol, 3.059 *N*) is slowly added via a 20.0-mL gastight syringe. The reaction is exothermic, and, depending on the particle size of the ReO_3 and on the *n*-BuLi concentration, addition of *n*-BuLi may cause boiling of the solvent. The product is purer (by X-ray powder diffraction) when boiling is avoided. The flask is capped with a serum stopper and stirred with a magnetic stirrer for ~24 hr at room temperature. The original red-bronze ReO_3 turns to a dark red-brown color upon completion of the reaction.

The reaction mixture is filtered in the glove box and the Li_2ReO_3 collected on a medium-porosity fritted-glass filter. The product is washed several times with hexane to ensure removal of any excess *n*-BuLi. The yield is quantitative. The filtrate and washings are then removed from the glove box after sufficient dilution with hexane and titrated for excess *n*-BuLi.

The titration consists of addition of a few milliliters of distilled water and an excess of standard HCl (0.1 *N*), followed by back-titration with standard NaOH solution (0.1 *N*). The lithium stoichiometry is calculated on the basis of the titration results. It has been shown that this total base titration gives the same results as are obtained with active lithium reagent and atomic absorption analysis.

Complete lithiation to the limiting lithium stoichiometry of $Li_{2.0}ReO_3$ may require more than one *n*-BuLi treatment. This can be due in part to dilution of *n*-BuLi as the reaction proceeds. Upon titrating the initial *n*-BuLi reaction solution, 10.491 mmol of Li remains from an original 3.059 *N* *n*-BuLi solution containing 12.5 mL (38.238 mmol) of *n*-BuLi in hexane and 4.080 g (17.421 mmol) of ReO_3. This indicates 1.59 mmol of Li per millimole of ReO_3. Further lithiation and subsequent titration results in $Li_{2.1}ReO_3$. X-ray powder diffraction data indicates the lithium composition, in excess of two Li per ReO_3, is due to impurities in the ReO_3.

Confirmation of the lithium stoichiometry is determined by an iodine reaction that yields the amount of lithium removed from the structure. A titration (described in Section A) performed after reaction of Li_xReO_3 with a standard iodine solution affords the stoichiometry $Li_{1.99}ReO_3$.

Properties

The compound Li_2ReO_3 is a dark red-brown solid that is reactive with atmospheric moisture, forming LiOH and H_2. The X-ray powder pattern data give the following hexagonal crystallographic parameters: $a = 4.977$ Å, $c = 14.793$ Å, $V = 52.88 \times 6$ Å^3.[10]

Neutron diffraction powder profile analysis establishes a rhombohedral structure in the space group $R3c$,[11] $a = 4.9711$ (1), $c = 14.788$ (1), $Z = 6$. Both Li and Re atoms occupy type (00z) positions, and oxygen atoms the general position (x,y,z) with six formula units per cell. The host lattice, ReO_3, undergoes a twist that creates two octahedral sites from the cubeoctahedral cavity of ReO_3. These are the sites occupied by lithium.[11]

2. Lithium Rhenium Trioxide

Procedure

In a 60.0-mL Schlenk filter within a helium glove box is placed 4.1891 g (17.37 mmol) of $Li_{2.0}ReO_3$. The filter is equipped with a serum cap and stir bar and removed from the box. The serum cap is secured with a twist of wire, and ~40 mL of absolute ethanol is transferred into the flask, under argon, using a double-edged transfer needle. Evolution of hydrogen is noted upon this addition, and the system is kept under argon flow using an oil bubbler. The mixture is stirred ~24 hr or until gas evolution has ceased, and it is then filtered, using the same Schlenk filter along with a 250-mL receiving flask. The product is rinsed three times with 5.0-mL aliquots of absolute ethanol under positive argon pressure using a transfer needle. Care must be taken so as not to overrinse, as more Li can be removed from the structure. The dark red product is isolated in quantitative yield by filtration and is then vacuum-dried. The filtrate containing lithium ethoxide is titrated using a standard acid/base method with phenolphthalein as the indicator.

Reaction of 4.189 g of Li_2ReO_3 together with excess absolute ethanol (~55.0 mL total) forms 19.45 mmol of lithium ethoxide. Therefore, 1.0 mmol of Li is removed from the structure, leaving $Li_{1.0}ReO_3$. Removal of all of the lithium using the standard iodine solution technique confirms the lithium stoichiometry.

Properties

The compound $LiReO_3$ is very similar in appearance and properties to Li_2ReO_3. Both are red-brown hygroscopic solids; however, $LiReO_3$ is less moisture-sensitive and can be exposed to air for brief periods without damage. The X-ray powder pattern shows that $LiReO_3$ is single-phase with hexagonal lattice parameters of $a = 5.096$ Å, $c = 13.400$ Å, and $V = 50.23 \times 6$ Å^3.[10,11] The structure of $LiReO_3$, according to neutron-diffraction powder profile analysis studies, shows $a = 5.0918$ (3), $c = 13.403$ (1), $z = 6$ in the $R3c$ space group.[11] The ReO_3 skeleton has undergone the same twist as in Li_2ReO_3. The lithium atoms order in half the octahedral sites. The compound is isostructural with $LiNbO_3$.

Both $LiReO_3$ and Li_2ReO_3 exhibit temperature-independent Pauli paramagnetism (0.8×10^{-6} and 3.3×10^{-6} emu/g, respectively).[10]

3. Lithium(0.2) Rhenium Trioxide

Procedure

As in the preparation of Li_2ReO_3, reaction and filtration operations take place in a helium-filled glove box. Rhenium(VI) oxide (10.381 g, 44.33 mmol) together with 70.0 mL of a 0.5 *M* LiI solution in acetonitrile are added to a flame-dried 100-mL round-bottomed flask equipped with a stirring bar. The flask is stoppered, and the reaction is monitored for completeness by periodic sampling of the solid using X-ray powder diffraction analysis. A total of three successive treatments with 70.0-mL aliquots of LiI in CH_3CN, with stirring over 7 days, are needed to obtain the homogeneous single-phase $Li_{0.2}ReO_3$ (smaller scale reactions[12] are complete with a single treatment). Each treatment is followed by filtration of the total solution and two washings with acetonitrile. The filtrates are removed from the glove box and titrated for iodine content using standard aqueous $Na_2S_2O_3$ solution as the titrant. Reaction of 10.381 g of ReO_3 with LiI forms 8.87 meq of iodine, indicating a final stoichiometry of $Li_{0.2}ReO_3$. Flame emission analysis of Li confirms this result.

References

1. D. W. Murphy and P. A. Christian, *Science,* **205**, 651 (1979).
2. R. R. Chianelli and M. B. Dines, *Inorg. Chem.*, **17,** 2758 (1978).
3. D. W. Murphy, C. Cros, F. J. DiSalvo, and J. V. Waszczak, *Inorg. Chem.*, **16,** 3027 (1977).
4. B. van Laar and D. J. W. Ijdo, *J. Solid State Chem.*, **3,** 590 (1971).
5. J. Galy and A. Hardy, *Bull. Soc. Chim. France,* **1964,** 2808.
6. M. S. Whittingham, *J. Electrochem. Soc.*, **123,** 315 (1976).
7. P. G. Dickens, S. J. French, A. T. Hight, and M. F. Pye, *Mater. Res. Bull.*, **14,** 1295 (1979).
8. D. W. Murphy, P. A. Christian, F. J. DiSalvo, and J. V. Waszczak, *Inorg. Chem.*, **18,** 2800 (1979).
9. G. Brauer (ed.), *Handbook of Preparative Inorganic Chemistry,* 2nd ed., Academic Press, New York, 1965, p. 1270.
10. D. W. Murphy, M. Greenblatt, R. J. Cava, and S. M. Zahurak, *Solid State Ionics,* **5,** 327 (1981).
11. R. J. Cava, A. J. Santoro, D. W. Murphy, S. M. Zahurak and R. S. Roth, *J. Solid State Chem.*, **42,** 251 (1982).
12. R. J. Cava, A. J. Santoro, D. W. Murphy, S. M. Zahurak, and R. S. Roth, *J. Solid State Chem.*, **50,** 121 (1983).

60. DINITROGEN COMPLEXES OF IRON(II) WITH (1,2-ETHANEDIYLDINITRILO)TETRAACETATE AND *trans*-(1,2-CYCLOHEXANEDIYLDINITRILO)TETRAACETATE

Submitted by M. GARCIA BASALLOTE,* J. M. LOPEZ ALCALA,* M. C. PUERTA VIZCAINO,* and F. GONZALEZ VILCHEZ*
Checked by CARRIE WOODCOCK† and DUWARD F. SHRIVER†

Since the initial report on the syntheses of the $[Ru(NH_3)_5N_2]X_2$ (X = halogen) dinitrogen complexes,[1] interest in these compounds has been very high, partly because of the apparent relationship between these substances and the nitrogen-fixation process. Several methods for the preparation of dinitrogen complexes were subsequently reported.[2–4]

Recently, Diamantis[5] prepared the complexes $[(Ru(edta))_2N_2]^{4-}$ and $[Ru(edta)N_2]^{2-}$ by bubbling N_2 gas into an aqueous solution of $[Ru(Hedta)H_2O]$ in the presence of hydrogen and platinum black. No analogous compounds that contain Fe(II) have been reported. As described here, dinitrogen complexes of Fe(II) with (1,2-ethanediyldinitrilo)tetraacetate(edta) and *trans*-(1,2-cyclohexanediyldinitrilo)tetraacetate(cdta) may be prepared by the reaction of $[Fe(HY)H_2O]$ (Y = edta, cdta) with sodium azide.

■ **Caution.** *All azides are potentially explosive and should be handled with care.*

A. AQUA[[(1,2-ETHANEDIYLDINITRILO)TETRAACETATO]-(3−)]IRON(III) HYDRATE

$$H_4Y + Fe(OH)_3 \rightarrow [Fe(HY)H_2O]\cdot H_2O + H_2O \qquad (Y = \text{edta, cdta})$$

Procedure

These compounds are prepared from modified published procedures.[6–8] Twenty grams of reagent grade iron(III) nitrate, $Fe(NO_3)_3\cdot 9H_2O$, is dissolved in 100 mL of water in a 250-mL beaker, and the solution is filtered. A solution of 50 mL of concentrated aqueous ammonia (28–30%) and 50 mL of water is added drop-wise to the solution of iron(III) nitrate. The $Fe(OH)_3$ that precipitates is recovered by vacuum filtration or centrifugation, using a 10-cm Büchner funnel, and is

*Department of Inorganic Chemistry, Faculty of Science, University of Cadiz, Cadiz, Spain.
†Department of Chemistry, Northwestern University, Evanston, IL 60201.

washed several times with water until NH_4^+ no longer appears in the washings. A 17.5 g sample of H_4edta (or 17.1 g of H_4cdta) is suspended in 40 mL of water, and the $Fe(OH)_3$ is added along with enough water to bring the total volume to 100 mL. The mixture is heated while stirring on a steam cone for ~2 hr. Any uncomplexed free acid that precipitates when the solution cools to room temperature is recovered by filtration. The reaction volume is then reduced to 60 mL, and 100 mL of acetone is added slowly. The solid complexes are collected and are recrystallized from 20 mL of H_2O/100 mL acetone. Yields: edta complex 6.5 g; cdta complex 4.0 g.

B. DISODIUM (DINITROGEN)[[(1,2-ETHANEDIYLDINITRILO)-TETRAACETATO](4—)]FERRATE(II) DIHYDRATE

$$[Fe(Hedta)H_2O]\cdot H_2O + 2NaN_3 \rightarrow Na_2[Fe(edta)N_2]\cdot 2H_2O + 2N_2$$

Procedure

A 0.50 g sample of $[Fe(Hedta)H_2O]\cdot H_2O$ is dissolved in 20 mL of N_2-degassed water in a 50-mL Pyrex flask. To this solution, 0.50 g (7.7 mmol) of NaN_3 is added with stirring. The color changes from yellow to red-orange immediately, and the pH increases to 5. The mixture is stirred for 1 hr (not more than 2 hr) at 70° while N_2 gas is bubbled into it (Fig. 1). The final pH of the solution should be about 7 and the corresponding volume about 10 mL. The solution is cooled to room temperature, and then 40–50 mL of ethanol is added to give a brown oil. The upper solution is decanted off, and the remaining oil is then solidified by the addition of 20 mL of ethanol while the mixture is stirred vigorously. An orange-brown powder is recovered. Attempts to recrystallize this product lead to regeneration of the starting material.[9] Yield: 0.40 g (67%).

Anal. Calcd. for $C_{10}FeH_{16}N_4Na_2O_{10}$: C, 26.45; N, 12.34; H, 3.55. Found: C, 26.6; N, 11.4; H, 3.6.

Properties

The complex forms as an orange-brown powder that decomposes slowly in air. A sharp intense band in the IR spectrum at 2040 cm^{-1} (KBr and Nujol) is assigned to $\nu_{N\equiv N}$ of coordinated nitrogen. The stretching vibrations for carboxylate groups appear at 1600 (asym) and 1380 (sym) cm^{-1}. The ESR spectrum indicates diamagnetic character. The UV/vis spectrum in aqueous solution has maxima at 51,000, 40,000, and 21,275 cm^{-1}. In a static air or argon atmosphere,

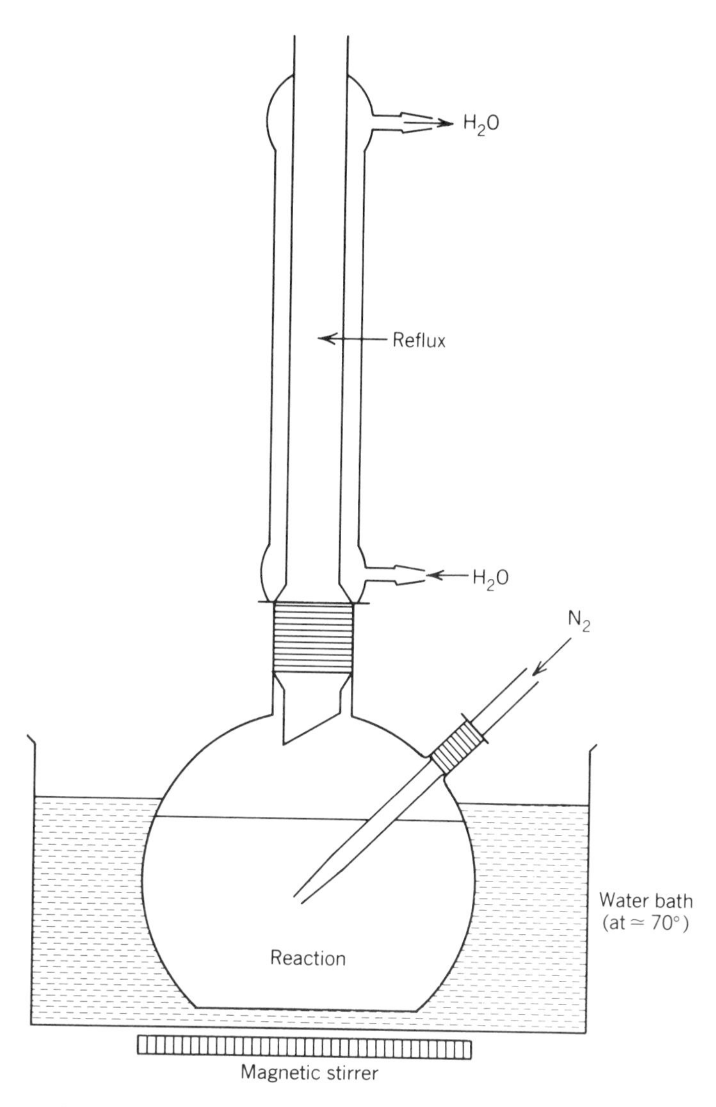

Fig. 1. Apparatus for preparation of dinitrogen complexes.

the weight loss observed in the TG study and the corresponding endothermic effect registered in DTA occur at ~220° and can be associated with the elimination of coordinated dinitrogen.

C. DISODIUM [[(1,2-CYCLOHEXANEDIYLDINITRILO)-TETRAACETATO](4−)](DINITROGEN)FERRATE(II) DIHYDRATE

$$[Fe(Hcdta)H_2O]\cdot H_2O + 2NaN_3 \rightarrow Na_2[Fe(cdta)N_2]\cdot 2H_2O + 2N_2$$

Procedure

A 0.50 g (7.7-mmol) sample of solid sodium azide is added to a solution of 0.50 g of $[Fe(Hcdta)H_2O]\cdot H_2O$ in 20 mL of water in a 50-mL Pyrex flask. The mixture is stirred while N_2 gas is bubbled into it (Fig. 1). The temperature is held at 70° for 2 hr. After this time, the brown-orange solution is cooled to room temperature. When the first crystals begin to appear, cold acetone is added to the solution to give a brown oil. The solution is decanted off the oil. The oil is dissolved in 5 mL of water, and 20 mL of acetone is added slowly. This results in a yellow-orange powder, which is recovered by filtration. The precipitate is washed with 10 mL of cold acetone and dried over P_4O_{10}. Yield: 0.40 g (68%).

Anal. Calcd. for $C_{14}FeH_{22}N_4Na_2O_{10}$: C, 33.09; N, 11.03; H, 4.36. Found: C, 33.2; N, 8.5; H, 4.6.

Properties

The complex crystallizes as yellow-orange microcrystals that decompose slowly in air. An intense sharp band at 2040 (KBr) or 2050 (Nujol) cm^{-1} in the IR spectrum is assigned to $\nu_{N\equiv N}$ of the coordinated dinitrogen. The stretching vibrations for carboxylate groups appear at 1620 (asym) and 1380 (sym) cm^{-1}. The ESR spectrum indicates diamagnetic character. The UV/vis spectrum in aqueous solution has maxima at 51,000, 37,040, and 21,700 cm^{-1}. In an atmosphere of static air or argon, weight loss in TG and an endothermic effect in DTA occur at ~215° due to the elimination of coordinated dinitrogen.

References

1. A. D. Allen and C. V. Senoff, *J. Chem. Soc., Chem. Commun.*, **1965,** 621.
2. L. A. P. Kane-Maguire, P. S. Sheridan, F. Basolo, and R. G. Pearson, *J. Am. Chem. Soc.*, **92,** 5865 (1970).
3. J. Chatt, *J. Organometal. Chem.*, **100,** 17 (1975).

4. G. Henrici-Olive and S. Olive, *Coordination and Catalysis (Monographs in Modern Chemistry*, Vol. 9), Verlag Chemie, New York, NY 1977.
5. A. A. Diamantis and J. V. Dubrawsky, *Inorg. Chem.*, **20,** 1142 (1981).
6. V. H. Brintzinger, H. Thiele, and U. Müller, *Z. Anorg. Chem.*, **251,** 289 (1943).
7. J. L. Lambert, C. E. Godsey, and L. M. Seitz, *Inorg. Chem.*, **2,** 127 (1963).
8. G. H. Cohen and J. L. Hoard, *J. Am. Chem. Soc.*, **88,** 3228 (1966).
9. J. M. Lopez-Alcala, M. C. Puerta-Vizcaino, F. Gonzalez Vilchez, E. N. Duesler, and R. E. Tapscott, *Acta Cryst.*, **C40,** 939 (1984).

61. POTASSIUM TETRAKIS[DIHYDROGEN DIPHOSPHITO(2 –)]DIPLATINATE(II)

$$2K_2[PtCl_4] + 8H_3PO_3 \rightarrow K_4[Pt_2(P_2O_5H_2)_4]\cdot 2H_2O + 8HCl + 2H_2O$$

Submitted by K. A. ALEXANDER,* S. A. BRYAN,* M. K. DICKSON,* D. HEDDEN,† and D. M. ROUNDHILL†
Checked by C.-M. CHE,‡ L. G. BUTLER,‡ and H. B. GRAY‡

Potassium tetrakis[dihydrogen diphosphito(2 –)]diplatinate(II) dihydrate was discovered because of its intense emission intensity in aqueous solution at room temperature.[1] The complex has strong absorption maxima at 367 nm (ϵ 33,500 $M^{-1}cm^{-1}$) and 452 nm (ϵ 120 $M^{-1}cm^{-1}$), and emissions at 403 and 515 nm.[2–4] The intense emission has been used as a basis for the quantitative determination of platinum.[5] The complex has been used as a source for the synthesis of binuclear platinum(III) complexes $K_4[Pt_2(P_2O_5H_2)_4X_2]$ (X = Cl, Br, I),[6] mixed valence binuclear complexes $[Pt_2(P_2O_5H_2)X]^{4-}$ (X = Cl, Br, I),[7] and higher condensation of oligomers of platinum(II).[8] The molecular structure of the complex shows an intermetallic separation of 2.925(1) Å.[7,9] A normal coordinate analysis of the vibrational Raman and IR spectra has shown that there is little intermetallic bonding in the ground state,[10] but an excited state Raman spectrum[11] has allowed direct comparison between ν_{Pt-Pt} in the two states. The preparative method described here is based on the previously reported synthesis of $K_4[Pt_2(P_2O_5H_2)_4]\cdot 2H_2O$.[1,9]

*Department of Chemistry, Washington State University, Pullman, WA 99164.
†Department of Chemistry, Tulane University, New Orleans, LA 70118.
‡Department of Chemistry, California Institute of Technology, Pasadena, CA 91125.

Procedure

Potassium tetrachloroplatinate(II) [Johnson Matthey] (0.500 g, 1.2 mmol), and phosphorous acid (2.3 g, 28.0 mmol) are dissolved in 10 ml of deionized water, and the solution is placed in a petri dish (1 cm deep × 10 cm diameter). If impure samples of $K_2[PtCl_4]$ are used, the yield will be considerably reduced. If necessary, the platinum salt can be purified by dissolving in water (~3 mL) followed by precipitation by addition of ethanol (~15 mL). The centrifuged $K_2[PtCl_4]$ is washed well with diethyl ether and air-dried prior to use. Crystalline samples of phosphorous acid are preferable, but very wet commercial samples can be dried over P_4O_{10} in a vacuum desiccator. The petri dish and contents are placed on a steam bath at 104° and heated for 3 hr. In order to prevent the solution from evaporating to dryness, the water content of the reaction solution is replenished at 40-min intervals. Alternatively, the dish can be covered with a watch glass, which can reduce the time of heating. During the heating process the color of the solution changes from red to brown, then to pale yellow. After the 3-hr heating period, the volume of the pale-colored solution is again replenished to 10 mL with water. The petri dish is then transferred to an oven at 110°, and the solution is allowed to evaporate to dryness from the open dish (approximately 4 hr). To the yellow-green residue is added 10 mL of methanol (reagent grade), and the resulting suspension is filtered with suction through a glass-frit filter (15 mL capacity, medium porosity). The solid is then washed successively with methanol (4 × 10 mL) and diethyl ether (2 × 10 mL) and is air-dried to give 0.615 g of product as a yellow-green powder. Yield: 88%.

Anal. Calcd. for $H_{12}K_4O_{22}P_8Pt_2$: Pt, 33.7. Found: Pt, 34.9.

The purity of the complex can be checked by aqueous solution ^{31}P NMR spectroscopy.

The checkers find that the pale green color is due to traces of oxidized binuclear platinum compounds.[7] In order to obtain ultrahigh-purity crystalline material for solid state measurements, further purification by recrystallization is necessary. Alternatively, cations other than potassium can be used, since the nature of the counterion affects the solid state trace impurities that are occluded.

Properties

Potassium tetrakis[dihydrogen diphosphito(2 −)]diplatinate(II) is a diamagnetic yellow-green solid that is air-stable. It is soluble in water but insoluble in common organic solvents. Aqueous solutions decompose over a 24-hr period, the stability being higher at low pH. The ^{31}P NMR spectrum of the complex in D_2O solvent shows a resonance at δ 66.5 ppm to high frequency of 85% H_3PO_4, the peak being flanked by satellites due to coupling with ^{195}Pt (33% abundance, $^1J_{PtP}$ =

3075 Hz). The complex can be recrystallized with significant loss from a concentrated aqueous solution by addition of methanol. The IR spectrum (Nujol mull) shows broad bands in the 1100–900 cm^{-1} region characteristic of ν(PO) in the dihydrogen diphosphite group.[10]

References

1. R. P. Sperline, M. K. Dickson, and D. M. Roundhill, *J. Chem. Soc., Chem. Commn.*, **1977,** 62.
2. W. A. Fordyce, J. G. Brummer, and G. A. Crosby, *J. Am. Chem. Soc.*, **103,** 7061 (1981).
3. C.-M. Che, L. G. Butler, and H. B. Gray, *J. Am. Chem. Soc.*, **103,** 7796 (1981).
4. S. F. Rice and H. B. Gray, *J. Am. Chem. Soc.*, **105,** 4571 (1983).
5. M. K. Dickson, S. K. Pettee, and D. M. Roundhill, *Anal. Chem.*, **53,** 2159 (1981).
6. C.-M. Che, W. P. Schaefer, H. B. Gray, M. K. Dickson, P. B. Stein, and D. M. Roundhill, *J. Am. Chem. Soc.*, **104,** 4253 (1982).
7. C.-M. Che, F. H. Herbstein, W. P. Schaefer, R. E. Marsh, and H. B. Gray, *J. Am. Chem. Soc.*, **105,** 4604 (1983).
8. M. K. Dickson, W. A. Fordyce, D. M. Appel, K. Alexander, P. Stein, and D. M. Roundhill, *Inorg. Chem.*, **21,** 3857 (1982).
9. M. A. Filomena Dos Remedios Pinto, P. J. Sadler, S. Neidle, M. R. Sanderson, and A. Subbiah, *J. Chem. Soc., Chem. Commun.*, **1980,** 13.
10. P. Stein, M. K. Dickson, and D. M. Roundhill, *J. Am. Chem. Soc.*, **105,** 3489 (1983).
11. C.-M. Che, L. G. Butler, H. B. Gray, R. M. Crooks, and W. H. Woodruff, *J. Am. Chem. Soc.*, **105,** 5492 (1983).

62. ETHENE COMPLEXES OF BIS(TRIALKYLPHOSPHINE)PLATINUM(0)

Submitted by R. A. HEAD*
Checked by D. M. ROUNDHILL† and D. HEDDEN†

A limited number of platinum(0) complexes that contain ethene and organophosphine ligands are known. The triphenylphosphine complex, $[Pt(C_2H_4)[P(C_6H_5)_3]_2]$, is readily prepared by the reaction of $[Pt[P(C_6H_5)_3]_2O_2]$ with ethene.[1] It is an excellent precursor to a range of platinum(0) complexes. Preparations of analogous compounds containing trialkylphosphine ligands are less convenient and include reaction of the very air-sensitive $[Pt(C_2H_4)_3]$ with trialkyl-

*New Science Group, Imperial Chemical Industries PLC, The Heath, Runcorn, Cheshire, England WA7 4QE.

†Department of Chemistry, Tulane University, New Orleans, LA 70118.

phosphines[2] and the thermal decomposition of $[Pt[P(C_2H_5)_3]_2(C_2H_5)_2]$.[3] The synthesis reported here is both simple and quick and gives stable solutions of these complexes, which should make them useful starting materials for new platinum(0) chemistry. Displacement of the coordinated ethene by alkenes and ketones that contain electron-withdrawing groups is facile, and monomeric formaldehyde effects a slow displacement reaction to give solutions of $[Pt(CH_2O)[P(C_2H_5)_3]_2]$.[4]

A. (ETHENE)BIS(TRIETHYLPHOSPHINE)PLATINUM(0)

$$PtCl_2[P(C_2H_5)_3]_2 + 2NaC_{10}H_8 + C_2H_4 \rightarrow Pt(C_2H_4)[P(C_2H_5)_3]_2 + 2\,NaCl + 2C_{10}H_8$$

Procedure

■ **Caution.** *All solvents should be dried thoroughly and air-free before use. The reactions must be performed under a rigorously oxygen-free atmosphere. Metallic sodium is exceptionally reactive, especially toward moisture, and should be used with utmost care. Excess sodium can be destroyed by careful reaction with 2-propanol.*

Freshly distilled oxygen-free tetrahydrofuran (120 mL) and naphthalene (2.2 g) are added to a 250-mL three-necked round-bottomed flask fitted with a nitrogen bubbler and containing a magnetic stirring bar. Sodium wire (0.5 g) is then added, and the solution is stirred vigorously for 5 hr, during which time the sodium dissolves to give an intense green solution of $NaC_{10}H_8$. The molarity of this solution is then determined by removing an accurately known volume (~5 mL) and quenching into water. Titration of this aqueous solution to neutrality with 0.1 *N* HCl, using phenolphthalein as the indicator, allows the concentration of $NaC_{10}H_8$ in tetrahydrofuran to be calculated. If the above procedure is observed, the concentration of $NaC_{10}H_8$ is approximately 0.11 *N*.

To a 250-mL round-bottomed flask fitted with an ethene inlet, a bubbler, and a rubber septum and containing a magnetic stirring bar is added dichlorobis(triethylphosphine)platinum(II)* (0.62 g, 1.2 mmol) and dry, oxygen-free tetrahydrofuran (50 mL). The reaction works equally well with both *cis* and *trans* isomers of the platinum complex. A tetrahydrofuran solution of $NaC_{10}H_8$ (24 mL, 0.1 *N*) prepared as above is then added over a 15–20-min period to the stirred suspension at room temperature by means of a gastight syringe (Fig. 1). On mixing the two solutions, the deep green color of $NaC_{10}H_8$ is rapidly destroyed,

*Syntheses of $PtCl_2(PR_3)_2$ complexes are best carried out using the procedure outlined for $PtCl_2(PPr_3^i)_2$ by Yoshida et al. in Reference 5.

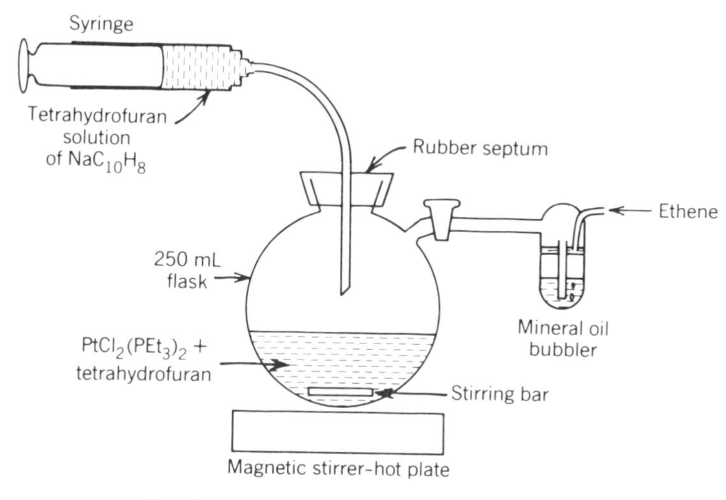

Fig. 1. Apparatus for platinum(0) complexes.

and the platinum complex gradually dissolves, eventually giving a slightly cloudy solution of (ethene)bis(triethylphosphine)platinum(0). Addition of excess reducing agent gives an intense red solution, but this is easily reversed by adding a crystal of $PtCl_2[P(C_2H_5)_3]_2$. The yield of product is essentially quantitative, as evidenced by ^{31}P NMR spectroscopy. A single line is observed at 20.4 ppm, relative to 85% H_3PO_4, with $^1J_{^{195}Pt-P} = 3520$ Hz.

B. (ETHENE)BIS(PHOSPHINE)PLATINUM(0)

$$PtCl_2(PR_3)_2 + 2NaC_{10}H_8 + C_2H_4 \rightarrow Pt(C_2H_4)(PR_3)_2 + 2NaCl + 2C_{10}H_8$$

Phosphine, PR_3* = tris(1-methylethyl)phosphine; diethylphenylphosphine; triphenylphosphine; ½[1,2-ethanediylbis(diphenylphosphine)]

Procedure

The method for preparation of the zero valent ethene complexes is identical to that described in Section A. For less soluble starting materials the reducing agent is best added over a longer period of time (up to 30 min). All complexes are

*All phosphines are available through Strem Chemicals, Inc., P.O. Box 108, Newburyport, MA 01950, who also supply certain Pt(II) complexes including $PtCl_2(PPh_3)_2$.

stable in solution. They are prepared in nearly quantitative yield as shown by ^{31}P NMR spectroscopy (see Table I).

TABLE I. ^{31}P NMR Data for $[Pt(C_2H_4)(PR_3)_2]$ Complexes

PR_3	Chemical shift (ppm)[a]	$^1J_{^{195}Pt-P}$ (Hz)
$P(C_2H_5)_3$	20.4	3520
$P(i\text{-}C_3H_7)_3$	53.4	3657
$P(C_2H_5)_2C_6H_5$	23.2	3574
$P(C_6H_5)_3$	32.0	3660
$\frac{1}{2}(C_6H_5)_2PCH_2CH_2P(C_6H_5)_2$	54.5	3300

[a]Relative to 85% H_3PO_4.

Properties

Solutions of the complexes are stable for several hours at room temperature under an ethene atmosphere. Decomposition takes place if the compound is heated ($>\sim 65°$). If the solvent is removed under reduced pressure, a red-brown oil of uncertain composition is formed.

References

1. F. R. Hartley, *The Chemistry of Platinum and Palladium with Particular Reference to Complexes of the Elements*, Halstead, New York, 1973.
2. M. Berry, J. A. K. Howard, and F. G. A. Stone, *J. Chem. Soc., Dalton Trans.*, **1980,** 1601, 1609; J. Spencer, *Inorg. Synth.*, **19,** 216 (1979).
3. R. G. Nuzzo, T. J. McCarthy, and G. M. Whitesides, *Inorg. Chem.*, **20,** 1312 (1981).
4. R. A. Head, *J. Chem. Soc., Dalton Trans.*, **1982,** 1637.
5. T. Yoshida, T. Matsuda, and S. Otsuka, *Inorg. Synth.*, **19,** 108 (1979).

63. NITROSYL HEXACHLOROPLATINATE(IV)

$$Pt + 6NOCl \rightarrow (NO)_2[PtCl_6] + 4NO \uparrow$$

Submitted by RICHARD T. MORAVEK* and GEORGE B. KAUFFMAN†
Checked by TARIQ MAHMOOD‡

Nitrosyl hexachloroplatinate(IV) was first prepared by the action of excess aqua regia on platinum[1,2] and has been reported since by several investigators using essentially the same method.[3,4] It can also be prepared by the action of fuming nitric acid on platinum(IV) chloride solution[5] or on hexachloroplatinic(IV) acid.[6] The yields obtained by the above methods are quite low, and extreme care is necessary in order to isolate the compound.

In addition to these preparations in aqueous solution, the compound has also been prepared in low yield by the action of nitrosyl chloride at 100° on platinum.[7] This low yield is due to incomplete reaction and the difficulty of separating the deliquescent powder from the unreacted metal.[8] It is possible to prepare the compound in higher yield by the action of nitrosyl chloride on platinum(II) or platinum(IV) chloride.[8] Neither platinum(II) nor platinum(IV) chloride is readily obtained commercially. Furthermore, platinum(IV) chloride is quite hygroscopic and is difficult both to prepare and to handle.

The following modification of the literature method[8] utilizes pure platinum metal. The manipulations involved are fairly simple, and the equipment used is kept to a minimum.

Procedure

■ **Caution.** *Nitrosyl chloride is a severe respiratory irritant. The operations must be performed behind a safety shield in a well-ventilated hood. Because of the danger of explosion a face shield should be worn at all times. Heavy gloves should be worn when handling glass vessels that contain gases at high pressure.*

Four grams (0.021 mole) of pure platinum wire (for example, 0.025-in. diameter platinum wire, 99.9+%) [Alfa Products] is cut into short lengths (~1 mm) with scissors. It is placed in a large ampule made from a 16-in. length

*Enjay Chemical Co., Baytown, TX 77520. Work performed at Research and Development Division, Humble Oil and Refining Co., Baytown, TX 77520.

†Department of Chemistry, California State University, Fresno, Fresno, CA 93740.

‡Department of Chemistry, University of Idaho, Moscow, ID 83843.

of ¾-in. o.d. Pyrex tubing with heavy walls (at least ⅛ in. thick) to withstand high pressure. The mouth of the ampule is constricted to fit a length of rubber pressure tubing. The vessel is cooled in a liquid nitrogen or Dry Ice/methanol bath.

About 15 mL of nitrosyl chloride (bp −5.5°) is then distilled at room temperature through a length of rubber pressure tubing (only slowly attacked by nitrosyl chloride) into the ampule, where it condenses as a yellow-orange solid on the platinum. A suitable mark is made on the ampule to show when the approximate volume of nitrosyl chloride has been added. The nitrosyl chloride can be obtained from a cylinder [Matheson] or may be prepared by the reaction of dry hydrogen chloride with nitrosylsulfuric acid[9] or sodium nitrite.[10] A corrosion-resistant nickel pressure regulator should be used to control the flow rate. Contrary to the work of previous investigators, no attempt is made to dry the gas, since a trace of moisture appears to catalyze the reaction.[11]*

After the required amount of nitrosyl chloride has been condensed in the ampule, which is still retained in the liquid nitrogen trap, the ampule is sealed off and *carefully annealed with a blowtorch.* (■ **Caution.** *Strains present in the glass may result in explosion of the ampule.*) It is removed from the cooling bath and allowed to warm to room temperature *behind a safety shield* in a well-ventilated hood. It is then placed upright in a tubular heating device consisting of a 1-in. i.d. stainless steel pipe closed at the bottom end, wrapped with electrical heating tape, and insulated with Fiberglas tape. The heater should be of such a height (15¾ in.) that the sealed tip of the ampule is just visible over the top. The temperature is controlled with a Powerstat variable transformer and can be measured with a thermocouple.

The temperature is slowly raised to ~100° over a period of 1 week and is maintained at this temperature for two additional weeks. (■ **Caution.** *Considerable pressure is built up, and the reaction should be carried out behind a safety shield.*) At the end of the required time the ampule is allowed to cool to room temperature and is then placed in a liquid nitrogen bath to condense nitrogen(II) oxide (mp −163.6°, bp −151.7°) produced in the reaction, together with excess nitrosyl chloride. If a Dry Ice/methanol bath is used, considerable pressure due to uncondensed nitrogen(II) oxide is to be expected on opening the tube.

The ampule is opened carefully with a blowtorch and is allowed to warm to room temperature in a well-ventilated hood, whereupon the nitrogen(II) oxide and excess nitrosyl chloride distill off and are absorbed in a beaker of dilute sodium hydroxide solution. Rubber pressure tubing is again used for the con-

*The checker carried out the synthesis by using standard vacuum line techniques, which are believed to be less hazardous. Heating was carried out inside a metal pipe in an ordinary laboratory oven.

nections, and a drying tube that contains P_4O_{10} is included to prevent the entrance of moisture into the ampule. An empty glass trap is included between the drying tube and the beaker to protect the product from any sodium hydroxide solution accidentally drawn back by suction. The residual red vapors that remain in the system can be removed by aspirator suction through a trap that contains P_4O_{10}. A sharp scratch is made around the body of the ampule, and together with the attached drying tube it is placed in a dry box. The ampule is broken, and the granular contents are removed and powdered in an agate mortar. The yield (~9.5 g; 20 mmol) is nearly quantitative.* The powder is readily separated from the few residual platinum granules by sifting through a square of 52-mesh platinum gauze. The product is dried under vacuum over P_4O_{10} and should be stored in vials that are kept in a desiccator. All transfers of the dry product are made in a dry box.

Anal. Calcd. for $(NO)_2PtCl_6$: Pt, 41.72; Cl, 45.47; N, 5.99; O, 6.84. Found: Pt, 41.72; Cl, 45.57; N, 5.90; O, 6.72.

Properties

Nitrosyl hexachloroplatinate(IV) is a deliquescent yellow-brown powder that reacts vigorously with water to form hexachloroplatinic(IV) acid and nitrogen(II) oxide. The substance exhibits no definite melting point but begins to decompose at 300°, as evidenced by blackening. It dissolves with effervescence in ethanol, methanol, isopropyl alcohol, acetone, and *N,N*-dimethylformamide. A freshly prepared concentrated acetone solution of the substance effervesces violently, evolving much heat and becoming dark brown. A black tarry residue that is incompletely soluble in aqua regia can be obtained by removal of solvent under reduced pressure. The exact nature of the reaction in these solvents is unknown. Nitrosyl hexachloroplatinate(IV) is insoluble in benzene, carbon disulfide, and carbon tetrachloride. The IR spectrum (KBr) has characteristic bands at 2210 (ν_{NO^+}) and 348 ($\nu_{PtCl_6^{2-}}$) cm^{-1}.

Although the empirical formula $PtCl_4 \cdot 2NOCl$ had been established by early investigators, it was not until the compound was shown to be diamagnetic[12] that the structure $(NO)_2[PtCl_6]$ could be assigned with any degree of certainty. An extensive review of the literature on inorganic nitrosyl compounds has been compiled by Moeller.[13]

References

1. H. D. Rogers and M. H. Boyé, *Phil. Mag.* [3] **17,** 397 (1840).
2. H. D. Rogers and M. H. Boyé, *Trans. Am. Phil. Soc.*, **7,** 59 (1841).

*The checker used one-fourth as much of the reactants as specified and obtained a yield of ~100%.

3. H. Precht, *Z. Anal. Chem.*, **18,** 509 (1879).
4. H. Bornträger, *Repertm. Anal. Chem.*, **7,** 741 (1887).
5. R. Weber, *Ann. Physik*, **131,** 441 (1867).
6. S. M. Jørgensen, *Kgl. Danske Videnskab. Selskabs Skrifter, Naturvidenskab. Math, Afdel.*, [5], **6,** 451 (1867).
7. J. J. Sudborough, *J. Chem. Soc.*, **59,** 655 (1891).
8. J. R. Partington and A. L. Whynes, *J. Chem. Soc.*, **1949,** 3135.
9. G.H. Coleman, G. A. Lillis, and G. E. Goheen, *Inorg. Synth.*, **1,** 55 (1939).
10. J. R. Morton and H. W. Wilcox, *Inorg. Synth.*, **4,** 48 (1953).
11. L. J. Beckham, W. A. Fessler, and M. A. Kise, *Chem. Rev.*, **48,** 345 (1951).
12. R. W. Asmussen, *Z. Anorg. Allgem. Chem.*, **243**, 127 (1939).
13. T. Moeller, *J. Chem. Educ.*, **23,** 441, 542 (1946).

64. *cis*-TETRAAMMINE AND *cis*-BIS (1,2-ETHANEDIAMINE) COMPLEXES OF RHODIUM(III)

Submitted by MARTIN HANCOCK,* BENTE NIELSEN,* and JOHAN SPRINGBORG*
Checked by ALAN FRIEDMAN,† P. C. FORD,† and MICHAEL J. SALIBY‡

Methods are described for the preparation and purification of the following mononuclear complexes: *cis*-$[Rh(NH_3)_4Cl_2]Cl \cdot \frac{1}{2}H_2O$, *cis*-$[Rh(NH_3)_4(H_2O)(OH)][S_2O_6]$, $[Rh(en)_2(C_2O_4)]ClO_4$, *cis*-$[Rh(en_2Cl_2]_2Cl(ClO_4)$, and *cis*-$[Rh(en)_2(H_2O)(OH)][S_2O_6]$. A simple and rapid method for the synthesis of $[Rh(NH_3)_5Cl]Cl_2$ in essentially quantitative yield is reported.

Procedures for the preparation and purification of the bromide salts of the dinuclear di-μ-hydroxo complex ions $[(NH_3)_4Rh(OH)_2Rh(NH_3)_4]^{4+}$ and Δ,Λ-$[(en)_2Rh(OH)_2Rh(en)_2]^{4+}$ (*diols*) from the crude dithionate salts are also given. The methods described here are thoroughly tested modifications of those reported in the literature.

The present series of syntheses is timely in that currently there is considerable interest in the thermal and the photochemical reactions of mono- and dinuclear rhodium(III) ammine and amine complexes.[1-9] The dinuclear complexes have also proved to be suitable starting materials for the synthesis of new types of dinuclear rhodium(III) complexes with other bridging ligands.[10]

In several of the preparations given below, the crude products are almost pure and are obtained in high yields. The crude salts are purified by reprecipitation, and the UV/visible absorption spectra as well as elemental analyses are used as

*Department of Chemistry, Royal Veterinary and Agricultural University, Thorvaldsensvej 40, DK-1871 Copenhagen V, Denmark.
†Department of Chemistry, University of California, Santa Barbara, CA 93106.
‡Department of Chemistry, University of New Haven, West Haven, CT 06516.

TABLE 1. UV/vis Absorption Spectra

Compound	Medium	$(\epsilon,\lambda)_{max}$[a]	$(\epsilon,\lambda)_{min}$[a]
cis-$[Rh(NH_3)_4Cl_2]Cl\cdot\frac{1}{2}H_2O$	H_2O	(127,360)(110,295)	(72,322)(82,275)
cis-$[Rh(NH_3)_4(H_2O)(OH)][S_2O_6]$	0.1 *M* $HClO_4$, 0.9 *M* $NaClO_4$	(104,326)(85,269)	(61,293)(8,227)
	0.1 *M* NaOH, 0.9 *M* $NaClO_4$	(122,335)(117,282)	(93,308)(28,245)
	1 *M* $NaClO_4$	(121,330)(100,281)	(96,299)(35,241)
$[(NH_3)_4Rh(OH)_2Rh(NH_3)_4]Br_4\cdot 4H_2O$	1 *M* $NaClO_4$	(308,336)	(203,304)
cis-$[Rh(en)_2Cl_2]_2Cl[ClO_4]$	H_2O	(200,352)(195,294)	(129,320)(124,272)
cis-$[Rh(en)_2(H_2O)OH)][S_2O_6]$	1 *M* $HClO_4$	(177,318)(139,271)	(119,289)(36,236)
	1 *M* $NaClO_4$	$(203,322)(144,280)_{sh}$	(63,242)
	1 *M* NaOH	(185,329)(177,279)	(141,302)(46,243)
$\Delta,\Lambda - [(en)_2Rh(OH)_2Rh(en)_2]Br_4\cdot 2H_2O$	1 *M* $NaClO_4$	(535,331)	(375,301)

[a]All ϵ values are per mole of complex cation.

criteria of purity. When it is stated that a sample is pure, this means that the positions of the maxima and minima of the spectrum remain constant upon further reprecipitation and that a deviation of less than 1% in the molar absorptivity (ϵ) (for both maxima and minima) is found for the two crops. For some of the compounds the spectrum changes with time, and linear extrapolation of the spectrum back to the time of dissolution is then made. The corrections are never greater than 1%. In Table I the (ϵ, λ) extremum values are collected and are those obtained for the sample reprecipitated one time more than the sample identified as pure. All ϵ values are per mole of complex cation.

■ **Caution.** *Mechanical handling and heating of perchlorates is potentially dangerous because of the reducing character of coordinated ammonia or amines. The common situation in which a glass rod is used on a sintered glass filter may create an exorbitant local pressure that is likely to act as a shock. Rods of soft polyethylene are therefore strongly recommended for the handling of perchlorates. Washing with alcohol or similar solvents should also be performed with caution.*

A. PENTAAMMINECHLORORHODIUM(III) CHLORIDE

$$RhCl_3 + 5NH_3 \xrightarrow[NH_4Cl]{EtOH} [Rh(NH_3)_5Cl]Cl_2$$

The preparation described here is a slight modification of that reported by Addison et al.[11] and is given here simply as a reliable alternative to the method described by Osborn et al.[12]

Procedure

Hydrated rhodium(III) chloride [Johnson Matthey] (0.094 mole*; ~23.5 g) is dissolved in 100 mL of water in a 400-mL beaker at room temperature. Ammonium chloride (20 g) is added, and the solution is stirred until it is essentially clear. The dark red solution is then heated to 50°, after which 25 mL of 95% ethanol is added with stirring. After the solution is allowed to stand for 2–3 min, it is cooled to ~40°, by agitating the flask in a running stream of cold water, for example. Any precipitation that occurs at this point may be ignored. Fifty milliliters of concentrated ammonia solution (density 0.88 g/mL, ~12 *M*) is added with vigorous stirring. (■ **Caution.** *Addition should be made in a well-ventilated fume hood.*) Immediate precipitation occurs, and the temperature of

*Checkers were able to obtain similar yields when 5–20% of the stated quantities of starting materials were used.

the mixture rises to ~65°. The mixture is allowed to cool with stirring for 1.5 hr and is stirred for a further 30 min in an ice bath.

The pale-yellow product is isolated by filtration on a medium-porosity 7-cm sintered glass funnel and washed with three 30-mL portions of 0.5 *M* hydrochloric acid, then thoroughly with 95% ethanol, and finally with diethyl ether. Drying in air gives 26.2 g (yield 95%). The product is sufficiently pure for use in the preparation of *cis*-$[Rh(NH_3)_4Cl_2]Cl \cdot \frac{1}{2}H_2O$.

Anal. Calcd. for $RhH_{15}N_5Cl_3$: H, 5.14; N, 23.79; Cl, 36.13. Found: H, 5.28; N, 22.90; Cl. 35.43.

B. *cis*-TETRAAMMINEDICHLORORHODIUM(III) CHLORIDE

$$2[Rh(NH_3)_5Cl]Cl_2 + Na_2C_2O_4 + H_2C_2O_4 \cdot 2H_2O \rightarrow 2[Rh(NH_3)_4(C_2O_4)]Cl + 2NH_4Cl + 2NaCl + 2H_2O$$

$$[Rh(NH_3)_4(C_2O_4)]Cl + HClO_4 + H_2O \rightarrow [Rh(NH_3)_4(C_2O_4)]ClO_4 \cdot H_2O + HCl$$

$$2[Rh(NH_3)_4(C_2O_4)]ClO_4 \cdot H_2O + 6HCl \rightarrow 2cis\text{-}[Rh(NH_3)_4Cl_2]Cl \cdot \tfrac{1}{2}H_2O + 2HClO_4 + 2H_2C_2O_4 + H_2O$$

The synthesis of the *cis*-tetraamminedichlororhodium(III) ion, using the reaction of tetraammine[oxalato]rhodium(III) perchlorate with boiling 6 *M* hydrochloric acid, was first reported in 1975.[13] The oxalato complex is formed by the reaction of pentaamminechlororhodium(III) chloride with oxalate and oxalic acid in water at 120°. The synthesis has been repeated on numerous occasions in our laboratory, and we have found that only in a minority of cases is it possible to isolate $[Rh(NH_3)_4(C_2O_4)]ClO_4 \cdot H_2O$ in sufficiently pure form to prepare the pure perchlorate by recrystallization. In a majority of cases the product that precipitates from the reaction mixture on addition of perchloric acid contains an appreciable quantity of an unidentified impurity that is not removed by recrystallization. The procedure given below permits the preparation of pure *cis*-$[Rh(NH_3)_4Cl_2]Cl \cdot \frac{1}{2}H_2O$ from the latter crude, impure oxalato complex.

Procedure

Pentaamminechlororhodium(III) chloride (5.0 g, 0.017 mole), disodium oxalate (2.275 g, 0.017 mole), and oxalic acid dihydrate (1.075 g, 0.0085 mole) are stirred together briefly in 190 mL of water in a stainless steel autoclave (volume ~300 mL). The autoclave is then closed tightly and kept at a temperature of 120° in an oven for 24 hr, after which it is cooled just sufficiently (under running water) to allow it to be handled without burning one's fingers. It is opened carefully and the contents are filtered immediately by suction through a fine-porosity, 4-cm sintered glass funnel into a 500-mL suction flask. The autoclave is rinsed with 25 mL of water, and the combined filtered solution is allowed to cool to room temperature. A 6.25-mL portion of 12 *M* perchloric acid is then added to the solution, whereupon crystals begin to form. (■ *Caution. Perchloric acid is a strong oxidizing agent and should not be permitted to contact organics or other oxidizable materials.*) After further cooling in an ice bath with very slow magnetic stirring for 10 min, the product is isolated by filtration on a 4-cm sintered glass funnel. It is washed with a little ice-cold 2 *M* perchloric acid, then methanol, and dried in air. (■ **Caution.** *Perchlorates may be shock-sensitive and potentially dangerous because of the reducing character of ammonia.*) The yield of crude, impure tetraammine[oxalato]rhodium(III) perchlorate is typically 3.2–3.8 g. Sometimes the oxalato complex separates in lower yield (2.6–2.8 g), but in this case the product is generally free of the impurity that is otherwise normally present. On subsequent boiling with 6 *M* hydrochloric acid, a clear solution is obtained that yields almost pure *cis*-$[Rh(NH_3)_4Cl_2]Cl\cdot\frac{1}{2}H_2O$ in essentially quantitative yield.

cis-Tetraamminedichlororhodium(III) chloride is obtained from the crude, impure product as follows: A portion of the latter product (2.40 g) is boiled for 1 min with 70 mL of 6 *M* hydrochloric acid in a 250-mL conical flask, and the solution is filtered quickly through a fine-porosity 3-cm sintered glass funnel. The clear bright-yellow solution is then allowed to cool spontaneously while the temperature is followed with a thermometer. If any precipitation occurs before the temperature falls to 50°, the solution is again filtered as before. The solution is cooled in an ice bath for 10 min, and methanol (70 mL) is added. After cooling for a further 10 min the bright-yellow crystals are isolated by filtration on a 3-cm sintered glass funnel of fine porosity, washed thoroughly with methanol, and dried in air. Yield: 1.2–1.5 g (72–90%). The crude product is sufficiently pure for use in the preparation of *cis*-$[Rh(NH_3)_4(H_2O)(OH)][S_2O_6]$ (see Section C).

The pure complex is obtained by dissolving the crude product in hot (~95°) 3 *M* hydrochloric acid (25 mL per gram of crude product), filtering the hot solution through a fine-porosity 3-cm sintered glass funnel, and allowing it to cool to room temperature. After the mixture has been kept at ~5° in a refrigerator overnight, the bright yellow needle crystals are isolated by filtration and washed and dried as before (recrystallization yield 91%).

Anal. Calcd. for $RhH_{13}N_4Cl_3O_{0.5}$: H, 4.58; N, 19.56; Cl (total), 37.14; Cl (ionic), 12.38. Found: H, 4.41; N, 19.50; Cl (total), 37.20; Cl (ionic), 12.66.

Properties

cis-Tetraamminedichlororhodium(III) chloride forms bright yellow needle crystals that are moderately soluble in water. It is the starting material for the synthesis of *cis*-$[Rh(NH_3)_4(H_2O)(OH)][S_2O_6]$ (Section C).

C. *cis*-TETRAAMMINEAQUAHYDROXORHODIUM(III) DITHIONATE

$$cis\text{-}[Rh(NH_3)_4Cl_2]Cl\cdot\tfrac{1}{2}H_2O + 3AgNO_3 + 1\tfrac{1}{2}H_2O \rightarrow cis\text{-}[Rh(NH_3)_4(H_2O)_2](NO_3)_3 + 3AgCl$$

$$cis\text{-}[Rh(NH_3)_4(H_2O)_2](NO_3)_3 + C_5H_5N \rightarrow cis\text{-}[Rh(NH_3)_4(H_2O)(OH)](NO_3)_2 + (C_5H_5NH)NO_3$$

$$cis\text{-}[Rh(NH_3)_4(H_2O)(OH)](NO_3)_2 + Na_2S_2O_6 \rightarrow cis\text{-}[Rh(NH_3)_4(H_2O)(OH)][S_2O_6] + 2NaNO_3$$

Procedure

cis-Tetraamminedichlororhodium(III) chloride (1.0 g, 0.00349 mole) and silver nitrate (1.77 g, 0.01042 mole) are heated together under reflux in 30 mL of water for 3.5 hr in a 100-mL round-bottomed flask that is wrapped in aluminum foil to exclude light. The mixture is allowed to cool to room temperature and is then filtered through a fine-porosity 3-cm sintered glass funnel into a 100-mL suction flask. The silver chloride residue is washed twice with 3 mL of water, and the washings are added to the pale yellow bulk filtrate. Solid sodium dithionate dihydrate (1.80 g) is added to the solution, which is then stirred at room temperature until the crystals are dissolved. Two milliliters of pyridine are added dropwise under vigorous magnetic stirring and cooling in an ice bath. After further stirring and cooling for 2 hr, the pale yellow crystalline product is isolated by filtration on a fine-porosity 3-cm sintered glass funnel and washed, first with a little ice-cold water, then 95% ethanol, and finally diethyl ether. Drying in air gives 0.98 g (yield 77%). The crude product is used in the preparation of the crude ammonia dihydroxo (*diol*) dithionate (see Section D).

The product is purified by dissolving the crude product at room temperature in slightly more than the calculated volume of 0.50 *M* hydrochloric acid in which sodium dithionate dihydrate has been dissolved (that is, 6.0 mL of 0.5 *M* HCl and 1.0 g of $Na_2S_2O_6\cdot 2H_2O$ per gram of complex). The stoichiometric quantity of 1.00 *M* sodium hydroxide solution is added to the filtered solution with stirring and cooling as before. Washing and drying are performed as above. The product when reprecipitated twice in this manner is pure.

Anal. Calcd. for $RhH_{15}N_4O_8S_2$: H, 4.13; N, 15.30; S, 17.51. Found: H, 4.38; N, 14.99; S, 17.42

Properties

cis-Tetraammineaqua(hydroxo)rhodium(III) dithionate is a pale yellow crystalline solid. It is sparingly soluble in water but readily soluble in strong acid and strong base, giving solutions of the *cis*-diaqua and *cis*-dihydroxo complex ions, respectively. At elevated temperatures the complex loses water, with formation of the dihydroxo compound (diol) (Section D). The acid dissociation constants for the *cis*-diaqua complex ion in 1.0 *M* sodium perchlorate medium at 25° have been determined[14]: $pK_{a1} = 6.40$; $pK_{a2} = 8.32$.

D. DI-μ-HYDROXO-BIS[TETRAAMMINERHODIUM(III)] BROMIDE

$$2cis\text{-}[Rh(NH_3)_4(H_2O)(OH)][S_2O_6] \rightarrow [(NH_3)_4Rh(OH)_2Rh(NH_3)_4][S_2O_6]_2 + 2H_2O$$

$$[(NH_3)_4Rh(OH)_2Rh(NH_3)_4][S_2O_6]_2 + 4NH_4Br + 4H_2O \rightarrow [(NH_3)_4Rh(OH)_2Rh(NH_3)_4]Br_4\cdot 4H_2O + 2(NH_4)_2[S_2O_6]$$

The starting material for the preparation of the bromide salt of di-μ-hydroxo-bis[tetraamminerhodium(III)] (*diol*) is the crude dithionate salt, which is obtained by heating *cis*-tetraammineaqua(hydroxo)rhodium(III) dithionate at 120°. In order to avoid contamination of the bromide with mono-hydroxo bridged (*monool*) species, it is necessary to keep dissolution times and, where possible, temperature to a minimum.

Procedure

Solid *cis*-$[Rh(NH_3)_4(H_2O)(OH)][S_2O_6]$ (4.76 g, 0.0130 mole) is heated at 120(±2)° for 20 hr in an oven. This gives crude di-μ-hydroxo (*diol*) dithionate (4.53 g), which is treated at room temperature with 50 mL of a saturated solution of ammonium bromide in a 100-mL conical flask. The suspension is stirred vigorously for 1 hr, and the resulting crude bromide salt is isolated by filtration through a fine-porosity 4-cm sintered glass funnel and washed with ice-cold 50% v/v ethanol/water, 95% ethanol, and then diethyl ether. The product is extracted on the filter with water (18–20°) in portions of ~40 mL within a total of 20 min (total extraction volume ~170 mL), and each successive portion of solution is filtered immediately into a single ice-cooled 500-mL conical flask. After each extraction 20 mL of saturated ammonium bromide solution is added to the flask with vigorous magnetic stirring (total added volume is 80 mL). A dense pale yellow precipitate is formed immediately, and the mixture is cooled with stirring for a further 20 min. The bromide salt is isolated as above. Drying in air gives 4.28 g (yield 86%) of pure di-μ-hydroxo (*diol*) bromide.

Anal. Calcd. for $Rh_2H_{34}N_8Br_4O_6$: H, 4.46; N, 14.60; Br, 41.63. Found: H, 4.44; N, 14.37; Br, 42.15.

Properties

See Section H.

E. BIS(1,2-ETHANEDIAMINE)(OXALATO)RHODIUM(III) PERCHLORATE

$$RhCl_3 + 2(NH_2CH_2CH_2NH_2)Cl_2 + 4NaOH \rightarrow \textit{cis/trans}\text{-}[Rh(en)_2Cl_2]Cl + 4NaCl + 4H_2O$$

$$\textit{cis/trans}\text{-}[Rh(en)_2Cl_2]Cl + NaClO_4 \rightarrow \textit{cis/trans}\text{-}[Rh(en)_2Cl_2]ClO_4 + NaCl$$

$$\textit{cis/trans}\text{-}[Rh(en)_2Cl_2]ClO_4 + Na_2C_2O_4 \xrightarrow{BH_4^-} [Rh(en)_2(C_2O_4)]ClO_4 + 2NaCl$$

The method of Johnson and Basolo[15] for the synthesis of $[Rh(en)_2Cl_2]^+$ gives a reasonably good yield of the *trans* isomer but a poor yield of the *cis* isomer. A number of modifications designed to improve the yield of *cis* isomer have subsequently been employed,[1,16–18] but with only limited success. The procedure described here gives a high yield (>70%) of a crude mixture of the perchlorate salts of *cis*- and *trans*- $[Rh(en)_2Cl_2)^+$, from which pure bis(1,2-ethanediamine)(oxalato)rhodium(III) perchlorate can be prepared in ~60% overall yield by the reductant-accelerated reaction with oxalate ion. The reductant used here is sodium tetrahydroborate (1—). The oxalato complex is an excellent starting material for the preparation of the *cis*-dichloro complex.[17–19]

Procedure

Hydrated rhodium(III) chloride (0.038 mole, ~10 g) and 1,2-ethanediamine dihydrochloride (10.1 g, 0.076 mole) are dissolved with gentle heating in 60 mL of water in a 250-mL round-bottomed flask fitted with a condenser. A copious red-brown precipitate is formed when 38 mL of 2.00 *M* sodium hydroxide solution is added. The mixture is heated to boiling under reflux. After boiling for ~1 min, 2.00 *M* sodium hydroxide (~45 mL) is added in small portions by means of the reflux condenser over a period of 10 min until the pH of the clear orange-red solution remains at 6.5. The solution is carefully evaporated to dryness on a vacuum rotary evaporator (water-bath temperature ~40°).

The flask containing the solid orange residue is kept in an oven at 170° for 24 hr, and the residue is then dissolved in 100 mL of 0.1 *M* hydrochloric acid by boiling. The solution is filtered while hot through a fine-porosity 3-cm sintered glass funnel into a 250-mL suction flask to remove a little insoluble dark brown residue. A hot solution of sodium perchlorate monohydrate (20 g) in 20 mL of water is added to the clear orange filtrate. On cooling slightly, copious precipitation sets in. After the mixture has been kept in a refrigerator (~5°) for 2 days, the product is isolated by filtration on a medium-porosity 7-cm sintered glass funnel and then washed with a little ice-cold water, 95% ethanol, and finally diethyl ether. Drying in air gives 11–11.5 g (~73–77%) of the crude dichloro complex.

Bis(1,2-ethanediamine)(oxalato)rhodium(III) perchlorate is then obtained as follows.

■ **Caution.** *Perchlorates may be shock-sensitive and potentially dangerous because of the reducing character of coordinated amines.*

Crude *cis*- and *trans*-dichlorobis(1,2-ethanediamine)rhodium(III) perchlorate (5.26 g, 0.0134 mole) and disodium oxalate (3.40 g, 0.0254 mole) are boiled together under reflux in 160 mL of water in a 250-mL round-bottomed flask fitted with a condenser until all the solid has dissolved. After a further 2 min or so, the reflux condenser is removed briefly, and a small speck (~1 mg) of sodium tetrahydroborate(1 −) is introduced into the boiling solution with a glass spatula, whereupon the solution immediately darkens. The addition of sodium tetrahydroborate(1 −) is repeated twice more at 5-min intervals, and the dark, cloudy solution is finally boiled under reflux for an additional 10 min.

The hot solution is then filtered quickly by suction through a fine-porosity 4-cm sintered glass funnel into a 500-mL suction flask. The flask and funnel are washed twice with 10 mL of boiling water. To the combined filtrate and washings is added a solution of sodium perchlorate monohydrate (10.0 g) in 10 mL of water. The mixture is then kept in a refrigerator for 24 hr.

The gray-tinged pale yellow crystals are isolated by filtration on a 4-cm sintered glass funnel, washed with a little ice-cold water, and then redissolved by boiling in 300 mL of water in a 500-mL conical flask. Activated charcoal (1 g) is added, and boiling is continued for ~1 min. The mixture is then filtered quickly by suction through a fine-porosity 4-cm sintered glass funnel into a 500-mL suction flask. The conical flask and funnel are washed with 25 mL of boiling water. The combined filtrate and washings are reheated to the boiling point, and a boiling, filtered solution of sodium perchlorate monohydrate (10.0 g) in 10 mL of water is added. On cooling, pale yellow crystals are formed. The mixture is left at room temperature for 24 hr and then in a refrigerator for a further 24 hr, and the product is then isolated by filtration on a fine-porosity 4-cm sintered glass funnel and washed with cold water, with 95% ethanol and finally with diethyl ether. Drying in air gives 4.3 g (yield 78%).

Anal. Calcd. for $RhH_{16}N_4C_6ClO_8$: C, 17.55; H, 3.93; N, 13.65; Cl, 8.63. Found: C, 17.13; H, 4.02; N, 13.41; Cl, 8.41.

Properties

Bis(1,2-ethanediamine)(oxalato)rhodium(III) perchlorate is a pale yellow crystalline solid that is sparingly soluble in water. It reacts with hot aqueous hydrochloric acid, with quantitative formation of *cis*-$[Rh(en)_2Cl_2]^+$ ion (see Section F).

F. *cis*-DICHLOROBIS(1,2-ETHANEDIAMINE)RHODIUM(III) CHLORIDE PERCHLORATE (2:1:1)

$$4[Rh(en)_2(C_2O_4)]ClO_4 + 10HCl \xrightarrow{\text{Boil}} 2[Rh(en)_2Cl_2]_2Cl(ClO_4) + 4H_2C_2O_4 + 2HClO_4$$

The synthesis of *cis*-$[Rh(en)_2Cl_2]Cl \cdot H_2O$ from the oxalato complex in 64% yield has been reported previously.[17,18] The procedure described here permits the isolation of the pure chloride perchlorate salt in ~80% yield.

Procedure

■ **Caution.** *Perchlorates may be shock-sensitive and potentially dangerous because of the reducing character of coordinated amines.*

To a 5.0-g (0.0122-mole) quantity of pure bis(1,2-ethanediamine)(oxalato) rhodium(III) perchlorate in a 250-mL conical flask is added 70 mL of 6 *M* hydrochloric acid. The mixture is boiled for 2 min. The clear, bright lemon-yellow solution that results is allowed to cool to room temperature and is then kept in a refrigerator for 24 hr. The large bright yellow crystals are isolated by filtration on a fine-porosity 4-cm sintered glass funnel, washed thoroughly with 95% ethanol and then diethyl ether, and dried in air. Yield: 3.94 g (90%). The unrecrystallized product is sufficiently pure for most purposes. The pure product is obtained by one recrystallization from boiling 6 *M* hydrochloric acid (15 mL per gram of crude product). The hot filtered solution is allowed to cool to room temperature, and the mixture is kept in a refrigerator for 2 days. The crystals are isolated and washed and dried as before (recrystallization yield 90–92%).

Anal. Calcd. for $Rh_2C_8H_{32}N_8Cl_6O_4$: C, 13.29; H, 4.46; N, 15.50; Cl (total), 29.42; Cl (ionic), 4.90. Found: C, 13.53; H, 5.05; N, 15.30; Cl (total), 29.27; Cl (ionic), 5.09.

Properties

cis-Dichlorobis(1,2-ethanediamine)rhodium(III) chloride perchlorate (2:1:1) is a bright yellow crystalline solid that is moderately soluble in water and can be repeatedly recrystallized unchanged from 6 *M* hydrochloric acid without the addition of perchlorate ion. As a result of this recrystallization behavior, it appears possible that the salt reported to be *cis*-$[Rh(en)_2Cl_2]Cl.H_2O$ obtained previously[16,17] by the reaction of bis(1,2-ethanediamine)(oxalato)rhodium(III) perchlorate with boiling dilute hydrochloric acid may in reality also be the chloride perchlorate.

G. *cis*-AQUABIS(1,2-ETHANEDIAMINE)HYDROXO-RHODIUM(III) DITHIONATE

$$cis\text{-}[Rh(en)_2Cl_2]^+ + 2Ag^+ + 2H_2O \rightarrow cis\text{-}[Rh(en)_2(H_2O)_2]^{3+} + 2AgCl$$

$$cis\text{-}[Rh(en)_2(H_2O)_2]^{3+} + OH^- + S_2O_6^{2-} \rightarrow cis\text{-}[Rh(en)_2(H_2O)(OH)][S_2O_6] + H_2O$$

The *cis*-aquabis(1,2-ethanediamine)(hydroxo)rhodium(III) cation has been isolated previously as the dithionate[6] and perchlorate[1] salts. The preparative procedure described here is essentially that described in Reference 6 for the dithionate.

Procedure

■ **Caution.** *Perchlorates may be shock-sensitive and potentially dangerous because of the reducing character of the coordinated amines.*

A mixture of *cis*-dichlorobis(1,2-ethanediamine)rhodium(III) chloride perchlorate (2:1:1) (6.0 g, 0.0166 mole rhodium) and silver nitrate (8.57 g, 0.05045 mole) in 50 mL of water in a 100-mL round-bottomed flask fitted with condenser is heated under reflux for 3 hr in the dark. The reaction mixture is left to stand overnight, and 11.0 mL of 1.00 *M* hydrochloric acid is then added with stirring. After the mixture has stood for a further 10 min, it is filtered through a fine-porosity 4-cm sintered glass funnel into a 250-mL suction flask. The reaction flask and the funnel are rinsed with three 10-mL portions of boiling water, and the filtered washings are added to the bulk filtrate. Finely powdered sodium dithionate dihydrate (6.0 g) is added to the latter and dissolved with stirring at room temperature, after which 24.0 mL of 1.00 *M* sodium hydroxide is added with stirring. A finely divided pale yellow precipitate is formed. After the mixture is allowed to stand for 1 hr at room temperature and then in an ice bath for 20 min, the product is isolated by filtration on a fine-porosity 4-cm sintered glass funnel and washed with ice-cold water, 95% ethanol, and finally diethyl ether. Drying in air gives 5.32 g (yield 77%).

The pure dithionate is obtained from the crude product by reprecipitation as follows: Crude dithionate (2.0 g, 0.00478 mole) and finely powdered sodium dithionate dihydrate (2.4 g) are stirred together in 12.0 mL of 0.50 *M* hydrochloric acid in a 50-mL conical flask at room temperature until all is dissolved. The solution is filtered through a fine-porosity 3-cm sintered glass funnel into a 100-mL suction flask, and 6.0 mL of 1.0 *M* sodium hydroxide is added dropwise with stirring. A pale yellow precipitate forms rapidly, and after stirring at room temperature for 15 min the mixture is kept in an ice bath for 45 min and the product is isolated as before. Yield: 1.88 g (94%). The product reprecipitated in this manner is pure.

Anal. Calcd. for $RhC_4H_{19}N_4O_8S_2$: C, 11.49; H, 4.58; N, 13.40. Found: C, 11.44; H, 4.50; N, 13.33.

Properties

cis-Aquabis(1,2-ethanediamine)hydroxorhodium(III) dithionate is a very pale yellow crystalline solid. It is sparingly soluble in water but readily soluble in strong acid and strong base, giving solutions of the *cis*-diaqua and *cis*-dihydroxo complex ions, respectively. At elevated temperatures the complex loses water, with the formation of the dinuclear di-μ-hydroxo compound (*diol*) (Section H). The acid dissociation constants for the *cis*-diaqua complex ion have been determined[1,6,20] for a range of ionic strengths in sodium perchlorate medium at 25°. For example, at ionic strength 0.5 *M*, $pK_{a1} = 6.09$ and $pK_{a2} = 8.08$[1]; at ionic strength 1.0 *M*, $pK_{a1} = 6.338(2)$ and $pK_{a2} = 8.244(2)$.[6]

H. DI-μ-HYDROXO-BIS[BIS(1,2-ETHANEDIAMINE)RHODIUM(III)] BROMIDE

$$2cis\text{-}[Rh(en)_2(H_2O)(OH)][S_2O_6] \rightarrow \Delta,\Lambda\text{-}[(en)_2Rh(OH)_2Rh(en)_2](S_2O_6)_2 + 2H_2O$$

$$\Delta,\Lambda\text{-}[(en)_2Rh(OH)_2Rh(en)_2](S_2O_6)_2 + 4NH_4Br + 2H_2O \rightarrow \Delta,\Lambda\text{-}[(en)_2Rh(OH)_2Rh(en)_2]Br_4\cdot 2H_2O + 2(NH_4)_2[S_2O_6]$$

The starting material for the preparation of the bromide salt of di-μ-hydroxo-bis[bis(1,2-ethanediamine)rhodium(III)] (*diol*) is the crude dithionate salt, which is obtained by heating *cis*-aquabis(1,2-ethanediamine)(hydroxo)rhodium(III) dithionate at 120°. In order to avoid contamination of the bromide with monohydroxo bridged (*monool*) species,[6] it is necessary to keep dissolution times and, where possible, temperature to a minimum.

Procedure

Crude *cis*-aquabis(1,2-ethanediamine)(hydroxo)rhodium(III) dithionate (5.32 g) is heated for 10–15 hr at 120(±2)° in an oven, giving an essentially quantitative yield of crude di-μ-hydroxo (*diol*) dithionate. Crude dithionate (4.81 g, 0.006 mole) is stirred vigorously in 50 mL of a saturated solution of ammonium bromide in a 100-mL conical flask at room temperature for 1 hr. The resulting crude bromide salt is then isolated by filtration on a fine-porosity, 4-cm sintered glass funnel and washed with ice-cold 50% v/v ethanol/water. The product is extracted on the filter with water (temp. 20°) in portions of ~50 mL within a total of 20 min (total extraction volume ~200 mL). Each successive portion of solution is immediately filtered into a single ice-cooled 500-mL suction flask. After the first extraction, 80 mL of saturated ammonium bromide solution is added with vigorous stirring, and stirring is continued during the subsequent extractions. A dense pale yellow precipitate is formed immediately. After the mixture is cooled with stirring for a further 20 min, the product is isolated by filtration on a fine-porosity 4-cm sintered glass funnel and is washed with ice-cold 50% v/v ethanol/water, 95% ethanol, and finally diethyl ether. Drying in air gives 3.85 g (yield 77%).

Anal. Calcd. for $Rh_2C_8H_{38}N_8Br_4O_4$: C, 11.49; H, 4.58; N, 13.41; Br, 38.24. Found: C, 11.28; H, 4.79; N, 13.37; Br, 38.12.

Properties

The di-μ-hydroxo bridged complexes, the preparations of which are described in Sections D and H, have several chemical and physical properties in common. The dithionates are only sparingly soluble in water, whereas the bromides are quite soluble. The dinuclear structures of the cations, as well as the Δ,Λ configuration of the 1,2-ethanediamine compounds have been established by the close similarity of the Guinier X-ray powder diffraction patterns of $[(NH_3)_4Rh(OH)_2Rh(NH_3)_4]Br_4 \cdot 4H_2O$ and $[(en)_2Rh(OH)_2Rh(en)_2](S_2O_6)_2$ to those of the well-characterized[21,22] cobalt(III) and chromium(III) analogs.

Both di-μ-hydroxo (*diol*) cations are stable in acidic, neutral, and basic aqueous solution with respect to hydrolysis of ammonia or 1,2-ethanediamine ligands. However, equilibrium between the dihydroxo compounds (*diols*) and the corresponding hydroxo compounds (*monools*) is established quite rapidly in aqueous solution; for example, the reaction

$$[(NH_3)_4Rh(OH)_2Rh(NH_3)_4]^{4+} + H_2O \rightarrow \textit{cis,cis}\text{-}[(H_2O)(NH_3)_4Rh(OH)Rh(NH_3)_4(OH)]^{4+}$$

has a half-life of ~3 hr at 25° in 1 *M* $NaClO_4$ (pH 5), and the equilibrium constant is 3.03.[23] The corresponding values for the 1,2-ethanediamine system are 25 min and 11.2, respectively.[6] The equilibration reaction is catalyzed by both acid and base; for example, both *diols* give *monools* quantitatively within minutes at 25° in 1 *M* $HClO_4$. The kinetics of *diol-monool* equilibration in both the ammonia and ethylenediamine systems have been studied recently,[6,23] and the results are qualitatively consistent with those found for the corresponding chromium(III) systems.[24–26]

References

1. D. A. Palmer, R. van Eldik, H. Kelm, and G. M. Harris, *Inorg. Chem.*, **19,** 1009 (1980).
2. L. H. Skibsted, D. Strauss, and P. C. Ford, *Inorg. Chem.*, **18,** 3171 (1979).
3. L.H. Skibsted and P. C. Ford, *Inorg. Chem.*, **19,** 1828 (1980).
4. S. F. Clark and J. D. Petersen, *Inorg. Chem.*, **19,** 2917 (1980).
5. K. Howland and L. H. Skibsted, *Acta Chem. Scand.*, **A37**, 647 (1983).
6. M. Hancock, B. Nielsen, and J. Springborg, *Acta Chem. Scand.*, **A36,** 313 (1982).
7. K. Wieghardt, W. Schmidt, B. Nuber, B. Prikner, and J. Weiss, *Chem. Ber.*, **113,** 36 (1980).
8. K. Wieghardt, W. Schmidt, R. van Eldik, B. Nuber, and J. Weiss, *Inorg. Chem.*, **19,** 2922 (1980).
9. K. Wieghardt, P. Chaudhuri, B. Nuber, and J. Weiss, *Inorg. Chem.*, **21,** 3086 (1982).
10. J. Springborg and M. Zehnder, *Helv. Chim. Acta*, **67,** 2218 (1984).
11. A. W. Addison, K. Dawson, R. D. Gillard, B. T. Heaton, and H. Shaw, *J. Chem. Soc., Dalton Trans.*, **1972**, 589.
12. J. A. Osborn, K. Thomas, and G. Wilkinson, *Inorg. Synth.*, **13,** 213 (1972).
13. M. P. Hancock, *Acta Chem. Scand.*, **A29,** 468 (1975).
14. L. H. Skibsted and P. C. Ford, *Acta Chem. Scand.*, **A34,** 109 (1980).
15. S. A. Johnson and F. Basolo, *Inorg. Chem.*, **1**, 925 (1962).
16. T. P. Dasgupta, R. M. Milburn, and L. Damrauer, *Inorg. Chem.*, **9,** 2789 (1970).
17. A. W. Addison, R. D. Gillard, P. S. Sheridan, and L. R. H. Tipping, *J. Chem. Soc., Dalton Trans.*, **1974,** 709.
18. R. D. Gillard, J. P. De Jesus, and P. S. Sheridan, *Inorg. Synth.*, **20,** 57 (1980).
19. M. P. Hancock, *Acta Chem. Scand.*, **A33**, 499 (1979).
20. U. Klabunde, Ph.D. thesis, Northwestern University, 1967.
21. J. Springborg and C. E. Schäffer, *Inorg. Synth.,*, **18,** 75 (1978) and references therein.
22. S. J. Cline, R. P. Scaringe, W. E. Hatfield, and D. J. Hodgson, *J. Chem. Soc., Dalton Trans.*, **1977,** 1662.
23. F. Christensson and J. Springborg, *Inorg. Chem.* **24,** 2129 (1985).
24. J. Springborg and H. Toftlund, *Acta Chem. Scand.*, **A30,** 171 (1976).
25. F. Christensson, H. Toftlund, and J. Springborg, *Acta Chem. Scand.*, **A34,** 317 (1980).
26. F. Christensson and J. Springborg, *Acta Chem. Scand.*, **A36,** 21 (1982).

65. μ_3-THIO-TRISILVER(1+) NITRATE

$$CS_2 + 6AgNO_3 + 2H_2O \rightarrow 2[Ag_3S]NO_3 + 4HNO_3 + CO_2\uparrow$$

Submitted by GEORGE B. KAUFFMAN* and GÜNTER BERGERHOFF†
Checked by JANE V. ZEILE KREVOR‡ and SANDRA I. BARBOUR‡

Whereas the great majority of coordination compounds contain a central metallic species surrounded by several nonmetallic species (ligands), a peculiar class of compounds, called metallo complexes, is known in which the reverse is true; the coordination center is a nonmetal (usually a large, easily polarizable anion such as I^-, Br^-, SCN^-, S^{2-}, Se^{2-}, Te^{2-}, or P^{3-}) surrounded by metal cations with both high electron affinities and large radii, such as Ag^+, Au^+, Cu^+, Hg^{2+}, Cd^{2+}, or Pb^{2+}.[1–6] Although the number of compounds that can be formulated as metallo complexes has grown steadily through the years, the syntheses of many are not reproducible, and actual structures are known for only relatively few. Ease and reproducibility of preparation make μ_3-thio-trisilver(1+) nitrate an excellent example of a metallo complex for which structural data are available.

μ_3-Thio-trisilver(1+) nitrate, $[Ag_3S]NO_3$, is formed by the action of silver ion in great excess on sulfide ion. It may be prepared by passing hydrogen sulfide through concentrated silver nitrate solution,[7] by the disproportionation of elemental sulfur in the presence of silver nitrate,[7] by dissolution of silver sulfide in molten silver nitrate, and by the hydrolysis of carbon disulfide in the presence of concentrated silver nitrate solution.[8] The hydrolysis of carbon disulfide proceeds only to a small extent and yields a very small concentration of sulfide ion, thus producing the metallo complex without the formation of very slightly soluble silver sulfide.

Procedure

■ **Caution.** *Carbon disulfide is very flammable, and the liquid and its vapors are toxic to the nervous, cardiovascular, and reproductive systems.*
A solution of 10.0 g (0.0588 mole) of silver nitrate in 10 mL of 2 *N* nitric acid is mechanically shaken vigorously with 1.70 mL (0.0282 mole) of carbon disulfide in a 25-mL cork-stoppered brown or actinic red flask for 24 hr. If more carbon disulfide than the amount specified is used to increase the yield, the

*Department of Chemistry, California State University, Fresno, Fresno, CA 93740.
†Anorganisch-Chemisches Institut der Universität Bonn, Gerhard-Domagk-Strasse 1, 53 Bonn, German Federal Republic.
‡Department of Chemistry, San Francisco State University, San Francisco, CA 94132.

concentration of silver nitrate is too low to form the complex, and insoluble silver sulfide is formed. The same result may be obtained if the reaction time is increased. If a stirrer is used instead of a shaking machine, more time is required for the desired reacton to take place.

Since the product is light-sensitive, all operations should be carried out in the dark or in subdued light. The resulting yellow precipitate is collected by suction filtration on a 30-mL sintered glass funnel (medium porosity) and washed with three 10-mL portions of 4 *N* nitric acid and three 10-mL portions of methanol. It is placed in an open amber bottle, which is allowed to stand overnight in a desiccator in the absence of light. The yield of product is 3.25 g (40%).

Analysis

The product is boiled with water, and the resulting precipitate of silver sulfide is collected by filtration, washed, dried, and weighed. Another sample of product is dissolved completely in concentrated nitric acid, any excess acid is removed by boiling, and the total silver ion is determined by titration with standard ammonium thiocyanate solution. Another sample of product is oxidized with concentrated nitric acid in a bomb tube,[8] and after removal of silver and nitrate the sulfur is determined by precipitation as barium sulfate by addition of barium chloride solution.

Anal. Calcd. for Ag_3SNO_3: Ag_2S, 59.33; Ag, 77.48; S, 7.68. Found: Ag_2S, 59.21; Ag, 77.55; S, 7.33.

Properties

The yellow product (density, 5.53 g/mL) is extremely light-sensitive, becoming green and eventually black on exposure to light. It decomposes on heating. It is decomposed into silver nitrate and silver sulfide by water and by all organic solvents that dissolve silver nitrate, such as acetonitrile, pyridine, or *N,N*-dimethylformamide. It is remarkably resistant to the action of nitric acid, even at high concentrations. It has been tested for use in photography, since sulfide ions are important in several steps of the photographic process.

The structure of μ_3-thio-trisilver(1+) nitrate has been determined from Patterson and Fourier projections.[9] The cubic structure contains approximately trigonal prismatic SAg_6 groups linked through common corners to form a three-dimensional framework. The ionically bonded NO_3^- groups are located in the cavities.

References

1. A. Werner, *Neuere Anschauungen auf dem Gebiete der anorganischen Chemie,* 2nd ed., Friedrich Vieweg und Sohn, Braunschweig, 1909, p. 246.

2. M. M. Jones, *Elementary Coordination Chemistry,* Prentice-Hall, Englewood Cliffs, NJ, 1965, p. 38.
3. G. Bergerhoff, *Ueber Metallkomplexe,* Habilitationsschrift, Universität Bonn, Bonn, 1962; *Angew. Chem.,* **76,** 697 (1964); *Angew. Chem., Int. Ed., Engl.,* **3,** 686 (1964).
4. K. B. Yatsimirskiĭ, *Dokl. Akad. Nauk S.S.S.R.,* **77,** 819 (1951); *Chem. Abstr.,* **45,** 7462a (1951).
5. K. H. Lieser, *Z. Anorg. Allgem. Chem.,* **292,** 114 (1957); *J. Inorg. Nucl. Chem.,* **26,** 1571 (1964).
6. G. B. Kauffman, M. Karbassi, and G. Bergerhoff, *J. Chem. Educ.,* **61,** 729 (1984).
7. T. Poleck and K. Thümmel, *Ber.,* **16,** 2435 (1883).
8. G. Bergerhoff, *Z. Anorg. Allgem. Chem.,* **299,** 328 (1959).
9. B. Wurzschmitt, *Mikrochem. Verein Mikrochem. Acta,* **36/37,** 769 (1951).

66. CARBONYLCHLOROGOLD(I)

$$2[H_3O]^+[AuCl_4]^- + 2SOCl_2 \rightarrow 2SO_2 + 6HCl + Au_2Cl_6$$
$$Au_2Cl_6 + 4CO \rightarrow 2Au(CO)Cl + 2COCl_2$$

Submitted by D. BELLI DELL'AMICO* and F. CALDERAZZO*
Checked by H. H. MURRAY† and J. P. FACKLER†

Carbonylchlorogold(I), a substance extremely reactive toward water, has been synthesized by the reaction of preformed anhydrous gold(III) chloride[1] with carbon monoxide in the solid state[2] or in a solvent at elevated temperature.[3] Also, gold(I) chloride is known[3] to be converted quantitatively to Au(CO)Cl by carbon monoxide in benzene as solvent. Gold(I) chloride, on the other hand, was prepared by thermal decomposition of $[H_3O]^+[AuCl_4]^-$ or of anhydrous gold(III) chloride.[4]

By the method described here,[5] anhydrous gold(III) chloride is formed by dehydration of commercially available hydrogen tetrachloraurate [Alfa Products] at room temperature and is caused to react with carbon monoxide at atmospheric pressure and room temperature to produce Au(CO)Cl. Sulfinyl chloride is used as the reaction medium.

Procedure

■ **Caution.** *The entire synthesis must be carried out in a well-ventilated hood. Extreme caution must be exercised with the gases that come off the reaction*

*Istituto Chimica Generale, University of Pisa, Via Risorgimento 35, 56100 Pisa, Italy.
†Department of Chemistry, Texas A&M University, College Station, TX 77843.

mixture, for they may be phosgene, sulfur dioxide, hydrogen chloride, carbon monoxide, and/or sulfinyl chloride. These gases are toxic.

Hydrogen tetrachloroaurate,‡ $HAuCl_4 \cdot 3H_2O$ (5.00 g; 12.7 mmol) is stirred with sulfinyl chloride (30 mL) in a 100-mL flask connected to a gas outlet through a Nujol bubbler to prevent moisture from contaminating the reaction mixture. Evolution of SO_2 and HCl is observed while the suspension is stirred. At this stage, a red precipitate of anhydrous gold(III) chloride is visible in the reaction flask. Carbon monoxide is now bubbled slowly through the $SOCl_2$ solution, which is being stirred rapidly with a magnetic stirrer. After being stirred for 15 hr under carbon monoxide at room temperature, the suspension of gold(III) chloride disappears, and the reaction mixture consists of colorless crystals of Au(CO)Cl suspended in an ivory-colored liquid. Absorption of carbon monoxide is observed due to the formation of Au(CO)Cl and $COCl_2$. The IR spectrum of the crude reaction mixture shows the presence of a band at 1804 cm^{-1} due to phosgene. During experiments carried out in closed systems—when fresh carbon monoxide is not supplied and the free volume of the reaction flask is small—the formation of a black precipitate is observed. This is due to the precipitation of Au_4Cl_8,[6–8] obtained according to the following stoichiometry:

$$2Au_2Cl_6 + 2CO \rightarrow Au_4Cl_8 + 2COCl_2$$

The presence of Au_4Cl_8, which can sometimes be the intermediate product under certain experimental conditions, is not detrimental to the yield. Introduction of fresh carbon monoxide readily converts the tetrameric gold(I)-gold(III) halo complex into Au(CO)Cl.

A mixture of 100 mL of heptane (pretreated with sulfuric acid, distilled over sodium and over $LiAlH_4$) and 5 mL of sulfinyl chloride is added to the solution of Au(CO)Cl to decrease the solubility of the chlorocarbonyl complex. The latter is collected by filtration under CO, dried quickly (about 15 min) under reduced pressure, and then sealed in vials under CO. Yield: 2.6 g (79%).

Anal. Calcd. for AuCClO: C, 4.6; H, 0.0; Cl, 13.6; CO, 10.8; Au, 75.6. Found: C, 4.8; H, 0.0; Cl, 13.7; CO, 10.0; Au, 76.3.

Properties

Carbonylchlorogold(I) is a colorless crystalline substance that decomposes rapidly in the presence of moisture to give gold, hydrogen chloride, CO, and CO_2[3]

‡The checkers suggest starting with pure fine 24 karat gold (99.99%) and converting it to hydrogen tetrachloroaurate according to Brauer.[1]

*The checkers obtained yields of 73 and 65% when the reaction was carried out on a comparable scale.

through the intermediacy of unidentified black-violet solids. The compound must be handled with great care in anhydrous and inert solvents, such as sulfinyl chloride or carbon tetrachloride, preferably under an atmosphere of carbon monoxide. The compound evolves carbon monoxide promptly when treated with pyridine, triphenylphosphine, or cyclohexylisocyanide at room temperature, even under an atmosphere of carbon monoxide. The carbonyl stretching vibration is slightly solvent-dependent: 2162 ($SOCl_2$ and CH_2Cl_2), 2156 ($CHBr_3$), 2152 (CCl_4), 2158 (tetrahydrofuran), and 2153 (benzene and toluene) cm^{-1}. The compound is monomeric in benzene, based on cryoscopic measurements, and in the solid state, according to a recent X-ray diffraction study.[9]

References

1. G. Brauer (ed.), *Handbook of Preparative Inorganic Chemistry*, 2nd ed., Vol. II, Academic Press, New York, 1965, pp. 1056–1057.
2. W. Manchot and H. Gall, *Chem. Ber.*, **58,** 2175 (1925).
3. M. S. Kharasch and H. S. Isbell, *J. Am. Chem. Soc.*, **52,** 2919 (1930).
4. G. Brauer (ed.), *Handbook of Preparative Inorganic Chemistry*, 2nd ed., Vol. II, Academic Press, New York, 1965, p. 1055.
5. D. Belli Dell'Amico and F. Calderazzo, *Gazz. Chim. Ital.*, **103,** 1099 (1973).
6. D. Belli Dell'Amico, F. Calderazzo, and F. Marchetti, *J. Chem. Soc., Dalton Trans.*, **1976,** 1829.
7. D. Belli Dell'Amico, F. Calderazzo, F. Marchetti, S. Merlino, and G. Perego, *J. Chem. Soc., Chem. Commun.*, **1977,** 31.
8. D. Belli Dell'Amico, F. Calderazzo, F. Marchetti, and S. Merlino, *J. Chem. Soc., Dalton Trans.*, **1982,** 2257.
9. P. G. Jones, *Z. Naturforsch.*, **37b,** 823 (1982).

67. PLATINUM MICROCRYSTALS

$$H_2[PtCl_6] + Na[BH_4] + 3H_2O \rightarrow Pt + H_3BO_3 + 5HCl + NaCl + 2H_2$$

Submitted by P. VAN RHEENEN,* M. McKELVY,† R. MARZKE,‡ and W. S. GLAUNSINGER§
Checked by R. KENT MURMANN¶

Small platinum particles are of considerable practical and fundamental importance. Finely divided platinum is a very active and commercially important

*Rohm and Haas Company, Philadelphia, PA 19105.
†Center for Solid State Science, Arizona State University, Tempe, AZ 85287.
‡Department of Physics, Arizona State University, Tempe, AZ 85287.
§Department of Chemistry, Arizona State University, Tempe, AZ 85287.
¶Department of Chemistry, University of Missouri, Columbia, MO 65211.

catalyst,[1,2] and the properties of such small particles may exhibit physical behavior that is entirely different from that of the bulk metal, due to the quantization of electronic energy levels within the particles or surface effects.[3–6] Therefore, the ability to synthesize and characterize platinum particles having a uniform size should be helpful in elucidating their chemical and physical properties.

Elemental platinum can be prepared in a highly dispersed state by aqueous reduction of its salts. Depending upon the reaction conditions and reducing agent used, one can obtain fairly homogeneous microcrystals.[7] In order to prevent coagulation, the particles must be protected with an organic protective agent. Ionic by-products can be removed by dialysis prior to drying the protected particles. Freeze-drying is the preferred method of water removal because it is a low-temperature technique that minimizes particle sintering. This chemical reduction technique has several advantages over other synthetic approaches, such as evaporation, irradiation, thermal decomposition, and particle-beam methods. With this method, relatively large amounts of sample can be prepared (50–100 mg) with a large weight percentage of platinum ($\approx$50 wt %) and a relatively narrow size distribution using standard, inexpensive equipment. Electron microscopy and powder X-ray diffraction can be used to characterize the microcrystals.

The method described here, which involves the aqueous reduction of dihydrogen hexachloroplatinate(2−) with sodium tetrahydroborate (1−), results in platinum particles having a size of 28 ± 11 Å.

Procedure

All water associated with the preparation should be deionized, preferably run through an organic filter, and then distilled in a glass still, because impurities in the water can greatly affect the growth and homogeneity of the microcrystals.[8,9] Stirring of the reaction mixtures is achieved with an overhead glass propeller-type stirrer. All glassware is cleaned with a cleaning solution composed of 45 mL of 48% HF, 165 mL of concentrated HNO_3, 200 mL of H_2O, and 10 g of Alconox detergent.

■ **Caution.** *Aqueous hydrogen fluoride solutions are highly corrosive and cause painful, long-lasting burns. Gloves should be worn, and the solution should be handled in a well-ventilated hood.*

The glassware is then rinsed 20 times with deionized water followed by five more rinses with glass-distilled water.

One liter of 0.0066 *M* $Na[BH_4]$ (99%) [Alfa Products] is prepared, and the pH of this solution is adjusted to 12 with 0.35 *N* NaOH, since $Na[BH_4]$ is stable in aqueous solutions at pH > 11 at ambient temperature. Then 0.02 g of the organic protective agent, poly(vinylpyrrolidone) (PVP; MW 10,000) [Sigma Chemical] is added to the pH-adjusted $Na[BH_4]$ solution. Next, an 8-mL sample of 0.015 *M* $H_2[PtCl_6]$ (99.9%) [Gallard-Schlesinger] is injected rapidly using a syringe, which causes instantaneous reduction of the $H_2[PtCl_6]$.

Ionic reaction by-products are removed by dialysis. The hydrosol is placed in 2-cm diameter cellulose dialysis tubing [VWR] that has been previously cleaned by dialysis. After addition of the hydrosol, the tubing is tied off at each end and placed in a large bath. The water in this bath is changed until its resistivity reaches that of pure, glass-distilled water ($\gtrsim 10^6\ \Omega-$cm). The hydrosol is then freeze-dried to remove water and obtain the final product.

Once the platinum microcrystals begin to form, they act as highly active catalysts for the hydrolysis of $Na[BH_4]$, so that a large excess of $Na[BH_4]$ is needed to maximize the reduction of $H_2[PtCl_6]$. However, below a certain $H_2[PtCl_6]$ concentration, the hydrolysis of $Na[BH_4]$ is predominant, and a concentration of about 4×10^{-5} *M* $H_2[PtCl_6]$, which was determined colorimetrically, is left after the reduction, regardless of initial $H_2[PtCl_6]$ concentration. The yield of platinum is $66 \pm 5\%$, so that about 15 mg of platinum is produced by this procedure.

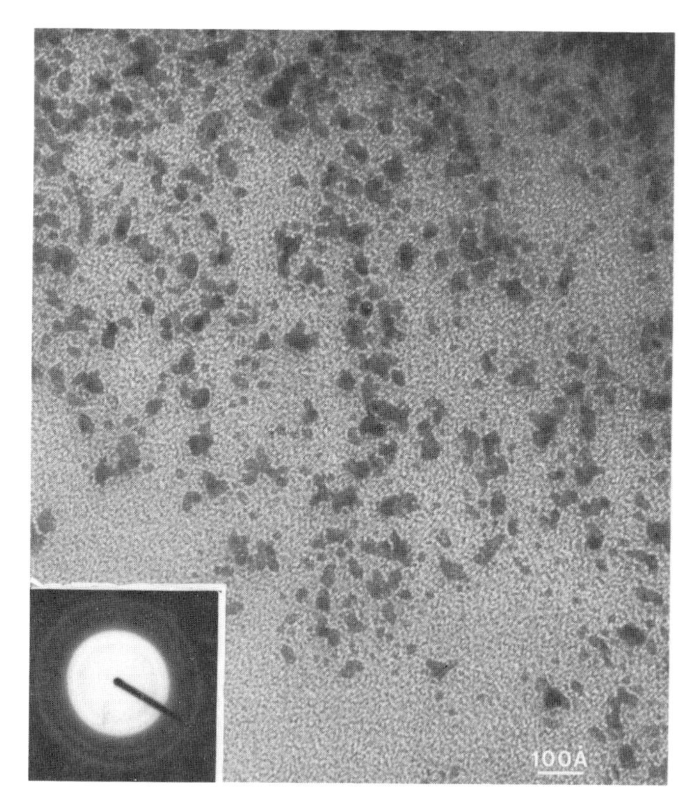

Fig. 1. Transmission electron micrograph of platinum microcrystals, with the corresponding Pt diffraction rings in the lower left corner. The granular background is the amorphous carbon film support.

The wt. % PVP was determined gravimetrically by heating 25 mg of sample to 800° for 8 hr in platinum-foil containers to remove the PVP. The wt. % PVP is 50.6%.

Powder X-ray diffraction patterns of the product exhibit line positions that are characteristic of platinum metal, but the lines are broader than those of the bulk metal. The X-ray line broadening can be used to estimate the crystallite size[10] and for this preparation was found to be 28 ± 11% Å.

The purity of the product is limited primarily by that of $H_2[PtCl_6]$, which contained the following impurities (in ppm by weight): Pd, 12; Au, 9; Fe, 191; Cu, 9; Si, 78; Mg, 25; Al, 85; Ni, 3; Cr, 6; Mn, 6; B $<$ 10; Co $<$ 1; Ca, 154; and Na, 70.

Properties

Samples were further characterized by transmission electron microscopy using a JEOL JEM100B electron microscope. The platinum hydrosol was freeze-dried onto carbon-coated copper microscope grids in order to simulate the final product. Figure 1 shows a micrograph of a typical preparation. The characteristic diffraction pattern of platinum was observed both by electron diffraction and by powder X-ray diffraction. Figure 2 depicts a histogram of the platinum microcrystal diameters. The mean diameter is 28 Å, and the standard deviation is

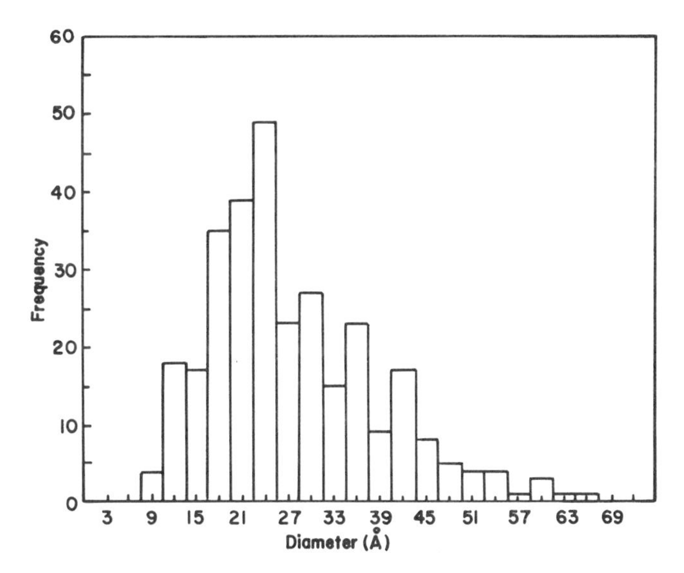

Fig. 2. Histogram of the diameters of the platinum microcrystals. The sample population is 300.

11 Å, which is in good agreement with the X-ray line-broadening results. However, the skewness of the distribution is characteristic of a lognormal[11] rather than a normal distribution. It is apparent from Fig. 1 that aggregation of individual particles has occurred to some extent. Part of the skewed tail on the histogram in Fig. 2 is due to measurements made on these larger particles.

The magnetic susceptibility of these microcrystals as well as commercial platinum foil (99.9%) [Johnson Matthey] has been measured at 300 K using a Faraday apparatus described elsewhere,[7] and their susceptibilities are equal within experimental error (1.00 ± 0.07 and 0.989 ± 0.004 cm^3/g, respectively). However, low-temperature susceptibility measurements on these microcrystals have revealed a Curie-Weiss law behavior below about 150 K, which has been attributed to their very small size.[7]

Acknowledgment

This work was supported by the Petroleum Research Fund.

References

1. H. C. Brown and C. A. Brown, *J. Am. Chem. Soc.*, **84,** 1495 (1962).
2. E. Borgarello, J. Kiwi, E. Pelizzetti, M. Visa, and M. Gratzel, *Nature,* **289,** 158 (1981).
3. R. Kubo, *J. Phys. Soc. Japan*, **17,** 975 (1962).
4. R. Denton, B. Mühlschlegel, and D. J. Scalapino, *Phys. Rev., B7*, **1973,** 3589.
5. R. F. Marzke, W. S. Glaunsinger, and M. Bayard, *Solid State Commun.*, **18,** 1025 (1976).
6. R. F. Marzke, *Catal. Rev. Sci. Eng.*, **19,** 43 (1979).
7. P. Van Rheenen, Ph.D. thesis, Arizona State University, Tempe, AZ, 1981.
8. H. B. Weiser and W. O. Milligan, *J. Phys. Chem.*, **36,** 1950 (1932).
9. S. W. Marshall, Ph.D. thesis, Tulane University, New Orleans, LA, 1962.
10. B. D. Cullity, *Elements of X-ray Diffraction*, Addison-Wesley, Reading, MA, 1978, pp. 284–285.
11. C. G. Granquist and R. A. Buhrman, *Solid State Commun.*, **18,** 123, (1976).

Chapter Five

TRIFLUOROMETHANESULFONATES AND TRIFLUOROMETHANESULFONATO-*O* COMPLEXES

68. INTRODUCTION TO TRIFLUOROMETHANESULFONATES AND TRIFLUOROMETHANESULFONATO-*O* COMPLEXES

NICHOLAS E. DIXON,* GEOFFREY A. LAWRANCE,† PETER A. LAY,‡ ALAN M. SARGESON,* and HENRY TAUBE§

Trifluoromethanesulfonic acid (triflic acid) and its organic esters have found considerable use in organic chemistry.[1,2] The desirable qualities of these compounds include a high thermal stability, excellent leaving-group properties of the $CF_3SO_3^-$ group, and the ease of purification of the acid and of synthesis of its esters. Its use in synthetic inorganic chemistry has been somewhat hampered by the lack of convenient routes for the preparation of the desired complexes. Despite this, the excellent leaving-group properties of the triflato ligand were established a decade ago by Scott and Taube,[3] who showed that the

*Research School of Chemistry, The Australian National University, G.P.O. Box 4, Canberra, A.C.T. 2601 Australia.

†Chemistry Department, The University of Newcastle, N.S.W. 2308 Australia.

‡Department of Chemistry, Stanford University, Stanford, CA 94305. CSIRO Postdoctoral Fellow.

§Department of Chemistry, Stanford University, Stanford, CA 94305.

$[Cr(OH_2)_5(OSO_2CF_3)]^{2+}$ and $[Cr(OH_2)_4(OSO_2CF_3)_2]^+$ complexes aquated rapidly. The preparative procedure utilized for these chromium complexes was to oxidize a Cr(II) solution in the presence of 6 *M* CF_3SO_3H followed by low-temperature cation-exchange chromatographic separation of the various species. Subsequent to this, the $[Co(NH_3)_5(OSO_2CF_3)](CF_3SO_3)_2$ complex was synthesized by way of the nitrosation of $[Co(NH_3)_5(N_3)](CF_3SO_3)_2$ by $[NO][CF_3SO_3]$ in nonaqueous solvents.[4,5] More recently, a method of general utility was found by Dixon et al.[6] that involved heating chloro complexes in neat triflic acid, followed by precipitation of the resultant triflato complexes by diethyl ether. This procedure, or modifications of the procedure, has been utilized for the synthesis of amine complexes of Co(III),[6–10] Rh(III),[8–9] Ir(III),[8,9] Cr(III),[8,9] Ru(III)[8,9,11] Os(III),[12,13] and Pt(IV).[8,9] In addition, polypyridyl triflato complexes of Ru(II), Ru(III), Os(II), and Os(III) have been synthesized.[14–16] Other useful techniques have appeared recently in the literature for the preparation of triflato complexes. These include heating triflate salts of aqua and other solvent complexes in the solid state under vacuum[17] and using trimethylsilyl triflate as a reagent instead of triflic acid in the above routes.[18]

Once prepared, the triflato complexes have many desirable properties. The most useful property in syntheses is the relatively high lability of the triflato group, which is substituted at a rate comparable to the best leaving groups known,[6] including the perchlorato[19] and fluoromethanesulfonato[20] ligands. This feature, combined with a high solubility in most polar organic solvents, a higher thermal stability, relatively low reactivity with atmospheric moisture, and simple and high-yielding preparative routes from readily available starting materials, makes these complexes extremely versatile synthetic intermediates en route to a large variety of important classes of transition metal complexes. Moreover, the use of the triflate anion for applications traditionally centered around the perchlorate anion is highly recommended. For instance, the explosive hazard, especially in organic solvents, is eliminated and the triflate anion often imparts higher solubility for the complexes in organic solvents.[21,22] Other commonly used anions that fall into the category of being poor nucleophiles, such as $[PF_6]^-$, $[BF_4]^-$, or $[BPh_4]^-$ do not have the thermal or photochemical stability exhibited by the $CF_3SO_3^-$ anion. Some chemical and physical properties of triflic acid are given in References 1, 2, and 7.

Procedures

■ **Caution.** *Triflic acid is one of the strongest known protic acids. It is necessary to take adequate precautions to prevent contact with the skin and eyes. Precautions should also be taken to minimize inhalation of the corrosive vapors given off from the acid. Reactions with the neat acid must be conducted in a well-ventilated fume hood.*

■ **Caution.** *Under no circumstances should perchlorate salts be used in any of the reactions involving neat triflic acid. The anhydrous hot perchloric acid thus produced represents an extremely explosive hazard, especially in contact with transition metal complexes. Addition of anhydrous diethyl ether to such solutions would represent an additional explosive hazard.*
Trifluoromethanesulfonic acid, its salts, and its complexes are extremely stable thermally, and no explosive hazards are known. However, consideration should be given to the thermal stability of other components of the complex before any new reactions are attempted at elevated temperatures.

Vacuum distillation of triflic acid [Aldrich Chemical] is performed as described previously.[6,7] Use of the acid as supplied does not appear to affect yields or purity of products markedly, although the complexes may be contaminated by highly colored impurities on occasion if the distillation procedure is not adopted.

The major synthetic methods utilized are described in the following sections. Methods for the syntheses of triflate salts are also described, because of their general use in nonaqueous chemistry in place of the perchlorate salts. Syntheses of specific trifluoromethanesulfonato complexes are described in the subsequent contributions to this chapter.

A. TRIFLATE SALTS FROM CHLORIDE SALTS

$$MCl_n + nCF_3SO_3H \rightarrow M(CF_3SO_3)_n + nHCl\uparrow$$

Procedure

This procedure may be applied to metal salts generally, and similar procedures have been described.[23–25]

To MCl_n (1 g) contained in a two-necked round-bottomed flask fitted with a nitrogen bubbler is carefully added anhydrous triflic acid (3–5 mL).

■ **Caution.** *Triflic acid is one of the strongest known protic acids, and gaseous hydrogen chloride is produced rapidly in the reaction. It is necessary to take adequate precautions to protect the skin and eyes and to prevent inhalation of the corrosive vapors. These manipulations must be performed in a well-ventilated fume hood. Because of the initial rapid evolution of HCl, care must be taken not to add the triflic acid too rapidly.*
A steady stream of nitrogen is passed through the solution while it is warmed to ~60°. After 0.5–1 hr, the heating is discontinued and the solution is cooled to ~0° in an ice bath while an N_2 flow is maintained. In order to precipitate the complex, anhydrous diethyl ether is added cautiously to the rapidly stirred solution in a dropwise fashion. (■ **Caution.** *This is a very exothermic addition and due care must be taken not to add the diethyl ether too quickly. Diethyl*

ether is toxic and very flammable. The addition must be performed in a well-ventilated fume hood.) The salt is filtered on a medium-porosity sintered glass funnel (15 mL) initially under gravity, and then the filtration is completed using a water aspirator. The ethereal solutions of triflic acid obtained at this stage may be kept for recovery of triflic acid as the sodium salt.[6] The powdery solid is washed with copious amounts of anhydrous diethyl ether (4×20 mL) and air-dried after each washing. At this stage further purification is generally not necessary, since the complexes are normally analytically pure. Yields are essentially quantitative except for mechanical losses.

In some of the precipitation processes, $Et_2O \cdot CF_3SO_3H$ may be coprecipitated but may be removed readily by boiling the solid in chloroform for ~0.5 hr after any solid lumps have been broken up using a mortar and pestle. (■ **Caution.** *Chloroform is toxic and a carcinogen; this procedure must be performed in a well-ventilated fume hood.*) The powder is collected on a medium-porosity frit and air-dried.

If crystalline material is required, most triflate salts are sparingly soluble in acetonitrile and may be recrystallized from hot solutions of this solvent. (■ **Caution.** *Acetonitrile is toxic and flammable. These crystallizations should be performed in a well-ventilated fume hood.*)

The same procedures may be applied to other salts such as other halides and pseudohalides, carbonates, and acetates.

■ **Caution.** *Under no circumstances should perchlorate salts be used in any of the reactions involving neat triflic acid. The anhydrous hot perchloric acid thus produced represents an extremely explosive hazard, especially in contact with transition metal complexes. Moreover, the addition of anhydrous diethyl ether to such solutions would represent an additional explosive hazard.*

B. TRIFLATE SALTS FROM SULFATE SALTS

$$M(SO_4)_n + nBa(CF_3SO_3)_2 \rightarrow M(CF_3SO_3)_{2n} + nBaSO_4 \downarrow$$

Procedure

This procedure is convenient for acid-sensitive complexes and is similar to that described elsewhere.[17] To a solution of $M(SO_4)_n$ (1 g) dissolved in water (~10 mL), $Ba(CF_3SO_3)_2$ (*n* equiv.) is added. (■ **Caution.** *Barium salts are extremely toxic. Avoid contact with skin.*) The solution is stirred, and the precipitated $BaSO_4$ is filtered using a medium-porosity sintered-glass filter. The solvent is removed by rotary evaporation to yield a powder, which may be recrystallized as before. For heat-sensitive complexes, the solvent is removed by freeze-drying techniques.

C. TRIFLATE SALTS USING SILVER TRIFLATE

$$MCl_n + nAgCF_3SO_3 \rightarrow M(CF_3SO_3)_n + nAgCl\downarrow$$

Procedure

This procedure[22] utilizes commercially available $AgCF_3SO_3$ [Alfa Products] and can also be used for other salts for which the AgX salts are very sparingly soluble. (■ **Caution.** *Silver salts are toxic and strong skin irritants. Avoid contact with skin.*) To MCl_n (1 g) dissolved in the minimum volume of water is added $AgCF_3SO_3$ (n equiv.). The solution is stirred rapidly for 5–10 min, and the precipitated AgCl is removed by vacuum filtration through a bed of Hyflo Supercel [Gallard Schlesinger] on a sintered glass funnel. Isolation and purification procedures are the same as described in Sections A and B.

D. TRIFLATO COMPLEXES FROM CHLORO COMPLEXES

$$[ML_xCl_y]Cl_n + (y + n)CF_3SO_3H \rightarrow [ML_x(OSO_2CF_3)_y](CF_3SO_3)_n + (y + n)HCl\uparrow$$

Procedure

This procedure is a general procedure applicable to inert transition metal complexes.[6–9] To $[ML_xCl_y]Cl_n$ (1 g) contained in a two-necked round-bottomed flask (25 mL) connected with a nitrogen bubbler is cautiously added anhydrous triflic acid (~5 mL).

■ **Caution.** *Triflic acid is one of the strongest known protic acids, and gaseous hydrogen chloride is produced rapidly in the reaction. It is necessary to take adequate precautions to protect the skin and eyes and to prevent inhalation of the corrosive vapors. These manipulations must be performed in a well-ventilated fume hood. Because of the initial rapid evolution of HCl, care must be taken not to add the triflic acid too quickly.*

■ **Caution.** *Under no circumstances should perchlorate salts be used in any of the reactions involving neat triflic acid. The anhydrous hot perchloric acid thus produced represents an extremely explosive hazard, especially in contact with transition metal complexes. Addition of anhydrous diethyl ether to such solutions would represent an additional explosive hazard.*

A steady stream of nitrogen is passed through the solution, which is then lowered into a silicone oil bath preheated to 100–120°. Lower temperatures may be required for certain complexes where heating may cause decomposition; lower temperatures require extended reaction times. Specific examples of reactions performed at lower temperature appear in the following sections.

Evolution of HCl gas is monitored by periodically passing the effluent gas through an $AgNO_3$ bubbler. After the HCl evolution has ceased (1–20 hr, depending on the complex and temperature), the flask is removed from the oil bath and allowed to cool to ~30° before cooling further in an ice bath. Ice cooling can be omitted for small-scale reactions, although boiling will occur on initial addition of diethyl ether. Use of a larger reaction vessel, vigorous mechanical stirring, and a well-ventilated fume hood are mandatory under these conditions. While the solution is rapidly stirred, diethyl ether (20 mL) is added carefully, in dropwise fashion, to precipitate the complex.

■ **Caution.** *This is a very exothermic addition, and due care must be exercised to avoid adding the diethyl ether too quickly. Diethyl ether is toxic and very flammable. Its addition must be performed in a well-ventilated fume hood.*

The complex is collected on a medium-porosity sintered glass funnel by initially allowing the solution to filter under gravity, then finally under vacuum. The filtrate may be saved for recovery of the triflic acid as $NaCF_3SO_3$.[6,7] The complex may be purified by boiling in chloroform as described in Section A. However, the complexes may not be recrystallized from acetonitrile, since the triflato ligands are readily substituted by acetonitrile molecules.

Similar procedures may be used for other complexes containing halo, pseudohalo, acetato, carbonato,[7] aqua, and many other ligands that are either relatively labile or are decomposed by strong acid.

E. TRIFLATO COMPLEXES FROM SOLID STATE REACTIONS OF TRIFLATO SALTS

$$[ML_x(\text{solvent})_y](CF_3SO_3)_n \rightarrow [ML_x(OSO_2CF_3)_y](CF_3SO_3)_{(n-y)} + y(\text{solvent})$$

Procedure

This procedure has been described for inert aqua complexes,[17] and is applicable to many other solvent species. For aqua complexes, $[ML_x(OH_2)_y](CF_3SO_3)_n$ (1 g) is placed in a vacuum oven that has been preheated to 100–180° (depending on the lability of the complex). The solid is kept under vacuum for from 8 to 24 hr depending on the complex, at which stage the substitution is complete. Once dehydrated, the triflato complexes are relatively stable to atmospheric moisture and generally may be manipulated without precautions to exclude air, unless otherwise stated.

For complexes containing solvent ligands that are both poorly coordinating and volatile, this procedure may be carried out at room temperature or even lower temperatures using vacuum line techniques. Such procedures are particularly useful for complexes that are thermally unstable toward isomerization or other chemical processes. Coordinated and ionic trifluoromethanesulfonate can usually be distinguished by infrared spectroscopy.[6,9]

F. REGENERATION OF TRIFLATO COMPLEXES

Procedure

Aged samples of triflato complexes may have undergone some aquation due to atmospheric moisture, although storage of the complexes in a desiccator over a suitable drying agent is generally sufficient to retain the integrity of the complexes for many months. The complexes are readily regenerated either by heating in the solid phase under vacuum (Section E) or by heating in neat triflic acid (Section D). Similar procedures may be used to regenerate the triflato complexes from a variety of product complexes.

References

1. J. B. Hendrickson, D. D. Sternbach, and K. W. Bair, *Acc. Chem. Res.*, **10,** 306 (1977).
2. R. D. Howells and J. D. McCown, *Chem. Rev.*, **71,** 69 (1977).
3. A. Scott and H. Taube, *Inorg. Chem.*, **10,** 62 (1971).
4. P. J. Cresswell, Ph.D. thesis, The Australian National University, 1974.
5. D. A. Buckingham, P. J. Cresswell, W. G. Jackson, and A. M. Sargeson, *Inorg. Chem.*, **20,** 1647 (1981).
6. N. E. Dixon, W. G. Jackson, M. J. Lancaster, G. A. Lawrance, and A. M. Sargeson, *Inorg. Chem.*, **20,** 470 (1981).
7. N. E. Dixon, W. G. Jackson, G. A. Lawrance, and A. M. Sargeson, *Inorg. Synth.*, **22,** 103 (1983).
8. N. E. Dixon, G. A. Lawrance, P. A. Lay, and A. M. Sargeson, *Inorg. Chem.*, **22,** 846 (1983).
9. N. E. Dixon, G. A. Lawrance, P. A. Lay, and A. M. Sargeson, *Inorg. Chem.*, **23,** 2940 (1984).
10. N. J. Curtis, K. S. Hagen, and A. M. Sargeson, *J. Chem. Soc., Chem. Commun.*, **1984,** 1571.
11. B. Anderes, S. T. Collins, and D. K. Lavallee, *Inorg. Chem.*, **23,** 2201 (1984).
12. R. H. Magnuson, P. A. Lay, and H. Taube, *J. Am. Chem. Soc.*, **105,** 2507 (1983).
13. P. A. Lay, R. H. Magnuson, J. Sen, and H. Taube, *J. Am. Chem. Soc.*, **104,** 7658 (1982).
14. P. A. Lay, A. M. Sargeson, and H. Taube, *Inorg. Synth.*, **24,** 291 (1986).
15. D. StC. Black, G. B. Deacon, and N. C. Thomas, *Trans. Metal Chem.*, **5,** 317 (1980).
16. D. StC. Black, G. B. Deacon, and N. C. Thomas, *Aust. J. Chem.*, **35,** 2445 (1982).
17. W. C. Kupferschmidt and R. B. Jordan, *Inorg. Chem.*, **21,** 2089 (1982).
18. M. R. Churchill, H. J. Wasserman, H. W. Turner, and R. R. Schrock, *J. Am. Chem. Soc.*, **104,** 1710 (1982).

19. J. MacB. Harrowfield, A. M. Sargeson, B. Singh, and J. C. Sullivan, *Inorg. Chem.*, **14,** 2864 (1975).
20. W. G. Jackson, and C. M. Begbie, *Inorg. Chem.*, **20,** 1654 (1981).
21. T. Fujinaga and I. Sakamoto, *Pure Appl. Chem.*, **52,** 1389 (1980).
22. W. G. Jackson, G. A. Lawrance, P. A. Lay, and A. M. Sargeson, *Aust. J. Chem.*, **35,** 1561 (1982).
23. J. S. Haynes, J. R. Sams, and R. C. Thompson, *Can. J. Chem.*, **59,** 669 (1981).
24. R. J. Batchelor, J. N. B. Ruddick, J. R. Sams, and F. Aubke, *Inorg. Chem.*, **16,** 1414 (1977).
25. A. M. Bond, G. A. Lawrance, P. A. Lay, and A. M. Sargeson, *Inorg. Chem.*, **22,** 2010 (1983).

69. PENTAAMMINE(TRIFLUOROMETHANESULFONATO-*O*)-CHROMIUM(III) TRIFLUOROMETHANESULFONATE AND BIS(1,2-ETHANEDIAMINE)-BIS(TRIFLUOROMETHANESULFONATO-*O*) CHROMIUM(III) TRIFLUOROMETHANESULFONATE

Submitted by GEOFFREY A. LAWRANCE* and ALAN M. SARGESON†
Checked by ANDREJA BAKAC‡ and JAMES H. ESPENSON‡

Despite the early synthesis of the labile pentaaquachromium(III) complex of the trifluoromethanesulfonate ligand,[1] only recently have aminechromium(III) complexes with coordinated trifluoromethanesulfonate been reported.[2,3] The relative lability of coordinated ammonia on chromium(III) compared to cobalt(III) precludes the usual approach adopted for $[Co(NH_3)_5(OSO_2CF_3)]^{2+}$ synthesis.[4] However, the labile $[Cr(NH_3)_5(OSO_2CF_3)](CF_3SO_3)_2$ is readily prepared from $[Cr(NH_3)_5Cl]Cl_2$[5] by a room-temperature reaction described below. The analogous reaction with *cis*-$[Cr(en)_2Cl_2]Cl$[6] (en = 1,2-ethanediamine) as precursor yields the *cis*-$[Cr(en)_2(OSO_2CF_3)_2]^+$ ion. Coordinated $[CF_3SO_3]^-$ is exceptionally labile [$t_{1/2}$ ~56 sec at 25° for the pentaamminechromium(III) complex in 0.1 M H^+][2] and may be readily substituted by other ligands such as OH_2, CH_3CN, urea, and *N,N*-dimethylformamide.[3,7]

■ **Caution.** *These reactions should be carried out in a well-ventilated fume hood.*

*Department of Chemistry, The University of Newcastle, N.S.W. 2308, Australia.
†Research School of Chemistry. The Australian National University, G.P.O. Box 4, Canberra 2601, Australia.
‡Ames Laboratory and Department of Chemistry, Iowa State University, Ames, IA 50011.

Procedure

A. PENTAAMMINE(TRIFLUOROMETHANESULFONATO-*O*)-CHROMIUM(III) TRIFLUOROMETHANESULFONATE, $[Cr(NH_3)_5(OSO_2CF_3)](CF_3SO_3)_2$

$$[Cr(NH_3)_5Cl]Cl_2 + 3CF_3SO_3H \rightarrow [Cr(NH_3)_5(OSO_2CF_3)](CF_3SO_3)_2 + 3HCl\uparrow$$

To pentaamminechlorochromium(III) dichloride (1 g) contained in a 50-mL two-necked round-bottomed flask fitted with a gas bubbler is added distilled anhydrous CF_3SO_3H (20 mL).

■ **Caution.** *Trifluoromethanesulfonic acid is a strong protic acid. Avoid contact with skin and eyes, and avoid breathing the corrosive vapors. Rapid evolution of HCl ensues; care must be taken not to add the acid too rapidly.* The solution is allowed to stand for 3 days at room temperature while a gentle stream of nitrogen gas is passed through the solution continuously. The gas flow is disconnected, and the solution is poured into a 0.5-L flask. Diethyl ether (200 mL) is added dropwise with vigorous mechanical stirring. (■ **Caution.** *This is a very exothermic addition, and care must be taken not to add the diethyl ether too rapidly.*) A fine pink precipitate is separated on a medium-porosity sintered frit, initially by gravity and then, after a bed of precipitate has formed, by suction. The precipitate is washed copiously with diethyl ether (5 × 20 mL) and dried under vacuum over P_4O_{10}. Yield: 1.75 g.

Anal. Calcd. for $C_3H_{15}N_5F_9O_9S_3Cr$: C, 6.17; H, 2.59; N, 11.98; S, 16.46. Found: C, 6.2; H, 2.5; N, 11.6; S, 16.2.

B. *cis*-BIS(1,2-ETHANEDIAMINE)BIS-(TRIFLUOROMETHANESULFONATO-*O*)CHROMIUM(III) TRIFLUOROMETHANESULFONATE, *cis*-$[Cr(en)_2(OSO_2CF_3)_2](CF_3SO_3)$

$$cis\text{-}[Cr(H_2NCH_2CH_2NH_2)_2Cl_2]Cl + 3CF_3SO_3H \rightarrow$$
$$cis\text{-}[Cr(H_2NCH_2CH_2NH_2)_2(OSO_2CF_3)_2](CF_3SO_3) + 3HCl\uparrow$$

cis-Dichlorobis(1,2-ethanediamine)chromium(III) chloride (1 g) and anhydrous CF_3SO_3H (20 mL) are allowed to react for 3 days at room temperature, and the product is isolated in exactly the manner described above for the pentaammine analog. Yield: 2.1 g.

*Anal.** Calcd. for $C_7H_{16}N_4F_9O_9S_3Cr$: C, 13.57; H, 2.60; N, 9.04; S, 15.53. Found: C, 13.45; H, 2.8; N, 8.8; S, 15.2.

Properties

Both complexes are isolated as pink powders that are air-stable if not subjected to prolonged exposure to atmospheric moisture. The complexes can be stored for several months in a desiccator, but exposure to strong light leads to slow decomposition. The pentaammine aquates (0.1 *M* CF_3SO_3H, 25°) with a rate constant of 1.24×10^{-2} sec^{-1}, which is 20 times faster than the pentaaquachromium(III) complex and only two times slower than the pentaamminecobalt(III) analog. Two consecutive steps for aquation of *cis*-$[Cr(en)_2(OSO_2CF_3)_2]^+$ under similar conditions are observed ($k_1 = 5.7 \times 10^{-3}$ sec^{-1}; $k_2 = 3.2 \times 10^{-3}$ sec^{-1}). The visible absorption spectra in anhydrous CF_3SO_3H exhibit maxima at 499 nm (ϵ 36.8 $M^{-1}cm^{-1}$) and 364 nm (ϵ31.7) for the pentaammine and at 497 nm (ϵ 69.8) and 380 nm (ϵ 42.7) for the bis(1,2-ethanediamine). In coordinating solvents such as water, acetonitrile, and *N,N*-dimethylformamide, solvent complexes are readily formed.[3,7] In poorly coordinating solvents such as tetrahydrothiophene-1,1-dioxide (sulfolane), substitution by ligands such as urea is facile.[7] The sensitivity of the aminetriflatochromium(III) complexes to prolonged heating limits their syntheses or regeneration by methods involving heating.

References

1. A. Scott and H. Taube, *Inorg. Chem.*, **10,** 62 (1971).
2. N. E. Dixon, G. A. Lawrance, P. A. Lay, and A. M. Sargeson, *Inorg. Chem.*, **22,** 846 (1983).
3. N. E. Dixon, G. A. Lawrance, P. A. Lay, and A. M. Sargeson, *Inorg. Chem.*, **23,** 2940 (1984).
4. N. E. Dixon, W. G. Jackson, M. A. Lancaster, G. A. Lawrance, and A. M. Sargeson, *Inorg. Chem.*, **20,** 470 (1981).
5. G. Schlessinger, *Inorg. Synth.*, **6,** 138 (1960).
6. C. L. Rollinson and J. C. Bailar, *Inorg. Synth.*, **2,** 200 (1946).
7. N. J. Curtis, G. A. Lawrance, and A. M. Sargeson, *Aust. J. Chem.*, **36,** 1495 (1983).

*This synthesis was not checked because of its similarity to that of Section A. The same cautionary procedures should be observed.

70. PENTAAMMINE(TRIFLUOROMETHANESULFONATO-*O*)RHODIUM(III) TRIFLUOROMETHANESULFONATE, PENTAAMMINEAQUARHODIUM(III) PERCHLORATE, AND HEXAAMMINERHODIUM(III) TRIFLUOROMETHANESULFONATE OR PERCHLORATE

Submitted by N. E. DIXON* and A. M. SARGESON*
Checked by H. ENGLEHARDT† and M. HERBERHOLD†

Complexes of trifluoromethanesulfonate anion with cobalt(III) are labile toward substitution under mild conditions,[1] and they have proved to be useful synthetic precursors to a variety of aminecobalt(III) complexes.[1,2] The pentaammine(trifluoromethanesulfonato-*O*)rhodium(III) ion, which is readily prepared from $[Rh(NH_3)_5Cl]Cl_2$ in hot CF_3SO_3H, is also versatile as a synthetic precursor.[3,4] Its synthesis and solvolysis to give essentially quantitative yields of the pentaammineaqua- and hexaamminerhodium(III) ions are described below. The aqua complex has previously been prepared by the base hydrolysis[5] or Ag^+-induced aquation[6] of $[Rh(NH_3)_5Cl]Cl_2$ in water, but the present method presents a cleaner and more rapid alternative. The methods for preparation of the $[Rh(NH_3)_6]^{3+}$ ion have evolved from the procedure of Jørgensen.[7] They involve prolonged reaction of $[Rh(NH_3)_5Cl]Cl_2$ with ammonia in a pressure vessel at elevated temperature. The solvolysis of $[Rh(NH_3)_5(OSO_2CF_3)](CF_3SO_3)_2$ in liquid ammonia is a simple, high-yield, and rapid alternative.

A. PENTAAMMINE(TRIFLUOROMETHANESULFONATO-*O*)-RHODIUM(III) TRIFLUOROMETHANESULFONATE

$$[Rh(NH_3)_5Cl]Cl_2 + 3CF_3SO_3H \rightarrow [Rh(NH_3)_5(OSO_2CF_3)](CF_3SO_3)_2 + 3HCl\uparrow$$

Procedure

To $[Rh(NH_3)_5Cl]Cl_2$[8] (25 g) in a 1-L three-necked round-bottomed flask fitted with a gas bubbler is cautiously added distilled anhydrous CF_3SO_3H (150 mL).

■ **Caution.** *Triflic acid is one of the strongest known protic acids. Gaseous*

*Research School of Chemistry, The Australian National University, G.P.O. Box 4, Canberra, A.C.T. 2601, Australia.

†Laboratorium für Anorganische Chemie, Universität Bayreuth, Universitätsstrasse 30, Bayreuth, FRG.

hydrogen chloride is produced rapidly in the reaction. It is necessary to take adequate precautions to protect the skin and eyes and to prevent inhalation of the corrosive vapors. These manipulations must be performed in a well-ventilated fume hood. Because of the initial rapid evolution of HCl, care must be taken not to add the triflic acid too rapidly.

A steady stream of argon or nitrogen is passed through the resulting solution while the flask is heated in an oil bath at 90–100°. After evolution of HCl has ceased (~2.5 hr), the flask is cooled to room temperature, then chilled in an ice-water bath to <5°. The gas flow is disconnected, and diethyl ether (500 mL) is added dropwise over ~30 min while the solution is stirred vigorously. (■ **Caution.** *This is a very exothermic addition, and due care must be exercised to avoid adding the diethyl ether too rapidly. Diethyl ether is toxic and very flammable. The addition must be performed in a well-ventilated fume hood.*) The pale yellow suspension is allowed to filter under gravity through a 10-cm sintered glass funnel (porosity grade 3) at room temperature. Gentle vacuum is applied as the residue is washed thoroughly with diethyl ether (5 × 50 mL) and allowed to dry in air. The product is ground in a mortar and thoroughly dried in a vacuum desiccator over P_4O_{10}. Yield: 54 g (100%).

Anal. Calcd. for $C_3H_{15}N_5F_9O_9S_3Rh$: C, 5.67; H, 2.38; N, 11.02; S, 15.14. Found: C, 5.6; H, 2.5; N, 11.3; S, 14.9.

Properties

Isolated as described in quantitative yield, $[Rh(NH_3)_5OSO_2CF_3](CF_3SO_3)_2$ is a free-flowing pale yellow powder. Thus prepared it contains small amounts of $[(CH_3CH_2)_2OH]^+[CF_3SO_3]^-$ that can be removed easily by boiling with chloroform.[2] It is stable to decomposition at room temperature over several months if precautions are taken to prevent contact with atmospheric moisture. Samples that have deteriorated through contact with moisture can be regenerated by treatment with hot CF_3SO_3H and isolated as above. The crude product is suitable for most preparative purposes. In 0.1 *M* CF_3SO_3H at 25°, the $[Rh(NH_3)_5(OSO_2CF_3)]^{2+}$ ion aquates ($t_{1/2}$ ≈40 sec) only a little more slowly than the corresponding Co(III) complex ($t_{1/2}$ ≈25 sec). It is susceptible to solvolysis in common organic solvents and to substitution by ligands dissolved in poorly coordinating solvents such as tetrahydrothiophene 1,1-dioxide (sulfolane).[3,4]

B. PENTAAMMINEAQUARHODIUM(III) PERCHLORATE

$$[Rh(NH_3)_5(OSO_2CF_3)](CF_3SO_3)_2 + H_2O \rightarrow [Rh(NH_3)_5(OH_2)](CF_3SO_3)_3$$

$$[Rh(NH_3)_5(OH_2)](CF_3SO_3)_3 + 3HClO_4 \rightarrow [Rh(NH_3)_5(OH_2)](ClO_4)_3 + 3CF_3SO_3H$$

Procedure

Pentaammine(trifluoromethanesulfonato-*O*)rhodium(III) trifluoromethanesulfonate (10 g) is dissolved in water (80 mL) at room temperature, and the solution is filtered. The solution is warmed to ~80° on a steam bath. Then $HClO_4$ (70% w/v) is added slowly until the first sign of persistent cloudiness (~8 mL). (■ **Caution.** *Perchloric acid is a very strong acid and highly oxidizing. Organic materials must be absent. Gloves and face shields should be used.*) Slow cooling of the solution gives pale yellow crystals, which are collected by filtration, washed extensively with ethanol and diethyl ether, and dried in air. Yield: 6.7 g (86%). A second crop (1.1 g) may be obtained by addition of more $HClO_4$ (~10 mL) to the filtrate, followed by cooling. The combined products may be recrystallized quantitatively from warm 0.5 *M* $HClO_4$ on slow cooling from 80°.

Anal. Calcd. for $H_{17}N_5O_{13}Cl_3Rh$: H, 3.40; N, 13.88; Cl, 21.09. Found: H, 3.4; N, 13.5; Cl, 21.0.

Absorption maxima [λ, nm (ε, $M^{-1}cm^{-1}$)]: 315 (104), 262 (90). Samples of $[Rh(NH_3)_5(OH_2)](CF_3SO_3)_3$ of acceptable analytical purity can be obtained from aqueous solutions of $[Rh(NH_3)_5(OSO_2CF_3)](CF_3SO_3)_2$ simply by evaporation of the solvent,[4] thus providing a convenient route for preparation of ^{17}O- or ^{18}O-labeled samples.

C. HEXAAMMINERHODIUM(III) TRIFLUOROMETHANESULFONATE AND HEXAAMMINERHODIUM(III) PERCHLORATE

$$[Rh(NH_3)_5(OSO_2CF_3)](CF_3SO_3)_2 + NH_3 \rightarrow [Rh(NH_3)_6](CF_3SO_3)_3$$

$$[Rh(NH_3)_6](CF_3SO_3)_3 + 3HClO_4 \rightarrow [Rh(NH_3)_6](ClO_4)_3 + 3CF_3SO_3H$$

Procedure

Pentaammine(trifluoromethanesulfonato-*O*)rhodium(III) trifluoromethanesulfonate (2 g) is dissolved in liquid ammonia[9] (−33°, ~50 mL) in a 250-mL rotary evaporator flask.

■ **Caution.** *Ammonia gas is toxic and corrosive. All manipulations should be performed in a well-ventilated fume hood, and adequate precautions should be taken to prevent contact with skin and eyes. If it is necessary to transport liquid ammonia, this should be performed using a loosely stoppered Dewar flask. Enclosed areas such as fire escapes should be avoided during transport*

The solvent is removed immediately at reduced pressure using a rotary evaporator. (■ **Caution.** *An efficient water aspirator is required. The temperature of the water bath should initially be ~20°, and the receiver flask should be half-*

filled with cold water.) A quantitative yield of a white crystalline product is obtained. The solid is collected, washed twice with diethyl ether (10 mL), and dried under vacuum over P_4O_{10}.

Anal. Calcd. for $C_3H_{18}N_6F_9O_9S_3Rh$: C, 5.55; H, 2.78; N, 12.89. Found: C, 5.8; H, 2.7; N, 12.5.

For preparation of the perchlorate salt, a filtered aqueous solution (150 mL) of the above crude product is treated with $HClO_4$ (70% w/v, 3 mL). (■ **Caution.** *Perchloric acid is a very strong acid and highly oxidizing. Organic materials must be absent. Gloves and face protection should be worn.*) The suspension is warmed to dissolve the product (~80°), then cooled slowly to give white crystals. These are collected by filtration, washed extensively with ethanol and diethyl ether, and dried in air. Yield: 1.5 g (97%). Absorption maxima [λ, nm (ϵ, $M^{-1}cm^{-1}$)]: 305 (131), 256 (94).

Anal. Calcd. for $H_{18}N_6O_{12}Cl_3Rh$: H, 3.60; N, 16.69; Cl, 21.13. Found: H, 3.6; N, 16.4; Cl, 20.9.

References

1. N. E. Dixon, W. G. Jackson, M. J. Lancaster, G. A. Lawrance, and A. M. Sargeson, *Inorg. Chem.*, **20,** 470 (1981).
2. N. E. Dixon, W. G. Jackson, G. A. Lawrance, and A. M. Sargeson, *Inorg. Synth.*, **22,** 103 (1983).
3. N. J. Curtis, N. E. Dixon, and A. M. Sargeson, *J. Am. Chem. Soc.*, **105,** 5347 (1983).
4. N. E. Dixon, G. A. Lawrance, P. A. Lay, and A. M. Sargeson, *Inorg. Chem.*, **22,** 846 (1983).
5. F. Basolo and G. S. Hammaker, *Inorg. Chem.*, **1,** 1 (1962).
6. R. D. Foust, Jr., and P. C. Ford, *Inorg. Chem.*, **11,** 899 (1972).
7. S. M. Jørgensen, *J. Prakt. Chem.*, **44,** 49 (1891).
8. J. A. Osborn, K. Thomas, and G. Wilkinson, *Inorg. Synth.*, **13,** 213 (1972).
9. D. Nicholls, *Inorganic Chemistry in Liquid Ammonia,* Elsevier, Amsterdam, 1979.

71. PENTAAMMINERUTHENIUM(III), PENTAAMMINERUTHENIUM(II), AND BINUCLEAR DECAAMMINEDIRUTHENIUM(III)/(II) COMPLEXES

Submitted by GEOFFREY A. LAWRANCE,* PETER A. LAY,†‡ ALAN M. SARGESON,§ and HENRY TAUBE‡
Checked by H. ENGLEHARDT¶ and M. HERBERHOLD¶

Pentaammineruthenium(II) and (III) complexes have an extensive literature of recent years.[1-7] This interest has stemmed from the chemical reversibility of the Ru(III)/Ru(II) couples, the strong π-backbonding between Ru(II) and π-acid ligands, and the interest in mixed valence chemistry of binuclear ions. The standard preparative route to these complexes involves the use of the labile $[Ru(NH_3)_5(OH_2)]^{2+}$ ion in either aqueous or nonaqueous media. This complex is generally prepared from $[Ru(NH_3)_5Cl]^{2+}$ by Zn(Hg) reduction. Recently, we reported the $[Ru(NH_3)_5(OSO_2CF_3)](CF_3SO_3)_2$ complex,[8,9] which like the other triflato complexes readily undergoes solvation. A limitation occurs for basic ligands, where reactions are complicated by base-catalyzed disproportionation to Ru(II) and Ru(IV).[10] However, the high solubility of $[Ru(NH_3)_5(OSO_2CF_3)](CF_3SO_3)_2$ in nonaqueous solvents and water allows the preparation of labile Ru(II) pentaammine complexes by Zn(Hg) reductions. This enables their use in a large variety of synthetic reactions, particularly for the preparation of mononuclear and binuclear complexes containing π-acid ligands. Such pathways are more facile, more convenient, and often higher yielding routes than those previously available using the reduction of $[Ru(NH_3)_5Cl]^{2+}$.[1-7]

The ready commercial availability of $[Ru(NH_3)_6]Cl_3$ as a starting material has made its reaction with 12 *M* HCl the preferred method for the preparation of $[Ru(NH_3)_5Cl]Cl_2$.[11] Other methods have used either $[Ru(NH_3)_6]Cl_2$[12] or $[Ru(NH_3)_5N_2]Cl_2$[13] (oxidation followed by treatment with HCl). The former reaction and the subsequent conversion of the chloro complex into the useful synthetic intermediate $[Ru(NH_3)_5(OSO_2CF_3)](CF_3SO_3)_2$ are described here. Simple substitution reactions are not reported, since they are similar to those described in other sections. Instead, examples are given of the preparation of mononuclear

*Chemistry Department, The University of Newcastle, N.S.W. 2308, Australia.
†C.S.I.R.O. Postdoctoral Fellow.
‡Chemistry Department, Stanford University, Stanford, CA 94305.
§Research School of Chemistry, The Australian National University, G.P.O. Box 4, Canberra, A.C.T. 2601, Australia.
¶Laboratorium für Anorganische Chemie, Universität Bayreuth, Universitätsstrasse 30, Bayreuth, FRG.

and binuclear Ru(II) complexes by reduction of the $[Ru(NH_3)_5(OSO_2CF_3)]^{2+}$ complex, followed by substitution in the labile $[Ru(NH_3)_5(solvent)]^{2+}$ species. Specifically, the preparation of $[Ru(NH_3)_5(pyrazine)]X_2$[6] and $[Ru(NH_3)_5$ (pyrazine)$Ru(NH_3)_5]I_5$[7] are reported as examples of the synthetic utility.

A. PENTAAMMINECHLORORUTHENIUM(III) CHLORIDE

$$[Ru(NH_3)_6]Cl_3 + HCl \rightarrow [Ru(NH_3)_5Cl]Cl_2 + NH_4Cl$$

Procedure

The time required for this procedure is ~5 hr. Hexaammineruthenium(III) chloride [Alfa Products] (10 g, 0.0323 mole) is dissolved in warm water (100 mL) contained in a 500-mL flask fitted with a reflux condenser. Concentrated HCl (36%, 100 mL) is added carefully, and the solution is heated at reflux for 4 hr. (■ **Caution.** *HCl is toxic and corrosive. This procedure must be performed in a well-ventilated fume hood.*) During this time, a copious yellow crystalline precipitate of $[Ru(NH_3)_5Cl]Cl_2$ forms. It is essential not to heat the reaction for too long, since this results in some conversion of $[Ru(NH_3)_5Cl]^{2+}$ to $[Ru(NH_3)_4Cl_2]^+$. After the solution is allowed to cool to room temperature, the crystals are collected on a medium-porosity sintered glass filter, washed with HCl (18%, 2 × 20 mL) and then methanol (2 × 20 mL), and dried under vacuum at room temperature. Yield: 9.0 g (95%).

Anal. Calcd. for $H_{15}N_5Cl_3Ru$: H, 5.17; N, 23.94; Cl, 36.35. Found: H, 5.3; N, 24.3; Cl, 36.6.

B. PENTAAMMINE(TRIFLUOROMETHANESULFONATO-*O*)-RUTHENIUM(III) TRIFLUOROMETHANESULFONATE

$$[Ru(NH_3)_5Cl]Cl_2 + 3CF_3SO_3H \rightarrow [Ru(NH_3)_5(OSO_2CF_3)](CF_3SO_3)_2 + 3HCl\uparrow$$

Procedure

This procedure requires variable amounts of time depending on the time taken for the filtration process. It generally requires ~4 hr; however, for larger quantities of compound, extra time should be allowed for the filtration process. To $[Ru(NH_3)_5Cl]Cl_2$ (1.0 g) in a two-necked round-bottomed flask connected with a nitrogen bubbler is cautiously added anhydrous CF_3SO_3H (8 mL).

■ **Caution.** *Triflic acid is one of the strongest known protic acids. Gaseous* *hydrogen chloride is produced rapidly in the reaction. It is necessary to take adequate precautions to protect the skin and eyes and to prevent inhalation of the corrosive vapors. These manipulations must be performed in a well-ventilated fume hood. Because of the initial rapid evolution of HCl, care must be taken not to add the triflic acid too quickly.*

A steady stream of nitrogen is passed through the solution, which is lowered into a silicone oil bath preheated to 100°. After the solution is heated for 2 hr, it is removed from the oil bath and allowed to cool to room temperature. It is then chilled in an ice bath. Care should be taken not to heat the solution for prolonged periods or at temperatures above 100°, since this may result in further loss of ammine ligands. Slow addition of diethyl ether (40 mL) with stirring precipitates the complex.

■ **Caution.** *This is a highly exothermic addition, and due care must be exercised to prevent addition of the diethyl ether too quickly. Diethyl ether is toxic and very flammable. The addition must be performed in a well-ventilated fume hood.*

The product is collected on a fine-porosity sintered glass funnel, washed with diethyl ether (3 × 20 mL), air-dried, and then dried in a vacuum desiccator over P_4O_{10}. Yield: 2.05 g.

Anal. Calcd. for $C_3H_{15}N_5F_9S_3Ru$: C, 5.69; H, 2.39; N, 11.06; S, 15.18; F, 27.00. Found: C, 5.7; H, 2.4; N, 10.8; S, 15.2; F, 26.5.

C. PENTAAMMINE(PYRAZINE)RUTHENIUM(II) CHLORIDE AND PENTAAMMINE(PYRAZINE)RUTHENIUM(II) TETRAFLUOROBORATE

$$2[Ru(NH_3)_5(OSO_2CF_3)](CF_3SO_3)_2 + Zn + 2(CH_3)_2CO \xrightarrow{Ar} 2[Ru(NH_3)_5(OC(CH_3)_2)](CF_3SO_3)_2 + Zn(CF_3SO_3)_2$$

$$[Ru(NH_3)_5(OC(CH_3)_2)](CF_3SO_3)_2 + \text{pyrazine} \xrightarrow{Ar} [Ru(NH_3)_5(\text{pyrazine})](CF_3SO_3)_2 + (CH_3)_2CO$$

$$[Ru(NH_3)_5(\text{pyrazine})](CF_3SO_3)_2 + 2Et_4NCl \xrightarrow{Ar} [Ru(NH_3)_5(\text{pyrazine})]Cl_2 + 2Et_4NCF_3SO_3$$

$$[Ru(NH_3)_5(\text{pyrazine})]Cl_2 + 2NaBF_4 \rightarrow [Ru(NH_3)_5(\text{pyrazine})](BF_4)_2 + 2NaCl$$

In the following manipulations in which argon is used, the gas is passed through a chromium(II) scrubber to remove O_2 and then concentrated H_2SO_4 to remove H_2O.

Procedure

A time of about 5 hr is required for the synthesis and recrystallization of $[Ru(NH_3)_5(pyrazine)]^{2+}$. The triflato complex (0.10 g, 0.16 mmol) is dissolved in degassed (30 min, Ar) acetone (AR grade, 10 mL) contained in a 25-mL bubble flask. Alternatively, a Zwickel flask[14] may be used for this procedure (Fig. 1). (■ **Caution.** *Acetone is highly flammable and toxic. Due care should be taken to prevent inhalation and contact with skin and eyes.*) The outlet is connected to a second bubble flask (100 mL) containing pyrazine (0.5 g, 6.2 mmol, AR grade) in degassed acetone (AR, 10 mL). After oxygen is removed from the solutions (~30 min), a piece of Zn(Hg) is added to reduce the $[Ru(NH_3)_5(OC(CH_3)_2)]^{3+}$ to the Ru(II) complex. The reduction is allowed to proceed for 30 min before transferring the Ru(II) solution into the flask containing the ligand solution. It is important to exclude oxygen from these reactions, since it reacts with Ru(II) to form the bright green oxo bridged dimer $[(NH_3)_5RuORu(NH_3)_5]^{5+}$.[15] If incomplete replacement of chloride occurs in the preparation of the triflato complex, an intense blue color develops due to the formation of intensely colored chloro-bridged complexes.[16] These complexes may also form from chloride ions adhered to the Zn(Hg). However, such species do not interfere with the reactions. This addition is made over ~15 min to minimize formation of the μ-pyrazine dimer. Almost immediately the solution turns a deep purple, but the reaction is allowed to proceed for an additional 30 min.

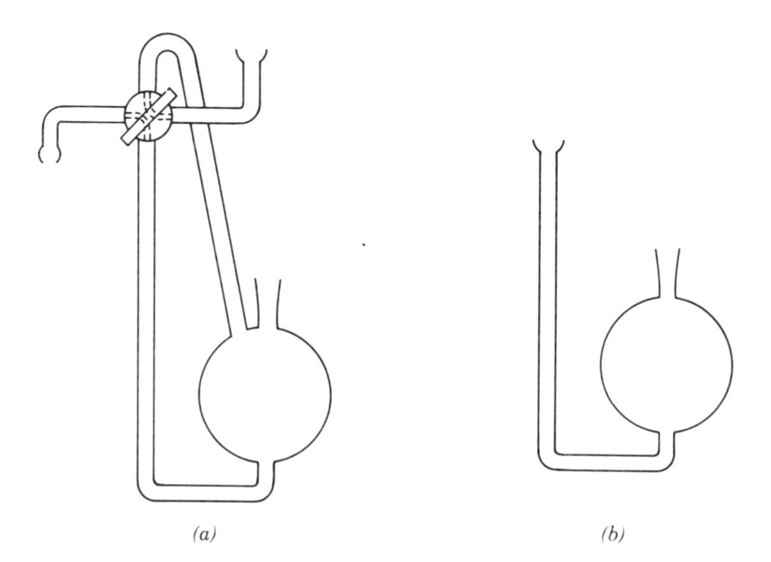

Fig. 1. a. Zwickel flask. b. Bubble flask.

To the resulting solution is added a degassed (Ar, 30 min) acetonitrile solution of Et_4NCl (0.5 g in 10 mL), with stirring. A precipitate of the $[Ru(NH_3)_5(pyrazine)]Cl_2$ forms immediately. The red-purple precipitate is collected on a sintered glass filter (under an Ar atmosphere) and is washed with degassed CH_3CN (5 mL) and degassed acetone (5 mL) and dried under vacuum. The yield is quantitative (0.054 g).

Anal. Calcd. for $C_4H_{19}N_7Cl_2Ru$: C, 14.25; H, 5.68; N, 29.08; Cl, 21.03. Found: C, 14.4; H, 5.4; N, 29.4; Cl, 21.4.

The crude product may be recrystallized as the $[BF_4]^-$ salt by dissolving $[Ru(NH_3)_5(pyrazine)]Cl_2$ (0.2 g) in degassed (Ar, 30 min) water (10 mL) and adding solid $Na[BF_4]$ (1 g) slowly with stirring.[17] After cooling to 5° for 2 hr (under Ar), the solid is collected and air-dried. Yield: 0.2 g (77%).

Anal. Calcd. for $C_4H_{19}N_7B_2F_8Ru$: C, 10.92; H, 4.35; N, 22.29. Found: C, 10.8; H, 4.4; N, 22.4.

The Ru(II) complex is stable in air in the solid state and is slowly oxidized in solution to the Ru(III) complex. The Ru(III) complex may be prepared by Ag^+ oxidation[17] or obtained directly from $[(NH_3)_5Ru(OSO_2CF_3)]^{2+}$ in a pyrazine melt. Procedures similar to that described allow the syntheses of a great variety of pentaammine complexes.

D. DECAAMMINE(μ-PYRAZINE)-DIRUTHENIUM(5+) IODIDE

$$2[Ru(NH_3)_5(OSO_2CF_3)](CF_3SO_3)_2 + Zn + 2(CH_3)_2CO \xrightarrow{Ar} 2[Ru(NH_3)_5(OC(CH_3)_2)](CF_3SO_3)_2 + Zn(CF_3SO_3)_2$$

$$2[Ru(NH_3)_5(OC(CH_3)_2)](CF_3SO_3)_2 + \text{pyrazine} \xrightarrow{Ar} \{[Ru(NH_3)_5]_2(\text{pyrazine})\}(CF_3SO_3)_4$$

$$\{[Ru(NH_3)_5]_2(\text{pyrazine})\}(CF_3SO_3)_4 + O_2 + CF_3SO_3^- \rightarrow \{[Ru(NH_3)_5]_2(\text{pyrazine})\}(CF_3SO_3)_5 + O_2^-$$

$$\{[Ru(NH_3)_5]_2(\text{pyrazine})\}(CF_3SO_3)_5 + 5\ n\text{-}Bu_4NI \rightarrow \{[Ru(NH_3)_5]_2(\text{pyrazine})\}I_5 + 5\ n\text{-}Bu_4NCF_3SO_3$$

Procedure

A reaction time of ~3 hr is required for the synthesis of the binuclear complex. Pentaammine(trifluoromethanesulfonato-*O*)ruthenium(III) trifluoromethanesulfonate (0.20 g, 0.31 mmol) is dissolved in degassed (30 min, Ar) acetone (AR, 10 mL) contained in a 50-mL bubble flask (Fig. 1) under a continuous stream of Ar. (■ **Caution.** *Acetone is highly flammable and toxic. Due care should* *be taken to prevent inhalation and contact with skin and eyes.*) A lump of freshly prepared Zn(Hg) is added, and the reduction is allowed to proceed for 30 min. Solid pyrazine (0.012 g, 0.15 mmol) is added. The solution immediately begins to become deep purple. The reaction is complete after ~30 min at room temperature [the air-sensitive diruthenium(4+) ion may be precipitated at this point using $[Bu_4N]I$], after which time the solution is filtered in air to remove the Zn(Hg). Air is bubbled through the filtrate for an additional 30 min, and a solution of $[Bu_4N]I$ (1.0 g) in acetone (10 mL) is added dropwise to precipitate the iodide salt of the mixed valence ion. The precipitate is collected, washed with diethyl ether (2 × 10 mL), and dried under vacuum. Yield: 0.16 g (95%).

Anal. Calcd. for $C_4H_{34}N_{12}I_5Ru_2$: C, 4.42; H, 3.15; N, 15.47. Found: C, 4.6; H, 3.2; N, 15.15.

This complex is interesting because of the observation of an intense transition in the near-infrared region at 1570 nm (ϵ 50,000 $M^{-1}cm^{-1}$).[7] It may be either oxidized to the 6+ ion or reduced to the 4+ ion. Similar techniques may be used for the syntheses of other binuclear decaammine complexes.

Properties

The $[Ru(NH_3)_5(OSO_2CF_3)](CF_3SO_3)_2$ complex undergoes rapid aquation in 0.1 *M* CF_3SO_3H to give the aqua complex ($k_{aq} = 9.3 \times 10^{-2}$ sec^{-1}, 25°). It is characterized by an electronic transition at 284 nm (ϵ 790 $M^{-1}cm^{-1}$) in the UV spectrum in neat CF_3SO_3H.[8,9] The colorless complex is moderately stable in the solid state, however, and can be stored for months in a desiccator without noticeable decomposition. The $[Ru(NH_3)_5Cl]Cl_2$ complex is characterized by a ligand-to-metal charge transfer at 328 nm (ϵ 1930 M^{-1} cm^{-1}) associated with the Cl^- ligand. It undergoes irreversible reduction to the Ru(II) complex with loss of chloride ligand.

Reduction of $[Ru(NH_3)_5(solvent)]^{3+}$ complexes in weakly coordinating solvents or solvents with poor π-acid properties results in the labile $[Ru(NH_3)_5(solvent)]^{2+}$ species, which are excellent synthetic intermediates. The $\{[Ru(NH_3)_5]_2(pyrazine)\}^{4+}$ ion obtained by using these intermediates is easily oxidized to the Ru(III) complex.

References

1. H. Taube, *Comments Inorg. Chem.*, **1,** 17 (1981).
2. H. Taube, *Pure Appl. Chem.*, **51,** 901 (1979).
3. P. C. Ford, *Coord. Chem. Rev.*, **5,** 75 (1970).
4. C. Creutz, *Prog. Inorg. Chem.*, **30,** 1 (1983).
5. P. Day, *Int. Rev. Phys. Chem.*, **1,** 149 (1981).
6. P. Ford, DeF. P. Rudd, R. Gaunder, and H. Taube, *J. Am. Chem. Soc.*, **90,** 1187 (1968).
7. C. Creutz and H. Taube, *J. Am. Chem. Soc.*, **95,** 1086 (1973).

8. N. E. Dixon, G. A. Lawrance, P. A. Lay, and A. M. Sargeson, *Inorg. Chem.*, **22,** 846 (1983).
9. N. E. Dixon, G. A. Lawrance, P. A. Lay, and A. M. Sargeson, *Inorg. Chem.*, **23,** 2940 (1984).
10. DeF. P. Rudd and H. Taube, *Inorg. Chem.*, **10,** 1543 (1971).
11. L. H. Vogt, Jr., J. L. Katz, and S. E. Wiberley, *Inorg. Chem.*, **4,** 1157 (1965).
12. J. E. Fergusson and J. L. Love, *Inorg. Synth.*, **13,** 208 (1972).
13. A. D. Allen, F. Bottomley, R. O. Harris, V. P. Reinsalu, and C. V. Senoff, *Inorg. Synth.*, **12,** 2 (1970).
14. C. Kuehn and H. Taube, *J. Am. Chem. Soc.*, **98,** 689 (1976).
15. J. A. Baumann and T. J. Meyer, *Inorg. Chem.*, **19,** 345 (1980).
16. F. A. Cotton and G. Wilkinson, *Advanced Inorganic Chemistry. A Comprehensive Text*, 4th ed., Wiley, New York, 1980, p. 919.
17. M. E. Gress, C. Creutz, and C. O. Quicksall, *Inorg. Chem.*, **20,** 1522 (1981).

72. PENTAAMMINEIRIDIUM(III) AND HEXAAMMINEIRIDIUM(III) COMPLEXES

Submitted by PETER A. LAY* and ALAN M. SARGESON*
Checked by FRODE GALSBØL,† SOLVEIG KALLESØE HANSEN,† and ANDREW W. ZANELLA‡

Pentaammineiridium(III) complexes are difficult to prepare because of the extreme kinetic inertness of the Ir(III)-ligand bonds. Such a property normally requires the use of quite harsh reaction conditions in order to prepare $[Ir(NH_3)_5X]^{n+}$ complexes. However, the use of the trifluoromethanesulfonato (triflato) complex as an intermediate allows the syntheses of a variety of complexes under relatively mild conditions. This has proved invaluable for the synthesis of both mononuclear and dinuclear species.[1-3] We report here the synthesis of $[Ir(NH_3)_5(OSO_2CF_3)](CF_3SO_3)_2$ from the readily prepared $[Ir(NH_3)_5Cl]Cl_2$[4,5] complex. The solvolysis reactions of the triflato complex to give quantitative yields of $[Ir(NH_3)_5(OH_2)]^{3+}$ and $[Ir(NH_3)_6]^{3+}$ in water and liquid ammonia, respectively, are also described. $[Ir(NH_3)_5(OH_2)]^{3+}$ has been prepared previously by the base hydrolysis[4,6,7] of $[Ir(NH_3)_5Cl]^{2+}$, while $[Ir(NH_3)_6]^{3+}$ has been prepared by the prolonged reaction of liquid ammonia on $Na_3[IrCl_6]\cdot xH_2O$ at high temperatures and pressures.[7,8] An alternative method involves the use of $[Ir(NH_3)_5N_3]^{2+}$ as a starting complex.[9] Such reactions are not as convenient, nor are such high yields obtained, as those reported here.

*Research School of Chemistry, The Australian National University, G.P.O. Box 4, Canberra, A.C.T. 2601, Australia.

†Chemistry Department I, H. C. Ørsted Institute, University of Copenhagen, Universitetsparken 5, DK-2100 Copenhagen Ø, Denmark.

‡Joint Science Department, Claremont Colleges, Claremont, CA 91711.

A. PENTAAMMINE(TRIFLUOROMETHANESULFONATO-*O*)-IRIDIUM(III) TRIFLUOROMETHANESULFONATE

$$[Ir(NH_3)_5Cl]Cl_2 + 3CF_3SO_3H \rightarrow [Ir(NH_3)_5(OSO_2CF_3)](CF_3SO_3)_2 + 3HCl\uparrow$$

Procedure

The total time required for this procedure is ~20–30 hr. As pointed out by the checkers, it is recommended that the purity of $[Ir(NH_3)_5Cl]Cl_2$ be checked prior to use. The electronic absorption spectrum reported by Schmidtke[4] appears to be that of a mixture of $[Ir(NH_3)_5Cl]Cl_2$ and *trans*-$[Ir(NH_3)_4Cl_2]Cl$. The correct absorption spectrum is the same as reported by Blanchard and Mason.[10] λ_{max} (nm) (ϵ_{max}, M^{-1} cm^{-1}): 226 (370), 287 (72), 362 (9.5) 0.01 *M* $HClO_4$. An alternative and more reliable method for preparation of $[Ir(NH_3)_5Cl]Cl_2$ is the reaction between $[IrCl_6]^{2-}$ or $[IrCl_6]^{3-}$ and concentrated ammonia solution in a bomb at 100°.[5,7]

To $[Ir(NH_3)_5Cl]Cl_2$ [Alfa Products]*[4,5] (1.0 g, 2.6 mmol) in a 50-mL two-necked round-bottomed flask is added carefully distilled trifluoromethanesulfonic (triflic) acid†[11] (5 mL, 57 mmol).

■ **Caution.** *Triflic acid is one of the strongest known protic acids, and gaseous hydrogen chloride is produced rapidly in the reaction. It is necessary to take adequate precautions to protect the skin and eyes from contact with both chemicals. Inhalation of the corrosive vapors should also be avoided, and the reaction must be performed in a well-ventilated fume hood.*

The stirred reaction mixture is heated to 120° while a flow of N_2 is maintained through the solution, as outlined in the general procedure of Section 68-D. After 1 hr, the gas bubbler is removed, the flask is lightly stoppered to prevent too much evaporation of CF_3SO_3H, and heating is continued for 15 hr at 120–123°. The nitrogen bubbler is replaced, and the solution is allowed to cool to room temperature under a stream of nitrogen. It is then further cooled to ~5° in ice. In order to precipitate the complex, anhydrous diethyl ether (40 mL) is cautiously added dropwise to the rapidly stirred solution.

■ **Caution.** *This addition leads to a very exothermic reaction, and due care should be exercised not to add the diethyl ether too quickly. Diethyl ether is toxic and highly flammable. The addition should be performed in a well-ventilated fume hood.*

The complex is filtered on a medium-porosity sintered-glass funnel, initially by

*The checkers are indebted to Johnson, Matthey and Co. for a loan of iridium chloride.

†The checkers used triflic acid as supplied by Alfa Products or Fluka Chemical Corporation.

gravity, and the filtration is completed using a water aspirator.* The white powder is washed with copious amounts of anhydrous diethyl ether (4 × 20 mL) and air-dried after each washing. Yield: 1.85 g (98%). The complex at this point is sufficiently pure for synthesis purposes. However, in some of the precipitation processes, $Et_2O{\cdot}CF_3SO_3H$ may be coprecipitated. This is removed by breaking up any lumps, and boiling the solid suspension in chloroform (30 mL) for 0.5 hr. (■ **Caution.** *Chloroform is toxic and a carcinogen; this procedure must be performed in a well-ventilated fume hood.*) The white powder is collected on a medium-porosity frit, washed with chloroform (2 × 5 mL), and air-dried. Yield: 1.81 g (95%).

Anal. Calcd. for $C_3H_{15}F_9IrN_5O_9S_3$: C, 4.97; H, 2.09; N, 9.67; S, 13.27; Cl, 0.00. Found: C, 5.13; H, 2.48; N, 9.58; S, 13.23; Cl, 0.00.†

B. PENTAAMMINEAQUAIRIDIUM(III) TRIFLUOROMETHANESULFONATE

$$[Ir(NH_3)_5(OSO_2CF_3)](CF_3SO_3)_2 + H_2O \rightarrow [Ir(NH_3)_5(OH_2)](CF_3SO_3)_3$$

Procedure

The time required for the preparation and isolation of the crude complex is ~1 hr. Crystallization of the complex from aqueous $NaCF_3SO_3$ solution requires ~4 hr. Pentaammine(trifluoromethanesulfonato-*O*)iridium(III) trifluoromethanesulfonate (0.50 g, 0.69 mmol) is added to distilled deionized water (5 mL) in a 20-mL round-bottomed flask, and the solution is heated to 80° for 10 min. The solution is filtered, and the solvent is removed by rotary evaporation (60°) to give $[Ir(NH_3)_5(OH_2)](CF_3SO_3)_3$ as a white powder. The white powder is washed from the flask with diethyl ether (3 × 10 mL) and then air-dried. Yield: 0.52 g (100%).

*The checkers use nitrogen pressure to aid the filtration.

†The checkers report that in one reaction conducted at 122–125° for 15 hr, the reaction was incomplete. *Anal.* Found: Cl, 2.80%. They report that a second reaction at 121° for 24 hr results in a yield of 1.76 g (93%). *Anal.* Found: C, 5.31; H, 2.18; N, 9.15; S, 13.82; Cl, trace. The authors point out that the figure of Cl 2.80% after 15 hr implies a formula of $[Ir(NH_3)_5(Cl)_{0.52}(OSO_2CF_3)_{0.48}](CF_3SO_3)_2$, that is, that the substitution reaction is only 50% complete. This further implies that the reaction should be only 75% complete after 24 hr (Cl, 1.3%), in disagreement with their second experiment. Conversely, the result after 24 hr implies that there can be only a trace of Cl after 15 hr, by the same arguments as above. The reasons for these discrepancies are unclear; however, it may be best to conduct the reaction for 24 hr. It should be pointed out that any *trans*-$[Ir(NH_3)_4Cl_2]Cl$ contaminating the starting material will be converted to *trans*-$[Ir(NH_3)_4(Cl)(OSO_2CF_3)](CF_3SO_3)$ under the reaction conditions. This would lead to chloride being present even after the reaction is complete.

Anal. Calcd. for $C_3H_{17}N_5F_9O_{10}S_3Ir$: C, 4.85; H, 2.31; N, 9.42; S, 12.95. Found: C, 4.7; H, 2.2; N, 9.3; S, 13.1.

If required, the complex may be crystallized from an aqueous $NaCF_3SO_3$ solution by the following procedure. Pentaammine(trifluoromethanesulfonato-*O*)iridium(III) trifluoromethanesulfonate (0.50 g, 0.69 mmol) is aquated in boiling aqueous $NaCF_3SO_3 \cdot H_2O$ [Aldrich Chemical][11] (1.0 g in 1.0 mL) for 10 min. The basicity of the $NaCF_3SO_3 \cdot H_2O$ to be used in this step should be checked prior to use, as many samples contain NaOH and/or Na_2SO_4. It should be recrystallized according to the published procedure[11] until solutions of the salt are near neutral (pH 6–8). If the $NaCF_3SO_3 \cdot H_2O$ contains the above impurities, this may lead to the formation of $[Ir(NH_3)_5OH]^{2+}$, $[Ir(NH_3)_5(OSO_3)]^+$, and other side products, causing contamination of the product. Sometimes the solution is turbid and needs to be filtered while still very hot. The solution is allowed to cool to room temperature, and then it is cooled in an ice bath for 2 hr. The colorless needles are filtered and air-dried. After washing with ethanol/diethyl ether (1:4, 4 × 5 mL) the crystals are again air-dried. Yield: 0.45 g (88%).

Anal. Found: C, 4.79; H, 2.55; N, 9.0.*

The filtrate containing the remaining pentaammine is normally retained for recycling of residues (Section 68-F) after column chromatographic recovery of pentaammine residues, as follows. Residues containing $[Ir(NH_3)_5(OH_2)]^{3+}$ and other pentaammineiridium(III) complexes containing acido ligands are dissolved in water to give solutions of ionic strength < 0.01 *M*. The solution is sorbed onto a column of Dowex 50W-X2, 200–400 mesh (H^+ form) cation exchange resin [Fluka Chemical] (a column of resin 3 cm in diameter and 10 cm high will be suitable for ~1 g of iridium residues), and the column is washed with water (1 L) and 0.7 *M* HCl (2 L). The complexes are eluted with 3 *M* HCl or 3 *M* CF_3CO_2H, with the latter being preferable in order to prevent the complexes precipitating on the column. The elution of the complexes may be followed by observing the movement of the white band down the brown resin. Alternatively, the eluant is monitored using a UV spectrometer (~250 nm) to follow the elution of the complexes. The eluate (or eluates) containing the iridium complexes is evaporated to dryness on a rotary evaporator, and converted to $[Ir(NH_3)_5(OSO_2CF_3)](CF_3SO_3)_2$ as described in Section A.

The first method described is convenient for the synthesis of isotopically labeled aqua complexes by use of vacuum-line techniques.

*The checkers report *Anal.* Found: C, 4.64; H, 2.5; N, 9.28; S, 12.70; Cl, 0.0.

C. HEXAAMMINEIRIDIUM(III) TRIFLUOROMETHANESULFONATE AND HEXAAMMINEIRIDIUM(III) CHLORIDE

$$[Ir(NH_3)_5(OSO_2CF_3)](CF_3SO_3)_2 + NH_3 \rightarrow [Ir(NH_3)_6](CF_3SO_3)_3$$

$$[Ir(NH_3)_6](CF_3SO_3)_3 + 3HCl \rightarrow [Ir(NH_3)_6]Cl_3 + 3CF_3SO_3H$$

Procedure

Preparation and recrystallization of the complex requires ~20 hr. Pentaammine-(trifluoromethanesulfonato-*O*)iridium(III) trifluoromethanesulfonate (0.50 g, 0.69 mmol) is dissolved in dry liquid ammonia[12] (20 mL) contained in a 50-mL round-bottomed flask that can be fitted with a mercury bubbler. The flask is initially cooled in a Dry Ice/acetone bath to prevent excessive splashing of the solvent and hence loss of complex.

■ **Caution.** *Ammonia gas is toxic and corrosive. All manipulations should be performed in a well-ventilated fume hood, and adequate precautions should be taken to prevent contact with skin and eyes. If it is necessary to transport liquid ammonia, this should be performed using a loosely stoppered Dewar flask, and enclosed areas such as fire escapes should be avoided during transport.*
The cooling bath is removed, and the open flask is warmed by a water bath (20°). This ensures rapid evaporation of some of the solvent, and the flow of gaseous ammonia produced minimizes condensaton of atmospheric moisture. After removal of the flask from the bath, the mercury bubbler is placed in position and arranged to maintain a positive pressure (~80 torr) of NH_3. The time of exposure of the solvent to the atmosphere should be minimized during these manipulations. The ammonia level is maintained periodically while the flask is allowed to stand at room temperature for 15 hr. After removal of the bubbler, the remaining liquid ammonia is boiled off by heating the solution in a water bath at 20°.

As an alternative to the addition of liquid ammonia to the flask, gaseous ammonia may be used in a closed system, as follows. To a 50-mL two-necked round-bottomed flask containing the complex is connected an ammonia gas inlet and an outlet consisting of a mercury bubbler. The flask is cooled in the Dry Ice/acetone bath as the ammonia (~20 mL) is condensed. The gas flow is stopped, the cooling bath is removed, and the reaction is allowed to proceed as before.

The off-white solid remaining is collected by washing the interior of the flask with diethyl ether (2 × 10 mL) and is collected on a medium-porosity sintered-glass frit. After air-drying, the product is dried under vacuum over P_4O_{10}. Yield; 0.49 g (96%).

Anal. Calcd. for $C_3H_{18}F_9N_6O_9S_3Ir$: C, 4.86; H, 2.45; N, 11.33; S, 12.97* Found: C, 5.0; H, 2.3; N, 11.0.

The chloride salt is obtained in the following manner. The crude triflate salt (0.50 g, 0.67 mmol) is dissolved in distilled water (10 mL) and passed down a column of Dowex 1 × 8 anion exchange resin, Cl^- form [Fluka] (3 cm diam × 10 cm high) that has been prewashed with distilled water. Additional distilled water (200 mL) is passed down the column, and the combined eluates are evaporated to dryness on a rotary evaporator. The residue of crude chloride salt is dissolved in water (2 mL) and is heated to 70°. Concentrated HCl (12 *M*) is added (0.5 mL), and the solution is allowed to cool to room temperature. It is cooled further in an ice bath for 2 hr, and then the solid is collected, washed with ethanol (2 × 5 mL) and then ether (2 × 5 mL), and air-dried. Yield: 0.22 g (81%).

Anal. Calcd. for $H_{18}N_6Cl_3Ir$: C, 0.0; H, 4.53; N, 20.97; Cl, 26.54. Found: C, 0.0; H, 4.52; N, 20.7; Cl, 26.39.

The remaining complex (0.04 g, 15%) precipitates from the filtrate when the ethanol and ether washings are added.†

Properties

Pentaammine(trifluoromethanesulfonato-*O*)iridium(III) trifluoromethanesulfonate is a white powder that is air-stable provided it is not subjected to prolonged exposure to atmospheric moisture. The complex may be kept for many months in a desiccator over a suitable drying agent (silica gel, $CaCl_2$) without any noticeable decomposition. The aquation rate constant for the complex is 2.6×10^{-4} sec^{-1} at 25° (0.1 *M* CF_3SO_3H), and at elevated temperatures (60–80°) the rate of substitution is quite rapid in both aqueous and nonaqueous solvents. When dissolved in poorly coordinating solvents such as acetone or tetrahydrothiophene 1,1-dioxide (sulfolane), the solvent-substituted species are themselves comparatively labile, and many substitution reactions may be performed in these solvents.[1–3]

The triflato complex may be regenerated from other pentaammine complexes as described in Section 68. Such a property is useful since it allows a regeneration of the starting materials from products. Characteristic spectral properties of the complex include the presence of bands due to coordinated triflate in the IR spectrum, and electronic [λ_{max} 315(sh), nm; ϵ_{max} 150 M^{-1} cm^{-1}, and 270 (220)]

*The checkers report C, 4.76; H, 2.50; N, 11.30; S, 12.66.

†The checkers report that direct use of the crude triflate salt results in a mixed salt of formulation $[Ir(NH_3)_6](CF_3SO_3)_{1.8}Cl_{1.2}$. Yield: 0.35 g (86% based on above formulation). *Anal.* Calcd. for $C_{1.8}H_{18}Cl_{1.2}F_{5.4}N_6O_{5.4}S_{1.8}Ir$: C, 3.57; H, 3.00; N, 13.89; S, 9.54; Cl, 7.03. Found: C, 3.51; H, 2.94; N, 13.38; S, 9.04; Cl, 7.63.

and ^1H NMR (δ 3.85 broad, versus NaTPS as internal standard) spectra in neat triflic acid.[1,2]

The $[Ir(NH_3)_6]^{3+}$ and $[Ir(NH_3)_5(OH_2)]^{3+}$ complexes are characterized by the following λ_{max} (ϵ_{max}) values in their Uv/vis spectra: [315 (14); 251 (92); 214 (160)][13] and [333 (12) (sh), 258 (86); 213 (128)],[14,15] respectively. These spectroscopic measurements were used to test the purity of the complexes in each instance.

References

1. N. E. Dixon, G. A. Lawrance, P. A. Lay, and A. M. Sargeson, *Inorg. Chem.*, **22,** 846 (1983).
2. N. E. Dixon, G. A. Lawrance, P. A. Lay, and A. M. Sargeson, *Inorg. Chem.*, **23,** 2940 (1984).
3. N. J. Curtis and A. M. Sargeson, *J. Am. Chem. Soc.*, **105,** 625 (1983).
4. H.-H. Schmidtke, *Inorg. Synth.*, **12,** 243 (1970).
5. D. N. Hendrickson and W. L. Jolly, *Inorg. Chem.*, **9,** 1197 (1970).
6. F. Basolo and G. S. Hammaker, *Inorg. Chem.*, **1,** 1 (1962).
7. W. Palmaer, *Z. Anorg. Allgem. Chem.*, **10,** 320 (1895).
8. G. W. Watt, E. P. Helvenston, and L. E. Sharif, *J. Inorg. Nucl. Chem.*, **24,** 1067 (1962).
9. B. C. Lane, J. W. McDonald, F. Basolo, and R. G. Pearson, *J. Am. Chem. Soc.*, **94,** 3786 (1972).
10. W. D. Blanchard and W. R. Mason, *Inorg. Chim. Acta,* **28,** 159 (1978).
11. N. E. Dixon, W. G. Jackson, G. A. Lawrance, and A. M. Sargeson, *Inorg. Synth.*, **22,** 103 (1983).
12. D. Nicholls, *Inorganic Chemistry in Liquid Ammonia,* Elsevier, Amsterdam, 1979.
13. H-H. Schmidtke, *J. Mol. Spectroscopy,* **11,** 483 (1963).
14. M. Talebinasab-Sarvari, A. W. Zanella, and P. C. Ford, *Inorg. Chem.*, **19,** 1835 (1980).
15. H.-H. Schmidtke, *Inorg. Chem.*, **5,** 1682 (1966).

73. PENTAAMMINEOSMIUM(III) AND HEXAAMMINEOSMIUM(III) COMPLEXES

Submitted by PETER A. LAY,*† ROY H. MAGNUSON,* and HENRY TAUBE*
Checked by ASBED VASSILIAN‡

Although pentaammineruthenium(III) chemistry has been studied extensively, pentaammineosmium(III) chemistry has received little attention, due largely to problems in synthesizing these compounds.[1,2] We have found[3] that the $[Os(NH_3)_5(OSO_2CF_3)](CF_3SO_3)_2$ complex is prepared readily and in quantitative

*Department of Chemistry, Stanford University, Stanford, CA 94305.
†CSIRO Postdoctoral Fellow.
‡Department of Chemistry, Rutgers University, New Brunswick, NJ 08901.

yield by the Br_2 oxidation of $[Os(NH_3)_5N_2]Cl_2$[4,5] in neat CF_3SO_3H. Like the other trifluoromethanesulfonato (triflato) complexes, it has proved to be an extremely useful intermediate in the preparation of a variety of mononuclear and binuclear osmium(III) complexes.[3,6] We report the syntheses of $[Os(NH_3)_5(OH_2)]^{3+}$, $[Os(NH_3)_6]^{3+}$, and $[Os(NH_3)_5(NCCH_3)]^{3+}$ as examples of typical reactions. Traditionally, pentaammine- and hexaammineosmium(III) complexes were prepared by the prolonged action of NH_3 on $(NH_4)_2[OsBr_6]$ or $(NH_4)_2[OsCl_6]$ at high temperatures and pressures.[7,8] More recently, the oxidation of the $[Os(NH_3)_5(N_2)]Cl_2$ complex has been used as a route into the $[Os(NH_3)_5L]^{n+}$ series.[2,4,5,9] The $[Os(NH_3)_5(OH_2)]^{3+}$ ion is normally prepared, in somewhat smaller yield than reported here, by the oxidation of the dinitrogen complex in aqueous media, followed by precipitation of the explosive perchlorate salt.[9] Hexaammine-osmium(III) compounds have been prepared by a variety of methods, but such methods have been plagued by synthesis difficulties, irreproducibility, and/or low yields.[5,7,8,10–13] The methods reported here are more straightforward and give higher yields than those reported previously.

A. PENTAAMMINE(DINITROGEN)OSMIUM(II) CHLORIDE

$$5(NH_4)_2[OsCl_6] + 13NH_2NH_2{\cdot}H_2O \rightarrow 4[Os(NH_3)_5(N_2)]Cl_2 + cis\text{-}[Os(NH_3)_4(N_2)_2]Cl_2 + 20HCl + 13H_2O$$

$$4cis\text{-}[Os(NH_3)_4(N_2)_2]Cl_2 + 3NH_2NH_2{\cdot}H_2O \rightarrow 4[Os(NH_3)_5N_2]Cl_2 + 5N_2 + 3H_2O$$

Procedure

The total time required for this procedure is 2–3 days depending on the purity of the complex after the second period of reflux. The dinitrogen complex is prepared in somewhat higher yield and purity by the use of several modifications to the original method of Allen and Stevens.[4,5,9]

■ **Caution.** *Hydrazine monohydrate is toxic, a suspected mutagenic agent, and potentially explosive. The reaction must be performed in a well-ventilated fume hood, preferably behind a safety screen, although to the authors' knowledge no explosions have occurred with these reactions.*

The compound $(NH_4)_2[OsCl_6]$[14] (10 g, 22.8 mmol) is added slowly over ~15 min to hydrazine monohydrate (90 mL) contained in a 250-mL round-bottomed flask connected with a reflux condenser while the mixture is stirred rapidly. It is essential not to add the compound too quickly and to grind up any lumps, since isolated high concentrations of $[OsCl_6]^{2-}$ result in the formation of the very stable complex $[(NH_3)_5OsNOs(NH_3)_5]Cl_5{\cdot}H_2O$. This nitrido-bridged species is not converted to the dinitrogen complexes and therefore reduces the yield. By the reverse addition of the hydrazine monohydrate to the solid, this product is obtained in yields of over 90%. The solution is refluxed for 10 hr, during which time the color changes from brown to golden yellow and a pale yellow precipitate forms. The solution is allowed to cool to room temperature, and the solid is collected on a medium-porosity frit. The yield is ~6–7 g. A further amount of compound (~1–1.5 g) is precipitated by the addition of anhydrous ethanol (~400 mL). The remainder of the complexes may be precipitated by the addition of more ethanol, but this material is a mixture consisting mainly of $[Os(NH_3)_5N_2]Cl_2$, *cis*-$[Os(NH_3)_4(N_2)_2]Cl_2$, and $[(NH_3)_5OsNOs(NH_3)_5]Cl_5$. These complexes have distinctive and strong stretching modes in their IR spectra at 2020 ($\nu_{N\equiv N}$), 2097 and 2168 ($\nu_{N\equiv N}$), and 1100 ($\nu_{Os\equiv N\equiv Os}$) cm^{-1}, respectively. These vibrations are a good diagnostic tool for assessing purity. The first two fractions are combined and heated at reflux with a second amount (90 mL) of $NH_2NH_2 \cdot H_2O$ as before. The precipitate obtained after this reflux is checked for impurities of *cis*-$[Os(NH_3)_4(N_2)_2]Cl_2$ by infrared spectroscopy. If the characteristic $\nu_{N\equiv N}$ bands for this complex are absent, this fraction is suitable for the following reactions. If this impurity is still present, a third reflux is necessary. Again more product is obtained by the addition of ethanol to the filtrate in these steps. These fractions are also examined for purity by IR spectroscopy. The overall yield of pure $[Os(NH_3)_5N_2]Cl_2$ is 6.8–7.0 g (80–85%). The complex is characterized by IR and UV/vis spectroscopy (see Table I).

Anal. Calcd. for $Cl_2H_{15}N_7Os$: Cl, 18.95; H, 4.04; N, 26.20. Found: Cl, 19.01; H, 4.20; N, 26.38.

B. PENTAAMMINE(TRIFLUOROMETHANESULFONATO-*O*)-OSMIUM(III) TRIFLUOROMETHANESULFONATE

$$2[Os(NH_3)_5N_2]Cl_2 + Br_2 + 6CF_3SO_3H \rightarrow 2[Os(NH_3)_5(OSO_2CF_3)](CF_3SO_3)_2 + 2N_2\uparrow + 2HBr\uparrow + 4HCl\uparrow$$

Procedure

The time required for this procedure is ~1 day but depends critically on the time taken for the filtration procedure. To pentaammine(dinitrogen)osmium(II) chloride, $[Os(NH_3)_5(N_2)]Cl_2$ (2.0 g, 5.3 mmol), contained in a 50-mL two-necked round-bottomed flask fitted with an Ar or N_2 gas bubbler, is added distilled[15,16] anhydrous CF_3SO_3H (10 mL).

■ **Caution.** *Triflic acid is one of the strongest known protic acids, and gaseous hydrogen chloride is produced rapidly in the reaction. It is necessary to take adequate precautions to protect the skin and eyes and to prevent inhalation* *of the corrosive vapors. These manipulations must be performed in a well-ventilated fume hood. Because of the initial rapid evolution of HCl, care must be taken not to add the triflic acid too quickly.*

Liquid Br_2 (0.5 mL, 10 mmol, fourfold excess) is added to the yellow or pale green solution, and N_2 immediately begins to evolve from the interface of the Br_2 and CF_3SO_3H layers. The pale green color that is sometimes observed results from an impurity of the mixed-valence species $[(NH_3)_5OsN_2Os(NH_3)_5]^{5+}$, but under the reaction conditions this compound is also oxidized to the $[Os(NH_3)_5(OSO_2CF_3)]^{2+}$ ion. (■ **Caution.** *Bromine is toxic and corrosive. Avoid contact with the skin and eyes. The reaction should be carried out in a well-ventilated hood.*) A constant flow of N_2 or Ar is commenced in order to stir the solution while the temperature is maintained around the boiling point of Br_2, ~50–60°. The gas flow is stopped periodically to determine if the evolution of N_2 gas has ceased from the interface of the two layers. When the evolution has ceased (~0.5–1 hr), the solution is heated to 110° in an oil bath, and a constant gas flow is maintained in order to drive off excess Br_2 and the HBr that has been produced in the reaction. (■ **Caution.** *Bromine and HBr are both corrosive and toxic.*) After the brown fumes have ceased and no Br_2 remains in the bottom of the flask, the solution is cooled to room temperature and is then cooled in an ice bath to <5°. Anhydrous diethyl ether (~30 mL) is added dropwise to the rapidly stirred solution.

■ **Caution.** *This is a very exothermic addition, and due care must be exercised to prevent addition of the diethyl ether too quickly. Diethyl ether is toxic and very flammable. The addition must be performed in a well-ventilated fume hood.*

A white to cream-colored precipitate begins to form immediately. The precipitate is collected on a medium-porosity frit and is air-dried. For larger scale reactions, it is best to let the precipitate settle and decant off most of the solvent through the frit. The filtrate may be kept for recovery of CF_3SO_3H as the Na^+ salt.[15,16] The solid is washed with copious quantities of anhydrous diethyl ether until the washings are colorless. After drying in air, the yield is 3.7 g (96%). At this point the compound is quite suitable for synthesis work, but it may be purified by boiling a suspension in chloroform as described in Section 68-A.

■ **Caution.** *Chloroform is toxic and a suspected carcinogen; this procedure must be performed in a well-ventilated fume hood.*

Anal. Calcd. for $C_3H_{15}F_9N_5O_9S_3Os$: C, 4.98; H, 2.09; N, 9.69; S, 13.31; F, 23.66. Found: C, 5.3; H, 2.1; N, 9.7; S, 13.1; F, 23.0.

The compound is stored in a desiccator over P_4O_{10} or silica gel to prevent absorption of atmospheric moisture, which will form $[Os(NH_3)_5(OH_2)](CF_3SO_3)_3$. However, this process is very slow, and the compound may be manipulated readily without the precautions necessary for working in a moisture-free atmosphere.

C. PENTAAMMINEAQUAOSMIUM(III) TRIFLUOROMETHANESULFONATE

$$[Os(NH_3)_5(OSO_2CF_3)](CF_3SO_3)_2 + H_2O \rightarrow [Os(NH_3)_5(OH_2)](CF_3SO_3)_3$$

Procedure

The total time required for this procedure is 3–4 hr. Pentaammine(trifluoromethanesulfonato-*O*)osmium(III) trifluoromethanesulfonate, $[Os(NH_3)_5(OSO_2CF_3)](CF_3SO_3)_2$ (0.30 g, 0.42 mmol), is added to a 0.1 *M* aqueous solution of CF_3SO_3H (1 mL) contained in a 15-mL beaker. The solution is boiled for 5 min. After being cooled to room temperature, the solution is cooled further to $<5°$ in an ice bath. Neat CF_3SO_3H (1 mL) is added slowly in dropwise fashion, while the solution is stirred rapidly.

■ **Caution.** *Triflic acid is one of the strongest known protic acids. It is necessary to take adequate precautions to protect the skin and eyes and to prevent inhalation of the corrosive vapors. These manipulations must be performed in a well-ventilated fume hood. The addition is very exothermic.*

A white precipitate of $[Os(NH_3)_5(OH_2)](CF_3SO_3)_3$ begins to form immediately. The solution is cooled in an ice bath for 1 hr, and the precipitate is collected on a medium-porosity sintered-glass funnel. After air-drying, the solid is washed thoroughly with anhydrous diethyl ether (4 × 10 mL) and air-dried. Yield: 0.2 g (65%).

Anal. Calcd. for $C_3H_{17}F_9N_5O_{10}S_3Os$: C, 4.87; H, 2.31; N, 9.46. Found: C, 5.0; H, 2.4; N, 8.95.

The remaining $[Os(NH_3)_5(OH_2)](CF_3SO_3)_3$ is precipitated by the addition of diethyl ether (6 mL). (■ **Caution.** *This is a very exothermic process, and the diethyl ether must be added slowly and with vigorous stirring of the solution at 0°.*) The precipitate is treated as above to give an essentially quantitative overall yield (0.30 g, 98%) of the complex.

D. HEXAAMMINEOSMIUM(III) TRIFLUOROMETHANESULFONATE AND HEXAAMMINEOSMIUM(III) CHLORIDE

$$[Os(NH_3)_5(OSO_2CF_3)](CF_3SO_3)_2 + NH_3 \rightarrow [Os(NH_3)_6](CF_3SO_3)_3$$

$$[Os(NH_3)_6](CF_3SO_3)_3 + 3HCl \rightarrow [Os(NH_3)_6]Cl_3 + 3CF_3SO_3H$$

Procedure

The total time required for the synthesis and recrystallization is ~1 day. Pentaammine(trifluoromethanesulfonato-*O*)osmium(III) trifluoromethanesulfonate, $[Os(NH_3)_5(OSO_2CF_3)](CF_3SO_3)_2$ (0.10 g, 0.14 mmol), is placed in a 100-mL two-necked round-bottomed flask, and the flask and solid are heated in a vacuum oven at 110°. A second 100-mL two-necked round-bottomed flask is connected to a KOH drying tower and placed in a dish containing Dry Ice. Liquid ammonia (~40 mL) is transferred to this flask, and the flask is stoppered.

■ **Caution.** *Ammonia gas is toxic and corrosive. All manipulations should be performed in a well-ventilated fume hood, and adequate precautions should be taken to prevent contact with skin and eyes. If it is necessary to transport liquid ammonia, this should be performed using a loosely stoppered Dewar flask, and enclosed areas such as fire escapes should be avoided during transport.*[17] The flask containing the osmium complex is removed from the vacuum oven with heat-resistant gloves while it is still hot. It is then connected to the outlet of the KOH drying tower, and the outlet from the flask is connected to a KOH drying tube. The Dry Ice dish is removed from the flask containing the liquid ammonia. By placing this flask in a water-bath for short periods of time (~15–20°), ~10 mL of the liquid is evaporated and allowed to pass through the flask containing the osmium complex. This procedure should take 15–30 min in order to prevent loss of compound by the gas flow. The outlet of the flask is connected to a mercury bubbler that is arranged to maintain a small positive pressure (~80 torr), and ammonia (~20 mL) is condensed into the flask containing the osmium complex by cooling the flask in a Dry Ice bath. The bath is removed and the solution is left at room temperature for 5–6 hr. After the mercury bubbler has been removed, the solvent is evaporated by warming in a water bath at ~20°. The off-white solid remaining is collected by washing the interior of the flask with diethyl ether (2 × 10 mL) and is collected on a medium-porosity sintered-glass frit. (■ **Caution.** *Diethyl ether is very flammable and toxic.*) After air-drying, the yield is 0.10 g (98%).

Anal. Calcd. for $C_3H_{18}F_9N_6O_9S_3Os$: C, 4.87; H, 2.45; N, 11.36. Found: C, 4.7; H, 2.6; N, 11.3.

Depending on the purity of the triflato complex used in this procedure, the product may be highly colored. The impurities giving rise to this coloration are not easily removed by recrystallization, and the complex is purified by cation exchange chromatography as follows. The crude triflate salt (0.10 g, 0.14 mmol) is dissolved in 0.01 *M* CF_3CO_2H (100 mL) and is sorbed onto a column of Dowex 50W-X2, 200–400 mesh, H^+ form [Fluka Chemical] (2 cm diameter × 10 cm). The column is washed with 0.01 *M* CF_3CO_2H (100 mL) and then 0.1 *M* HCl (1 L), and the colorless complex is eluted with 3 *M* HCl (~500 mL). The movement of the band down the column may be monitored visually by the movement of the white band against the colored background of the resin. Alternatively, it may be monitored using UV detection at 220 nm. The eluate containing the complex is evaporated to dryness on a rotary evaporator (water bath 50–60°). The crude chloride residue is dissolved in 0.001 *M* CF_3CO_2H (1 mL) and is heated to 70–80°. Concentrated HCl (12 *M*, 0.5 mL) is added, and the solution is allowed to cool to room temperature and then cooled further in an ice bath for 2 hr. The white precipitate is collected on a sintered-glass micro filter, washed with ethanol (2 × 5 mL) and diethyl ether (2 × 5 mL), and air-dried. Yield: 0.046 g (85%).

Anal. Calcd. for $H_{18}N_6Cl_3Os$: C, 0.00; H, 4.55; N, 21.09; Cl, 26.67. Found: C, 0.00; H, 4.47; N, 20.89; Cl, 26.63.

E. (ACETONITRILE)PENTAAMMINEOSMIUM(III) TRIFLUOROMETHANESULFONATE

$$[Os(NH_3)_5(OSO_2CF_3)](CF_3SO_3)_2 + CH_3CN \rightarrow [Os(NH_3)_5(NCCH_3)](CF_3SO_3)_3$$

Procedure

This procedure requires ~1 day. Pentaammine(trifluoromethanesulfonato-*O*)osmium(III) trifluoromethanesulfonate, $[Os(NH_3)_5(OSO_2CF_3)](CF_3SO_3)_2$ (0.5 g, 0.69 mmol), is added to spectrophotometric grade acetonitrile [Aldrich Chemical] (2 mL) containing several drops of trifluoromethanesulfonic anhydride (triflic anhydride) all contained in a 10-mL test tube.

■ **Caution.** *Acetonitrile is toxic and flammable. Manipulations involving this solvent should be performed in a fume hood. Trifluoromethanesulfonic anhydride is toxic and a very efficient dehydrating agent. Due care must be taken to avoid contact with the skin and eyes and inhalation of the vapors.*

The use of triflic anhydride can be avoided by rigorously drying the acetonitrile over freshly regenerated 4-Å molecular sieves overnight or by passing the solvent through an alumina column prior to the reaction with the triflato complex. The suspension of the triflato complex is rapidly stirred using a 10 × 3 mm magnetic stirring bar, and the solvent is protected from atmospheric moisture by sealing with an appropriate serum cap. After allowing the mixture to stir overnight, the white precipitate is removed with diethyl ether (4 × 5 mL) and is filtered on a medium-porosity frit. (■ **Caution.** *Diethyl ether is toxic and very flammable. The addition should be performed in a well-ventilated fume hood.*) The precipitate is washed two more times with anhydrous diethyl ether and air-dried. Yield: 0.5 g (90%).

Anal. Calcd. for $C_5H_{18}F_9N_6O_9S_3Os$: C, 7.86; H, 2.38; N, 11.00. Found: C, 7.8; H, 2.5; N, 10.7.

Properties

Pentaammine(trifluoromethanesulfonato-*O*)osmium(III) trifluoromethanesulfonate, $[Os(NH_3)_5(OSO_2CF_3)](CF_3SO_3)_2$, is a white powder (in some reactions it is pale yellow or pale brown due to the presence of highly colored minor impurities) that is moderately air-stable and may be stored for months in a desiccator over a suitable drying agent without noticeable decomposition. It undergoes aquation readily for an Os(III) complex, with a first-order rate constant of 1.4×10^{-3} sec^{-1} at 25° (0.1 *M* CF_3SO_3H). The complex is characterized by the IR stretching vibrations of the coordinated triflato ligand, where all bands normally attributed to the triflate anion have now been doubled. In particular, new vibrations assigned as the asymmetric S=O stretch occur at 1300–1400 cm^{-1}. Moreover, the peak at 3400–3500 cm^{-1} due to the coordinated water molecule of the $[Os(NH_3)_5(OH_2)](CF_3SO_3)_3$ complex is absent. The complex is readily regenerated from $[Os(NH_3)_5(OH_2)](CF_3SO_3)_3$, $[Os(NH_3)_5Cl]Cl_2$, and many other pentaammine complexes by the procedures described in Section 68. For the synthesis of pentaammine complexes containing basic ligands it is often necessary to add triflic acid to the solvent to prevent base-catalyzed disproportionation to Os(II) and Os(IV) and subsequent reactions.[3]

Characteristic electronic spectral and redox couples of the osmium(II) and osmium(III) complexes are contained in Table I. All Os(III) complexes also exhibit transitions in the near-infrared (1500–2100 nm),[2,3] due to the effects of spin-orbit coupling on the ground electronic state.

TABLE I. Electronic Spectral and Redox Couples of Os(II) and Os(III) Complexes

Compound	UV[a]/vis (nm)	$E_{1/2}$ (mV)[b]
$[Os(NH_3)_5(N_2)]Cl_2$	208 (25,100)	+580
$[Os(NH_3)_5(OSO_2CF_3)](CF_3SO_3)_2$	235.5 (843), 290 (sh) (108), 447 (55)	—
$[Os(NH_3)_5(OH_2)](CF_3SO_3)_3$	220 (sh) (1100)	−730
$[Os(NH_3)_6](CF_3SO_3)_3$	221 (760)	−780
$[Os(NH_3)_5(NCCH_3)](CF_3SO_3)_3$	~225 (sh) (1700), 250 (sh) (790) 320 (sh) (44), 450 (4)	−250

[a]Extinction coefficients in parentheses, $M^{-1}cm^{-1}$

[b]Formal potentials (vs. NHE).

References

1. H. Taube, *Pure Appl. Chem.*, **51,** 901 (1979).
2. J. Sen and H. Taube, *Acta Chem. Scand., Ser. A.*, **33,** 125 (1979).
3. P. A. Lay, R. H. Magnuson, J. Sen, and H. Taube, *J. Am. Chem. Soc.*, **104,** 7658 (1982).
4. A. D. Allen and J. R. Stevens, *Can. J. Chem.*, **50,** 3093 (1972).
5. F. Bottomley and S.-B. Tong, *Inorg. Synth.*, **16,** 9 (1976).
6. R. H. Magnuson, P. A. Lay, and H. Taube, *J. Am. Chem. Soc.*, **105,** 2507 (1983).

7. F. P. Dwyer and J. H. Hogarth, *J. Proc. Roy. Soc., N.S.W.*, **84,** 117 (1951); **85,** 113 (1952).
8. G. W. Watt and L. Vaska, *J. Inorg. Nucl. Chem.*, **5,** 304 (1958); **5,** 308 (1958); **6,** 246 (1958); **7,** 66 (1958).
9. J. D. Buhr and H. Taube, *Inorg. Chem.*, **18,** 2208 (1979).
10. F. Bottomley and S.-B. Tong, *J. Chem. Soc., Dalton Trans.*, **1973,** 217.
11. F. Bottomley and S.-B. Tong, *Inorg. Chem.*, **13,** 243 (1974).
12. I. P. Evans, G. W. Everett, and A. M. Sargeson, *J. Am. Chem. Soc.*, **98,** 8041 (1976).
13. J. D. Buhr, J. R. Winkler, and H. Taube, *Inorg. Chem.*, **19,** 2416 (1980).
14. F. P. Dwyer and J. W. Hogarth, *Inorg. Synth.*, **5,** 206 (1957).
15. N. E. Dixon, W. G. Jackson, M. J. Lancaster, G. A. Lawrance, and A. M. Sargeson, *Inorg. Chem.*, **20,** 470 (1981).
16. N. E. Dixon, W. G. Jackson, G. A. Lawrance, and A. M. Sargeson, *Inorg. Synth.*, **22,** 103 (1983).
17. D. Nicholls, *Inorganic Chemistry in Liquid Ammonia*, Elsevier, Amsterdam, 1979.

74. PENTAAMMINEPLATINUM(IV) COMPLEXES

Submitted by NEVILLE J. CURTIS,* GEOFFREY A. LAWRANCE,† and ALAN M. SARGESON*
Checked by RONALD C. JOHNSON‡

The substitution reactions of amine platinum(IV) complexes is appreciably slower than those of most other inert metal amines, at least in acidic solutions. Consequently, the availability of a relatively labile leaving group, such as trifluoromethanesulfonate, may have advantages where substitution is required at the sixth site about the pentaammineplatinum(IV) ion. In parallel with reports of other second- and third-row complexes in this chapter, the synthesis of $[Pt(NH_3)_5(OSO_2CF_3)]^{3+}$ from the $[Pt(NH_3)_5Cl]Cl_3$ precursor is readily achieved.[1] Both are described below.

A. PENTAAMMINECHLOROPLATINUM(IV) CHLORIDE, $[Pt(NH_3)_5Cl]Cl_3$

$$K_2[PtCl_6] + 5NH_3 + Na_2HPO_4 \rightarrow [Pt(NH_3)_5Cl](PO_4) + 2KCl + 2NaCl + HCl$$

$$[Pt(NH_3)_5Cl](PO_4) + 3HCl \rightarrow [Pt(NH_3)_5Cl]Cl_3 + H_3PO_4$$

*Research School of Chemistry, The Australian National University, G.P.O. Box 4, Canberra 2601, Australia.
†Department of Chemistry, The University of Newcastle, N.S.W. 2308, Australia.
‡Department of Chemistry, Emory University, Atlanta, GA 30322.

Procedure

This complex is prepared by a variation of the published method.[2] To potassium hexachloroplatinate(IV) [Aldrich Chemical] (4 g) and Na_2HPO_4 (8 g) in a 500-mL flask fitted with a condenser are added concentrated aqueous ammonia (25%, 75 mL) and water (120 mL). The mixture is stirred and heated to reflux, and the temperature is maintained until the suspension turns white (~10 min). Then the mixture is cooled to or below room temperature. The white precipitate is collected on a frit, washed with methanol (2 × 25 mL), and air-dried. The precipitate is dissolved in hot (~70°) 0.1 *M* HCl (~200 mL). The mixture is filtered, and concentrated HCl (50 mL) is added. The colorless solution is reduced in volume to ~40 mL on a rotary evaporator, during which time a white precipitate forms. After cooling, this is collected by filtration on a frit, washed with ice-cold 3 *M* HCl (20 mL), methanol (3 × 25 mL), and diethyl ether (25 mL), and dried in a vacuum desiccator over P_4O_{10}. Yield: 3.0 g (90%). Absorption maximum in water: 286 nm (ϵ 141 $M^{-1}cm^{-1}$).

Anal. Calcd. for $H_{15}N_5Cl_4Pt$: H, 3.58; N, 16.59; Cl, 33.60. Found: H, 3.7; N, 16.3; Cl, 33.8.

B. PENTAAMMINE(TRIFLUOROMETHANESULFONATO-*O*)-PLATINUM(IV) TRIFLUOROMETHANESULFONATE, $[Pt(NH_3)_5(OSO_2CF_3)](CF_3SO_3)_3$

$$[Pt(NH_3)_5Cl]Cl_3 + 4CF_3SO_3H \rightarrow [Pt(NH_3)_5(OSO_2CF_3)](CF_3SO_3)_3 + 4HCl\uparrow$$

Procedure

To pentaamminechloroplatinum(IV) chloride (0.6 g) contained in a 50-mL two-necked round-bottomed flask fitted with a gas bubbler is added distilled anhydrous CF_3SO_3H (5 mL).

■ **Caution.** *Trifluoromethanesulfonic acid is a strong protic acid. Avoid contact with skin and eyes, and avoid breathing the corrosive vapors. Rapid evolution of HCl ensues. An efficient fume hood is required, and the acid should be added slowly.*

A steady stream of argon or nitrogen is passed through the resulting solution while the flask is heated in an oil bath at ~110° for 16 hr. The heat source is removed, and the flask is cooled to ~20° while nitrogen continues to pass through it. The gas flow is discontinued, and diethyl ether (30 mL) is added slowly with vigorous mechanical stirring. (■ **Caution.** *This is a very exothermic addition, and care must be taken not to add the diethyl ether too rapidly.*) The white precipitate is removed by filtering through a fine-porosity sintered-glass funnel, initially under gravity and then, after a bed of solid has formed, under suction.

The precipitate is washed with diethyl ether (2 × 10 mL) and air-dried. As isolated, the product may be contaminated with $[(CH_3CH_2)_2OH]^+CF_3SO_3^-$. For further purification, the product is ground in a mortar, boiled in chloroform for ~10 min, and collected as above. It is washed with hot chloroform (20 mL) and diethyl ether (2 × 10 mL), air-dried, and then dried thoroughly over P_4O_{10} in a vacuum desiccator. Yield: 1.1 g (90%).

Anal. Calcd. for $C_4H_{15}N_5F_{12}O_{12}S_4Pt$: C, 5.48; H, 1.73; N, 8.00; S, 14.63. Found: C, 5.2; H, 2.0; N, 7.4; S, 14.4.

Properties

The trifluoromethanesulfonato product is a white powder that is air-stable for long periods in a closed tube and may be stored in a desiccator for months. The electronic spectrum shows a maximum in the near ultraviolet in neat CF_3SO_3H (λ_{max} 299, ϵ 186 M^{-1} cm^{-1}). In the ^1H NMR spectrum in neat CF_3SO_3H, a single broad peak is observed at δ 4.7 versus sodium (trimethylsilyl)propionate, with side bands observed due to $^2J_{^{195}Pt-^1H}$ coupling (~35 Hz) and $^1J_{^{14}N-^1H}$ coupling (~55 Hz). Following the general reactivity patterns for coordinated trifluoromethanesulfonates, $^-OSO_2CF_3$ may be substituted by a range of other ligands. The synthesis described here for the pentaammineplatinum(IV) complex may be applied to a range of other platinum(IV) amine and even platinum(II) amine complexes.

References

1. N. E. Dixon, G. A. Lawrance, P. A. Lay, and A. M. Sargeson, *Inorg. Chem.*, **23**, 2940 (1984).
2. A. Bakac, T. D. Hand, and A. G. Sykes, *Inorg. Chem.*, **14**, 2540 (1975).

75. PENTAKIS(METHANAMINE)-(TRIFLUOROMETHANESULFONATO-*O*) COMPLEXES OF CHROMIUM(III), COBALT(III), AND RHODIUM(III)

Submitted by GEOFFREY A. LAWRANCE* and ALAN M. SARGESON†
Checked by GEORGE R. BRUBAKER,‡ DAVID W. JOHNSON,‡ and RICHARD A. PETERSON‡

*Department of Chemistry, The University of Newcastle, N.S.W. 2308, Australia.
†Research School of Chemistry, The Australian National University, G.P.O. Box 4, Canberra 2601, Australia.
‡Department of Chemistry, Illinois Institute of Technology, Chicago, IL 60616.

Syntheses reported for the pentaammine(trifluoromethanesulfonato-*O*) complexes can be readily adapted for other amine or multidentate amine analogs.[1] Syntheses of colbalt(III) complexes with 1,2-ethanediamine or *N*-ethyl-1,2-ethanediamine ligands have been reported earlier in this series.[2] To exemplify the procedures further, trifluoromethanesulfonato-*O* complexes of cobalt(III), chromium(III), and rhodium(III) with unidentate methylamine ligands based on the readily prepared[3–5] $[M(NH_3)_5Cl]Cl_2$ precursors are reported here. The following sections report syntheses of 1,2-ethanediamine complexes of Rh(III) and Ir(III) and of Ru(II) and Os(II) diimines with trifluoromethanesulfonato ligands. Such syntheses indicate the diversity of the synthesis technique, and the complexes described are excellent precursors for other compounds.

■ **Caution.** *All reactions should be carried out in a well-ventilated fume hood.*

A. PENTAKIS(METHANAMINE)-(TRIFLUOROMETHANESULFONATO-*O*)CHROMIUM(III) TRIFLUOROMETHANESULFONATE, $[Cr(NH_2CH_3)_5(OSO_2CF_3)](CF_3SO_3)_2$

$$[Cr(NH_2CH_3)_5Cl]Cl_2 + 3CF_3SO_3H \rightarrow [Cr(NH_2CH_3)_5(OSO_2CF_3)](CF_3SO_3)_2 + 3\,HCl\uparrow$$

Procedure

To chloropentakis(methanamine)chromium(III) chloride (1.0 g) in a 50-mL two-necked round-bottomed flask fitted with a gas bubbler is added distilled anhydrous CF_3SO_3H [Aldrich Chemical] (20 mL). (■ **Caution.** *Trifluoromethanesulfonic acid is a strong protic acid. Avoid contact with skin and eyes, and avoid breathing the corrosive vapors. Rapid evolution of HCl ensues; care must be taken not to add the acid too rapidly.*) The solution is allowed to stand for 3 days at room temperature while a gentle stream of nitrogen gas is passed through the flask. The gas flow is disconnected, and the solution is poured into a 0.5-L flask. Anhydrous diethyl ether (200 mL) is added dropwise with vigorous mechanical stirring. (■ **Caution.** *This is a very exothermic addition and care must be taken not to add the diethyl ether too rapidly.*) A fine precipitate is separated on a medium-porosity frit, initially by gravity and then, after a bed of precipitate has formed, by suction. The precipitate is washed copiously with anhydrous diethyl ether (4×20 mL) and dried under vacuum over P_4O_{10}. Yield: 1.9 g (95%).

Anal. Calcd. for $C_8H_{25}N_5F_9O_9S_3Cr$: C, 14.68; H, 3.85; N, 10.70; S, 14.70. Found: C, 15.0; H, 3.9; N, 10.4; S, 14.45.

B. PENTAKIS(METHANAMINE)-(TRIFLUOROMETHANESULFONATO-*O*)COBALT(III) TRIFLUOROMETHANESULFONATE, $[Co(NH_2CH_3)_5(OSO_2CF_3)](CF_3SO_3)_2$

$$[Co(NH_2CH_3)_5Cl]Cl_2 + 3CF_3SO_3H \rightarrow [Co(NH_2CH_3)_5(OSO_2CF_3)](CF_3SO_3)_2 + 3HCl\uparrow$$

Procedure

To chloropentakis(methanamine)cobalt(III) chloride (0.8 g) in a 50-mL two-necked round-bottomed flask fitted with a gas bubbler is added distilled anhydrous CF_3SO_3H [Aldrich Chemical] (15 mL). (■ **Caution.** *Trifluoromethanesulfonic acid is a strong protic acid. Avoid contact with skin and eyes, and avoid breathing the corrosive vapors. Rapid evolution of HCl ensues; care must be taken not to add the acid too quickly.*) The solution is allowed to react, and the product is isolated exactly as described for the chromium(III) analog above. Yield: 1.5 g (94%).

Anal. Calcd. for $C_8H_{25}N_5F_9O_9S_3Co$: C, 14.53; H, 3.81; N, 10.59; S, 14.54. Found: C, 14.2; H, 3.8; N, 10.3; S, 14.7.

C. PENTAKIS(METHANAMINE)-(TRIFLUOROMETHANESULFONATO-*O*)RHODIUM(III) TRIFLUOROMETHANESULFONATE, $[Rh(NH_2CH_3)_5(OSO_2CF_3)](CF_3SO_3)_2$

$$[Rh(NH_2CH_3)_5Cl]Cl_2 + 3\ CF_3SO_3H \rightarrow [Rh(NH_2CH_3)_5(OSO_2CF_3)](CF_3SO_3)_2 + 3HCl\uparrow$$

Procedure

To chloropentakis(methanamine)rhodium(III) chloride (0.65 g) in a 150-mL two-necked round-bottomed flask fitted with a gas bubbler is added distilled anhydrous CF_3SO_3H [Aldrich Chemical] (15 mL). (■ **Caution.** *Trifluoromethanesulfonic acid is a strong protic acid. Avoid contact with skin and eyes, and avoid breathing the corrosive vapors. Rapid evolution of HCl ensues; care must be taken not to add the acid too quickly.*) A gentle stream of nitrogen is passed through the solution, which is heated in an oil bath at 110° for 3 hr. The flask is removed from the bath, and the nitrogen flow is continued until the solution has cooled to room temperature. The gas flow is disconnected, and anhydrous diethyl ether

(100 mL) is added dropwise with vigorous mechanical stirring. (■ **Caution.** *This is a very exothermic addition, and care must be taken not to add the diethyl ether too rapidly.*) The white precipitate is collected on a fine-porosity sintered frit, first by gravity and then, after a bed of precipitate has formed, by aspiration. The precipitate is washed well with diethyl ether (4 × 25 mL), and dried under vacuum over P_4O_{10}. Yield: 1.2 g (95%).

Anal. Calcd. for $C_8H_{25}N_5F_9O_9S_3Rh$: C, 13.62; H, 3.57; N, 9.93; S, 13.63. Found: C, 13.6; H, 3.5; N, 9.6; S, 13.8.

Properties

The complexes are air-stable powders and can be stored for months in a desiccator. Both the cobalt(III) and chromium(III) complexes are purple, and the rhodium(III) compound is off-white. Aquation of the cobalt(III) complex (k_{aq} = 0.11 sec^{-1} at 25°, 0.1 *M* CF_3SO_3H) is faster than that of the pentaammine analog (k_{aq} = 0.027 sec^{-1}). The pentakis(methanamine)chromium(III) ion aquates appreciably more slowly than the pentaammine analog ($k_{aq} = 6.3 \times 10^{-4}$ sec^{-1} versus 1.24×10^{-2} sec^{-1}) while the rate of aquation of the pentakis(methanamine)rhodium(III) ion ($k_{aq} = 3.28 \times 10^{-2}$ sec^{-1}) is similar to that of the pentaammine analog ($k_{aq} = 1.87 \times 10^{-2}$ sec^{-1}). Electronic absorption maxima of the pentakis(amine) complexes are shifted to lower energies compared with the pentaammine analogs. Maxima [λ, nm (ε, M^{-1} cm^{-1})] in CF_3SO_3H are observed for Co at 542 (58.9), 490 (sh) (51), and 362 (67.3), for Rh at 335 (142) and 275 (144), and for Cr at 506 (49.1) and 374 (45.8). The diamagnetic Co(III) and rhodium(III) complexes exhibit characteristic ^1H NMR for the methyl groups; for Co, singlets are observed at δ 1.24 (3H, *trans*-CH_3) and δ 1.87 (12H, *cis*-CH_3), measured versus sodium (trimethylsilyl)propionate in neat CF_3SO_3H, whereas for Rh the corresponding resonances are at δ 1.80 and δ 2.10. Further, amine resonances are observed for Co at δ 3.08 (2H, *trans*-NH_2) and δ 3.76 (8H, *cis*-NH_2) but are unresolved for Rh with a broad resonance at δ 3.56 ppm. The pentakis(methanamine) complexes are readily solvated in coordinating solvents, and in poorly coordinating solvents such as acetone or sulfolane they react with ligands such as urea. Syntheses described here for the methylamine complexes are valid for a range of alkylamines.

References

1. N. E. Dixon, G. A. Lawrance, P. A. Lay, and A. M. Sargeson, *Inorg. Chem.*, **23,** 2940 (1984).
2. N. E. Dixon, W. G. Jackson, G. A. Lawrance, and A. M. Sargeson, *Inorg. Synth.*, **22,** 103 (1983).
3. R. Mithner, P. Blankenburg, and W. Depkat, *Z. Chem.*, **9,** 68 (1969).
4. A. Rodgers and P. J. Staples, *J. Chem. Soc.*, **1965,** 6834.
5. T. W. Swaddle, *Can. J. Chem.*, **55,** 3166 (1977).

76. *cis*- AND *trans*-BIS(1,2-ETHANEDIAMINE)-RHODIUM(III) COMPLEXES

Submitted by PETER A. LAY* and ALAN M. SARGESON*
Checked by LIANGSHIU LEE† and JOHN D. PETERSEN†

The chemistry of bis(1,2-ethanediamine)cobalt(III) has a rich and important place in the understanding of substitution reactions.[1,2] One of the chief reasons for this has been the ease with which these complexes are prepared. Those of rhodium(III) are more difficult to prepare, so less is understood about substitution at this metal center. The chief entry into the $[Rh(en)_2XY]^{n+}$ complexes has been the preparation of the $[Rh(en)_2Cl_2]^+$ isomers from $RhCl_3 \cdot 3H_2O$. A variety of methods exist for performing this reaction,[3–8] but the best is that of Hancock.[3] We have modified his procedure by using cation-exchange chromatography to separate the *cis* and *trans* isomers.[9] Upon treatment with neat CF_3SO_3H, the *cis*- and *trans*-chloro complexes undergo ligand substitution to give *cis*-$[Rh(en)_2(OSO_2CF_3)_2]$ (CF_3SO_3) and *trans*-$[Rh(en)_2(OSO_2CF_3)Cl](CF_3SO_3)$, respectively, with full retention of configuration.[9] The *cis*-$[Rh(en)_2(OSO_2CF_3)Cl](CF_3SO_3)$ complex may be prepared by treatment of the dichloro complex with a stoichiometric amount of $AgCF_3SO_3$. In all these complexes the triflato ligand is labile and undergoes substitution with retention of stereochemistry, which makes them useful intermediates in syntheses.[9]

A. *cis*-DICHLOROBIS(1,2-ETHANEDIAMINE)RHODIUM(III) CHLORIDE HYDRATE, *cis*-$[Rh(en)_2Cl_2]Cl \cdot H_2O$, AND *trans*-DICHLOROBIS(1,2-ETHANEDIAMINE)RHODIUM(III) CHLORIDE HYDROCHLORIDE DIHYDRATE, *trans*-$[Rh(en)_2Cl_2]Cl \cdot HCl \cdot 2H_2O$

$$RhCl_3 \cdot 3H_2O + 2en \cdot 2HCl \rightarrow cis\text{-}[Rh(en)_2Cl_2]Cl \cdot H_2O + 2H_2O + 4HCl$$

$$RhCl_3 \cdot 3H_2O + 2en \cdot 2HCl \rightarrow trans\text{-}[Rh(en)_2Cl_2]Cl \cdot HCl \cdot 2H_2O + H_2O + 3HCl$$

Procedure

The total time taken for the syntheses and separation of isomers is 2–3 days. These compounds are prepared by a modification of the method of Hancock.[3]

*Research School of Chemistry, The Australian National University, G.P.O. Box 4, Canberra, A.C.T. 2601, Australia.

†Department of Chemistry, Clemson University, Clemson, SC 29631.

Rhodium(III) chloride trihydrate [Alfa Products] (10.0 g, 38.0 mmol) and en·2HCl (10.1 g, 76 mmol) are dissolved in water (60 mL) contained in a 250-mL round-bottomed flask that is fitted with a reflux condenser and a magnetic stirring bar. (■ **Caution.** *1,2-Ethanediamine(en) dihydrochloride is a skin irritant.*) The mixture is warmed gently until all the solid has dissolved, giving a deep red solution. Aqueous NaOH (2 *M*, 38 mL) is added, and a brick red precipitate is formed immediately. The solution is stirred and allowed to reflux until the suspended solid dissolves to give a cherry red solution. More aqueous NaOH (2 *M*, 38 mL) is added dropwise through the reflux condenser until the pH of the boiling solution remains at ~7. If the solution is allowed to become too basic, the yields are drastically reduced. The solution is transferred to a 500-mL Buchi flask and evaporated to dryness on a rotary evaporator (80°). The solid residue is heated at 170° for 24 hr. The solid is dissolved in aqueous HCl (0.02 *M*, 1.5 L) and sorbed on a column of Dowex 50W-X2* cation exchange resin (H^+ form, 200–400 mesh, 50 × 4 cm). The column is washed with 0.05 *M* HCl (2 L), and the complexes are immediately eluted with 0.7 *M* HCl to give two clearly defined yellow bands. A red complex passes through the column during sorption of the complexes. This is a small amount of $[Rh(en)CL_4]^-$. If the complexes are left on the column for too long, some aquation may occur, thus complicating chromatographic separations. Each band is collected, and the solution is evaporated to dryness under reduced pressure to yield the *trans* isomer (first band eluted) and the *cis* isomer. The compounds at this point are suitable for conversion to the triflato complexes.

Yield of the *trans* isomer: 6.0 g (39%)† *Anal.* Calcd. for $C_4H_{21}N_4Cl_4O_2Rh$: C, 11.95; H, 5.27; N, 13.94; O, 7.96; Cl, 35.28. Found: C, 11.9; H, 5.3; N, 13.7; O, 8.2; Cl, 35.3.

Yield of the *cis* isomer: 6.2 g (47%). *Anal.* Calcd. for $C_4H_{18}N_4Cl_3ORh$: C, 13.83; H, 5.22; N, 16.13. Found: C, 14.0; H, 5.9; N, 15.7.

Total yield ~85%,‡ based on $RhCl_3·3H_2O$.

The *cis* isomer may be recrystallized by the following procedure. The crude product (4.0 g) is dissolved in boiling 6 *M* HCl (60 mL) contained in a 200-mL conical flask. (■ **Caution.** *Hydrochloric acid is corrosive and toxic. This procedure must be performed in a well-ventilated fume hood.*) After the solution is allowed to cool to room temperature, it is cooled further to ~3° in a refrigerator for 2 days. The large bright yellow crystals are collected on a medium-porosity frit, washed with ethanol (5 × 20 mL), and air-dried. Yield: 3.8 g (94%).

Anal. Calcd. for $C_4H_{18}N_4Cl_3ORh$: C, 13.83; H, 5.22; N, 16.13. Found: C, 13.7; H, 5.3; N, 16.0.

*Amberlite resin CG-120 (H) Type 1 100–200 mesh [Fluka Chemical] may also be used with 0.60 *M* HCl as eluent.

†On a one-tenth scale, the yield was 20%.

‡On a one-tenth scale, the overall yield was 43%.

This complex was first thought to be the sesquihydrate,[3] but this conclusion was due to the precipitation of the double salt *cis*-$[Rh(en)_2Cl_2]_2Cl(ClO_4)$ in the original isolation procedure using the crude perchlorate salt.[10] The *trans* isomer may be recrystallized according to published procedures.[5,6] If only the *cis* isomer is required, the *trans* isomer may be converted to the *cis* isomer by way of the oxalato complex $[Rh(en)_2(C_2O_4)]^+$.[3,5] Alternatively, the *cis* isomer may be converted to the *trans* isomer in the presence of $[BH_4]^-$ and excess HCl.[5]

B. *trans*-CHLOROBIS(1,2-ETHANEDIAMINE)-(TRIFLUOROMETHANESULFONATO-*O*)RHODIUM(III) TRIFLUOROMETHANESULFONATE, *trans*-$[Rh(en)_2(OSO_2CF_3)Cl](CF_3SO_3)$

$$trans\text{-}[Rh(en)_2Cl_2]Cl \cdot HCl \cdot 2H_2O + 2CF_3SO_3H \rightarrow trans\text{-}[Rh(en)_2(OSO_2CF_3)Cl](CF_3SO_3) + 3HCl\uparrow + 2H_2O$$

Procedure

The time taken for this procedure is ~1 day. This pale yellow powder is prepared in quantitative yield from *trans*-$[Rh(en)_2Cl_2]Cl \cdot HCl \cdot 2H_2O$ by the standard method described in Section 68-D, using a temperature of 110° and a reaction time of 4 hr.

Anal. Calcd. for $C_6H_{16}N_4ClF_6O_6S_2Rh$: C, 12.94; H, 2.89; N, 10.06; S, 11.52; F, 20.48; Cl, 6.37. Found: C, 13.0; H, 3.2; N, 9.8; S, 11.6; F, 20.2; Cl, 6.1.

C. *cis*-BIS(1,2-ETHANEDIAMINE)BIS-(TRIFLUOROMETHANESULFONATO-*O*)RHODIUM(III) TRIFLUOROMETHANESULFONATE, *cis*-$[Rh(en)_2(OSO_2CF_3)_2](CF_3SO_3)$

$$cis\text{-}[Rh(en)_2Cl_2]Cl \cdot H_2O + 3CF_3SO_3H \rightarrow cis\text{-}[Rh(en)_2(OSO_2CF_3)_2](CF_3SO_3) + 3HCl\uparrow + H_2O$$

Procedure

The time taken for this procedure is ~1 day. The compound is prepared and isolated in quantitative yields by the method described in Section 68-D. A reaction time of 4 hr and a temperature of 110° are used.

Anal. Calcd. for $C_7H_{16}N_4F_9O_9S_3Rh$: C, 12.54; H, 2.14; N, 8.36; S, 14.35; F, 25.51. Found: C, 12.4; H, 2.5; N, 8.0; S, 13.8; F, 25.0.

Properties

The rhodium(III) triflato complexes undergo substitution with retention of stereochemistry to give the respective solvent complexes.[9] They are pale yellow to white powders (depending on whether or not Cl^- is in the coordination sphere) and may be handled for short periods in air without any noticeable decomposition. They may be stored for months in a desiccator over P_4O_{10} but are hygroscopic and undergo hydrolysis to the aqua complexes in air. The *trans* isomers are characterized by a single CH_2 resonance in their ^{13}C NMR spectra, while the *cis* isomers have two resonances.[11] Thus, this is a convenient method for determining the stereo course of a reaction *in situ*. UV/vis spectroscopic data are presented in Table I.

TABLE I. UV/vis Spectral Data for Rh(III) Complexes

Complex	λ (nm)[a]
cis-$[Rh(en)_2Cl_2]^+$	352 (203), 295(205)[b]
trans-$[Rh(en)_2Cl_2]^+$	406(83), 286(134), 240(1350), 206(38900)[b]
cis-$[Rh(en)_2(OSO_2CF_3)_2]^+$	341(193), 271(142)[c]
trans-$[Rh(en)_2(OSO_2CF_3)Cl]^+$	407(43), 270sh(127)[c]

[a]$(M^{-1}\ cm^{-1})$ in parentheses.
[b]0.1 *M* HCl.
[c]Neat CF_3SO_3H.

cis-$[Rh(en)_2(OSO_2CF_3)Cl]^+$ may be prepared by reaction of *cis*-$[Rh(en)_2Cl_2]^+$ with $AgCF_3SO_3$. In addition, reactions similar to those described within may also be performed for other bis(diamine) complexes and those containing macrocyclic ligands.

References

1. W. G. Jackson and A. M. Sargeson, in *Rearrangements in Ground and Excited States*, Vol. II, P. de Mayo (ed.), Academic Press, New York, 1980, Chap. 11.
2. F. Basolo and R. G. Pearson, *Mechanisms of Inorganic Reactions*, 2nd ed., Wiley, New York, 1967.
3. M. P. Hancock, *Acta Chem. Scand., Ser. A.*, **33**, 15 (1979).
4. S. A. Johnson and F. Basolo, *Inorg. Chem.*, **1**, 925 (1962).
5. R. D. Gillard, J. Pedroso De Jesus, and P. S. Sheridan, *Inorg. Synth.*, **20**, 57 (1980).
6. S. N. Anderson and F. Basolo, *Inorg. Synth.*, **7**, 214 (1963).
7. A. W. Addison, R. D. Gillard, P. S. Sheridan, and L. R. H. Tipping, *J. Chem. Soc., Dalton Trans.*, **1974**, 709.
8. I. B. Baranovskii and G. Ya Mazo, *Russ. J. Inorg. Chem.*, **20**, 244 (1975).
9. N. E. Dixon, G. A. Lawrance, P. A. Lay, and A. M. Sargeson, *Inorg. Chem.*, **23**, 2940 (1984).

10. M. Hancock, B. Nielson, and J. Springborg, *Acta Chem. Scand., Ser. A.*, **A36,** 314 (1982).
11. F. P. Jakse, J. V. Paukstelis, and J. D. Petersen, *Inorg. Chim. Acta*, **27,** 225 (1978).

77. *cis*- and *trans*-BIS(1,2-ETHANEDIAMINE)-IRIDIUM(III) COMPLEXES

Submitted by PETER A. LAY,*† ALAN M. SARGESON,‡ and HENRY TAUBE*
Checked by LIANGSHIU LEE§ and JOHN D. PETERSEN§

Of the bis(1,2-ethanediamine) complexes of the cobalt triad in oxidation state III, those of iridium(III) have been least studied. This has been due chiefly to a lack of suitable methods for preparing the $[Ir(en)_2Cl_2]Cl$ isomers. Several methods are described in the literature for preparing these complexes,[1-3] but until very recently[4] these reactions were unreliable or gave poor yields. Reported here is a slight modification of the recent method of Galsbøl and Rasmussen[4] for preparing the $[Ir(en)_2Cl_2]Cl$ isomers and subsequent conversion of these complexes into the corresponding triflato complexes.[5] Again, the triflato complexes prove to be very useful intermediates for further syntheses.

A. *cis*-DICHLOROBIS(1,2-ETHANEDIAMINE)IRIDIUM(III) CHLORIDE HYDRATE, *cis*-$[Ir(en)_2Cl_2]Cl\cdot H_2O$, and *trans*-DICHLOROBIS(1,2-ETHANEDIAMINE)IRIDIUM(III) CHLORIDE HYDROCHLORIDE DIHYDRATE, *trans*-$[Ir(en)_2Cl_2]Cl\cdot HCl\cdot 2H_2O$

$$IrCl_3\cdot 6H_2O + 2\ en\cdot 2CH_3CO_2H \rightarrow cis\text{-}[Ir(en)_2Cl_2]Cl\cdot H_2O + 2CH_3CO_2H + 5H_2O$$

$$IrCl_3\cdot 6H_2O + 2en\cdot 2CH_3CO_2H \rightarrow trans\text{-}[Ir(en)_2Cl_2]Cl\cdot HCl\cdot 2H_2O + 2CH_3CO_2H + 4H_2O$$

Procedure

The total time required for the syntheses and separation of isomers is 2–3 days. The procedure is identical to that of Rasmussen and Galsbøl, except that the

*Department of Chemistry, Stanford University, Stanford, CA 94305.
†CSIRO Postdoctoral Fellow.
‡Research School of Chemistry, The Australian National University, G.P.O. Box 4, Canberra, A.C.T. 2601, Australia.
§Department of Chemistry, Clemson University, Clemson, SC 29631.

isomers are separated by chromatography instead of using fractional crystallization.[4] Iridium(III) chloride hexahydrate [Alfa Products] (6.2 g, 15.2 mmol) is dissolved with heating in a mixture of acetic acid (0.9 mL, 16 mmol) and water (17 mL) contained in a 50-mL round-bottomed flask fitted with a condenser. The amount of water in $IrCl_3 \cdot xH_2O$ varies among commercial samples. Appropriate allowances should be made for this variation. While heating the iridium(III) solution at reflux, a solution of 1,2-ethanediamine (3.1 mL, 46 mmol) in water (total volume 6 mL) is added in portions as follows: Initially, 1 mL is added and then 0.5 mL every ½ hr (5 hr total). (■ **Caution.** *1,2-Ethanediamine is toxic and a strong irritant. Manipulations should be performed in a well-ventilated fume hood, and adequate protection for the eyes and skin should be worn.*) As the 1,2-ethanediamine is added, the solution changes from dark green to orange and then to yellow. After the last addition, the solution is heated at reflux for an additional 5 hr and then is evaporated to dryness in a rotary evaporator (final bath temperature ~90°). The resultant brown residue is heated to 170° for 24 hr, and the solid is dissolved in water (1 L). Dowex 50W-X2 cation exchange resin [Fluka Chemical] (200 × 400 mesh, H^+ form, 0.5 g) is added, and the mixture is stirred for 20 min to sorb any highly charged species. The suspension is filtered through a medium-porosity sintered-glass filter, and the resin is washed with 0.1 *M* HCl (4 × 25 mL) by gravity filtration. The combined filtrates are evaporated to dryness, the residue is dissolved in water (1 L) and sorbed onto a column of Dowex 50W-X2, 200–400 Mesh, H^+ form (25 × 4 cm).† A red solution sometimes passes through the column, which is washed with water (500 mL). The initial effluent and the washing may be used to recover $K[Ir(en)Cl_4]$.[4] The column is washed with 0.02 *M* LiOH (500 mL) to remove any *mer*-$[Ir(en)(en^*)Cl_3]$, water (100 mL), and 0.1 *M* HCl (2 L). The yellow complex removed by 0.02 *M* LiOH contains monodentate en (denoted by en*) and the *mer*-$[Ir(en)(enH)Cl_3]Cl \cdot H_2O$ complex may be recovered from this solution.[4] Finally, the yellow $[Ir(en)_2Cl_2]Cl$ isomers are removed with 0.7 *M* HCl and separated into two bands, with the *trans* isomer moving more rapidly. The separate eluates are evaporated to dryness on a rotary evaporator. The residues may be used directly for the subsequent preparations. The *trans* isomer (first eluted) is dissolved in boiling water (10 mL), and 12 *M* HCl (20 mL) is carefully added. (■ **Caution.** *HCl is toxic and corrosive. The reaction should be carried out in a well-ventilated fume hood.*) The mixture is allowed to cool to room temperature for 1 hr, and then it is cooled in ice for 3 hr, giving large yellow needles of the required product. These are collected on a medium-porosity

†Amberlite Resin CG-120 (H) Type 1 100–200 mesh [Fluka] may also be used with 0.60 *M* HCl as eluent.

filter, washed with ice-cold 6 *M* HCl (2 × 3 mL), and dried in air. Yield: 3.0 g (40%) of *trans*-$[Ir(en)_2Cl_2]Cl \cdot HCl \cdot 2H_2O$.†

Anal. Calcd. for $C_4H_{21}N_4O_2Cl_4Ir$: C, 9.78; H, 4.31; N, 11.41. Found: C, 10.1; H, 4.4; N, 11.35.

Instead of being dried in air, the complex may be washed with ethanol (2 × 10 mL) and diethyl ether (2 × 10 mL), and then dried at 135° for 8 hr to yield the anhydrous salt, *trans*-$[Ir(en)_2Cl_2]Cl$.

Anal. Calcd. for $C_4H_{16}N_4Cl_3Ir$: C, 11.47; H, 3.85; N, 13.38. Found: C, 11.6; H, 3.9; N, 13.3.

The *cis* isomer (second fraction) is recrystallized by dissolution in a boiling solution of NaCl (0.5 g, 10 mL), followed by cooling for 2 hr at room temperature and then for a day at ~3° in a refrigerator. The crystal cake is broken up, and the mixture is cooled in ice for 3 hr before the crystals are collected on a medium-porosity sintered-glass funnel. After being washed with ice water (2 × 1 mL), the crystals are dried in air. Yield of *cis*-$[Ir(en)_2Cl_2]Cl \cdot H_2O$: 2.6 g (39%).‡

Anal. Calcd. for $C_4H_{18}N_4OCl_3Ir$: C, 10.99; H, 4.15; N, 12.83. Found: C, 11.07; H, 4.29; N, 12.61.

The *cis* isomer may be resolved into its optical isomers using either (+)-α-bromocamphor-π-sulfonic acid[3] or $(+)_{546}$-(ethylenediaminetetracetato)cobaltate-(III).[6]

B. *trans*-CHLOROBIS(1,2-ETHANEDIAMINE)-(TRIFLUOROMETHANESULFONATO-*O*)IRIDIUM(III) TRIFLUOROMETHANESULFONATE, *trans*-$[Ir(en)_2(OSO_2CF_3)Cl](CF_3SO_3)$

$$trans\text{-}[Ir(en)_2Cl_2]Cl \cdot HCl \cdot 2H_2O + 2CF_3SO_3H \rightarrow trans\text{-}[Ir(en)_2(OSO_2CF_3)Cl](CF_3SO_3) + 3HCl\uparrow + 2H_2O$$

Procedure

The reaction time is ~2 days. This compound is prepared in quantitative yield by the method described for the Rh analog, using a reaction time of 24 hr and a temperature of 120°.[7]

Anal. Calcd. for $C_6H_{16}N_4ClF_6O_6S_2Ir$: C, 11.16; H, 2.50; N, 8.67; S, 9.93; Cl, 5.49. Found: C, 11.2; H, 2.3; N, 8.6; S, 9.8; Cl, 5.6.

†At one-fifth scale, the checkers obtained a 35% yield.
‡At one-fifth scale, the checkers obtained 15% yield.

C. *cis*-BIS(1,2-ETHANEDIAMINE)BIS-(TRIFLUOROMETHANESULFONATO-*O*)IRIDIUM(III) TRIFLUOROMETHANESULFONATE, *cis*-$[Ir(en)_2(OSO_2CF_3)_2](CF_3SO_3)$

$$cis\text{-}[Ir(en)_2Cl_2]Cl\cdot H_2O + 3CF_3SO_3H \rightarrow cis\text{-}[Ir(en)_2(OSO_2CF_3)_2](CF_3SO_3) + 3HCl\uparrow + H_2O$$

Procedure

The reaction time is ~2 days. This compound is prepared (120° and 15 hr) as described in Section 68-D as a quantitative yield of pale yellow powder.

Anal. Calcd. for $C_7H_{16}N_4F_9O_9S_3Ir$: C, 11.07; H, 2.12; N, 7.38; S, 12.66. Found: C, 11.2; H, 2.0; N, 7.4; S, 12.8.

Properties

The dichloro species may be characterized from their UV/vis spectra (Table I), their NMR spectra,[4,5] and their chromatographic behavior. The presence of molecules of water and hydrogen chloride of crystallization has been confirmed by thermogravimetric analysis of these complexes[4] in addition to microanalytical data.

Conversion into the triflato complexes proceeds with full retention of configuration, and the compounds undergo solvolysis reactions with retention of configuration (as confirmed by ^1H and ^{13}C NMR spectroscopy).[5] The triflato complexes are white ($[Ir(en)_2(OSO_2CF_3)_2]^+$) to pale yellow ($[Ir(en)_2(OSO_2CF_3)Cl]^+$) powders that are stable in air and react only very slowly with atmospheric moisture. They are characterized by UV/vis (Table I) and ^1H and ^{13}C NMR spectroscopy.[5]

TABLE I UV/vis Spectroscopic Properties

Compound	λ (nm)[a]
cis-$[Ir(en)_2Cl_2]^+$	377(19.4) (sh), 320(100) (sh), 292(145), 254(162) (sh), 227(642) (sh)[b]
trans-$[Ir(en)_2Cl_2]^+$	428(6.8), 345(45.9), 275 (sh) (30)[b]
trans-$[Ir(en)_2(OSO_2CF_3)Cl]^+$	502 (sh) (16), 428 (sh) (25), 334(51), 275 (sh) (~60), 241(110)[c]
cis-$[Ir(en)_2(OSO_2CF_3)_2]^+$	469 (sh) (10), 368 (sh) (34), 296 (sh) (142), 272(184), 243(156)[c]

[a] (M^{-1} cm^{-1}) in parentheses.
[b] H_2O, Ref. 4.
[c] Neat CF_3SO_3H, Ref. 5.

The presence of coordinated triflate is also evident in the IR spectra, where all the triflate bands are split due to the presence of ionic and coordinated triflate.[5] The complexes aquate relatively slowly for triflato complexes (0.1 *M* CF_3SO_3H) and have half-lives of aquation of ~1 hr at 25°.[5] The stereospecificity of their syntheses and subsequent reactions makes them useful intermediates in syntheses.

Reactions similar to those described within may be performed with other diamine ligands, thus extending the synthetic utility of such reactions.

References

1. I. B. Baranovskii, G. S. Kovalenko, and A. V. Babaeva, *Russ. J. Inorg. Chem.*, **13,** 1708 (1968).
2. S. Kida, *Bull. Chem. Soc. Japan,* **39,** 2415 (1966).
3. R. A. Bauer and F. Basolo, *Inorg. Chem.*, **8,** 2231 (1969).
4. F. Galsbøl and B. S. Rasmussen, *Acta Chem. Scand., Ser. A,* **36,** 439 (1982).
5. N. E. Dixon, G. A. Lawrance, P. A. Lay, and A. M. Sargeson, *Inorg. Chem.*, **23,** 2940 (1984).
6. H. Ogino and J. C. Bailar, *Inorg. Chem.*, **17,** 1118 (1978).
7. P. A. Lay and A. M. Sargeson, *Inorg. Synth.*, **24,** 285 (1986).

78. *cis*-BIS(2,2′-BIPYRIDINE-*N,N′*) COMPLEXES OF RUTHENIUM(III)/(II) AND OSMIUM(III)/(II)

Submitted by PETER A. LAY,*·† ALAN M. SARGESON,* and HENRY TAUBE‡
Checked by MEI H. CHOU§ and CAROL CREUTZ§

Osmium and ruthenium polypyridine complexes initially received much attention from Dwyer and coworkers because the M(II), M(III), and M(IV) oxidation states are substitution inert.[1–6] Interest in them has been renewed because of their photochemical reactions[7,8] and the role they play in the study of reactions of coordinated ligands[9–11] and of mixed valence ions[12] and in the preparation of electroactive polymer films.[13] The aqua complexes[14–19] also have important potential applications in the selective oxidation of organic molecules[14,15] and water.[19] We found that trifluoromethanesulfonato (triflato) complexes are convenient synthetic intermediates in the preparation of aqua and oxo species,[20] and we describe the syntheses of the *cis*-bis(2,2′-bipyridine) complexes here.

*Research School of Chemistry, The Australian National University, GPO Box 4, Canberra, A. C. T. 2601, Australia.
†CSIRO Postdoctoral Fellow
‡Department of Chemistry, Stanford University, Stanford, CA 94305.
§Department of Chemistry, Brookhaven National Laboratories, Upton, Long Island, NY 11973.

cis-Bis(2,2′-bipyridine-*N*,*N*′)dichlororuthenium(II) was first prepared by the pyrolysis of (bpyH)[Ru(bpy)Cl_4][3] and subsequently by the reaction of HCl with [Ru$(bpy)_2$(ox)][21] or *cis*-[Ru$(bpy)_2(py)_2$]Cl_2.[22] We here outline a more convenient method that is based on the reaction of $RuCl_3 \cdot xH_2O$ and bpy in *N*,*N*-dimethylformamide (DMF)[23,24] and, in addition, its oxidation to [Ru$(bpy)_2Cl_2$]Cl by a modification of the methods of Liu, Liu, and Bailar.[21] *cis*-Bis(2,2′-bipyridine-*N*,*N*′)dichloroosmium(III) chloride was prepared by the reaction of $K_2[OsCl_6]$ with bpy in DMF and was reduced with $[S_2O_4]^{2-}$ to give [Os$(bpy)_2Cl_2$].[6] Modifications of both of these reactions are also described. The chloro ligands are readily substituted in hot trifluoromethanesulfonic acid to give the required triflato complexes.[20] These in turn react with water to produce the aqua complexes,[20] which are converted to a variety of oxo species.[18] Previous methods for preparing the aqua complexes include spontaneous and Ag^+-induced substitution of chloro ligands, substitution of pyridine ligands, azide-induced aquations of nitrosyl complexes, and acid-catalyzed aquation of carbonato complexes.[5,16,25]

A. *cis*-BIS(2,2′-BIPYRIDINE-*N*,*N*′)DICHLORORUTHENIUM(II) DIHYDRATE, *cis*-Ru$(bpy)_2Cl_2 \cdot 2H_2O$

$$RuCl_3 \cdot 3H_2O + 2bpy + DMF \rightarrow Ru(bpy)_2Cl_2 \cdot 2H_2O + H_2O + Cl^- + ?$$

Procedure

The total time required for the preparation and isolation of product is ~24 hr. It is prepared essentially by the published method of Meyer and coworkers.[23] Ruthenium(III) chloride trihydrate (10 g, 38.2 mmol), 2,2′-bipyridine (12.0 g, 76.9 mmol), and LiCl (11 g) are dissolved in reagent grade *N*,*N*-dimethylformamide (60 mL) contained in a 100-mL round-bottomed flask fitted with a reflux condenser. (■ **Caution.** *N,N-Dimethylformamide is toxic and flammable, and 2,2′-bipyridine is a skin irritant. Avoid contact with skin for both chemicals. This reaction is best performed in a fume hood.*) While the mixture is being stirred magnetically, it is refluxed for 8 hr by heating in an oil bath. After it cools to room temperature, the solution is poured into rapidly stirred acetone (200 mL). (■ **Caution.** *Acetone is toxic and highly flammable. This procedure is best performed in a fume hood.*) The round-bottomed flask is washed with two further portions of acetone (2 × 50 mL), and the combined mixtures are allowed to stand at 0° overnight. The resultant dark green microcrystalline material is collected on a medium-porosity sintered-glass filter and is washed with water (3 × 25 mL) and diethyl ether (3 × 25 mL). (■ **Caution.** *Diethyl ether is toxic and highly flammable. Avoid breathing vapors.*) Yield: 13.9 g (70%).

Anal. Calcd. for $C_{20}Cl_2H_{20}N_4O_2Ru$: C, 46.16; H, 3.87; N, 10.77. Found: C, 46.03; H, 3.76; N, 10.89.

On adding 1g $NaClO_4$ to the mother liquor remaining, the complex *cis*-$[Ru(bpy)_2(CO)Cl][ClO_4]$ can be obtained in up to 40% yield. It may form from the reduction of formic acid, which is present as an impurity due to hydrolysis of DMF.

B. *cis*-BIS(2,2'-BIPYRIDINE-*N,N'*)DICHLORORUTHENIUM(III) CHLORIDE DIHYDRATE, *cis*-$[Ru(bpy)_2Cl_2]Cl \cdot 2H_2O$

$$2[Ru(bpy)_2Cl_2] \cdot 2H_2O + Cl_2 \rightarrow 2[Ru(bpy)_2Cl_2]Cl \cdot 2H_2O$$

Procedure

The total time required for the synthesis and isolation of the product is ~4–5 hr. It is prepared by a modification of the method of Liu, Liu, and Bailar.[21] A suspension of $[Ru(bpy)_2Cl_2] \cdot 2H_2O$ (0.5 g, 0.46 mmol) in 2 *M* HCl (50 mL) contained in a 250-mL conical flask is warmed to 80° on a steam bath in a well-ventilated fume hood. Chlorine gas is bubbled through the solution via a Pasteur pipette. (■ **Caution.** *Chlorine gas is toxic and corrosive. This operation must be performed in a well-ventilated fume hood.*) The bubbling is continued until all the solid dissolves leaving a red solution. Concentrated aqueous HCl (36%, 15 mL) is added cautiously, and the solution is allowed to cool for 1 hr. (■ **Caution.** *Hydrogen chloride is toxic and corrosive, and therefore the operation is performed in a fume hood.*) After further cooling in an ice bath for 2 hr, the crystals are collected on a medium-porosity filter, washed with a little cold HCl (5 *M*, 5 mL), dried under vacuum, and then washed with diethyl ether (3 × 10 mL). Yield: 0.48 g (90%).

Anal. Calcd. for $C_{20}Cl_3H_{20}N_4O_2Ru$: C, 43.22; H, 3.63; N, 10.08. Found: C, 43.01; H, 3.73; N, 9.85.

C. *cis*-BIS(2,2'-BIPYRIDINE-*N,N'*)DICHLOROOSMIUM(III) CHLORIDE DIHYDRATE, *cis*-$[Os(bpy)_2Cl_2]Cl \cdot 2H_2O$ AND *cis*-BIS(2,2'-BIPYRIDINE-*N,N'*)DICHLOROOSMIUM(III) CHLORIDE, *cis*-$[Os(bpy)_2Cl_2]Cl$

$$K_2[OsCl_6] + 2bpy + 2H_2O \xrightarrow{DMF} [Os(bpy)_2Cl_2]Cl \cdot 2H_2O + 2KCl + Cl^- + ?$$

$$[Os(bpy)_2Cl_2]Cl \cdot 2H_2O \xrightarrow{heat} [Os(bpy)_2Cl_2]Cl + 2H_2O$$

These are prepared by a modification of the method described previously.[6] Although $K_2[OsCl_6]$ is available commercially [Alfa Products] it may be prepared by the method described by Dwyer and Hogarth[27] (except that KCl is used to precipitate the complex).

Procedure

The total time required for the preparation and isolation of the product is 4–5 hr. Potassium hexachloroosmate(IV) [Alfa Products] (1.9 g, 3.95 mmol) and 2,2′-bipyridine (1.3 g, 8.3 mmol) are suspended in DMF (40 mL) contained in a 100-mL round-bottomed flask fitted with a reflux condenser. Recently, it was claimed that the use of ethylene glycol as solvent improves the yield,[28] but no experimental details have been given. (■ **Caution.** *N,N-Dimethylformamide is toxic and flammable, and 2,2′-bipyridine is a skin irritant. Avoid contact with skin for both chemicals. This reaction is best performed in a fume hood.*) The flask is immersed in an oil bath, and the mixture is heated to reflux for 1 hr while the solution is stirred magnetically. After ~15 min, crystals of KCl begin to form and the solution darkens. After heating at reflux, the solution is allowed to cool to room temperature (1 hr). The KCl is removed from the solution by filtration, and ethanol (20 mL) is added to the filtrate contained in a 1-L beaker. The complex is precipitated by the slow addition of diethyl ether (500 mL) while the solution is stirred rapidly. (■ **Caution.** *Diethyl ether is toxic and highly flammable. This procedure should be performed in a well-ventilated fume hood.*) After the oily precipitate crystallizes, it is collected on a medium-porosity sintered-glass funnel and air-dried. During the crystallization process, the complexes absorb two molecules of water from moisture in the solvents. Yield: 2.2–2.45 g (86–96%).

Anal. Calcd. for $C_{20}Cl_3H_{20}N_4O_2Os$: C, 37.24; H, 3.13; N, 8.69. Found: C, 36.99; H, 3.27; N, 9.00.

If an oily precipitate remains on the frit, it is dried in a vacuum oven at 100° to give the anhydrous salt.

Anal. Calcd. for $C_{20}Cl_3H_{16}N_4Os$: C, 39.45; H, 2.65; N, 9.20. Found: C, 39.19; H, 2.71; N, 9.12.

D. *cis*-BIS(2,2′-BIPYRIDINE-*N,N′*)DICHLOROOSMIUM(II), *cis*-[Os(bpy)$_2$Cl$_2$]

$$2[Os(bpy)_2Cl_2]Cl\cdot 2H_2O + Na_2S_2O_4 \rightarrow 2[Os(bpy)_2Cl_2] + 2H_2SO_3 + 2H_2O$$

Procedure

The total time required for the synthesis and isolation of the product is ~2 hr. Bis(2,2'-bipyridine-*N,N'*)dichloroosmium(III) chloride dihydrate (1.0 g, 1.55 mmol) is dissolved in a mixture of DMF (20 mL) and MeOH (10 mL) contained in a 500-mL beaker. A dilute aqueous solution of sodium dithionite (2.0 g, 1.1 mmol in 200 mL) is added slowly with stirring over 0.5 hr. The solution containing the dark purple oily suspension of the complex is cooled in an ice bath, and the walls of the beaker are scratched with a glass rod until the complex crystallizes. It is then collected on a medium-porosity filter and washed with water (2 × 10 mL), methanol (2 × 10 mL), and diethyl ether (2 × 10 mL). Yield: 0.86 g (97%).

Anal. Calcd. for $C_{20}H_{16}N_4Cl_2Os$: C, 41.89; H, 2.81; N, 9.77. Found: C, 41.58; H, 3.07; N, 9.56.

E. *cis*-BIS(2,2'-BIPYRIDINE-*N,N'*)BIS(TRIFLUOROMETHANESULFONATO-*O*)-RUTHENIUM(III) TRIFLUOROMETHANESULFONATE, *cis*-$[Ru(bpy)_2(OSO_2CF_3)_2](CF_3SO_3)$, AND *cis*-BIS(2,2'-BIPYRIDINE-*N,N'*)BIS(TRIFLUOROMETHANESULFONATO-*O*)-OSMIUM(III) TRIFLUOROMETHANESULFONATE, *cis*-$[Os(bpy)_2(OSO_2CF_3)_2](CF_3SO_3)$

$$[M(bpy)_2Cl_2]Cl \cdot 2H_2O + 3CF_3SO_3H \rightarrow [M(bpy)_2(OSO_2CF_3)_2](CF_3SO_3) + 3HCl\uparrow + 2H_2O$$

$$M = Ru, Os$$

Procedure

The time required for the synthesis, isolation, and drying of each of the complexes is approximately 2 days. To $[M(bpy)_2Cl_2]Cl \cdot 2H_2O$ (1.0 g, 1.55 mmol for Os, 1.80 mmol for Ru) in a 50-mL two-necked round-bottomed flask, distilled trifluoromethanesulfonic acid[29] (triflic) (7 mL, 80 mmol) is added carefully and dropwise.

■ **Caution.** *Triflic acid is a powerful protic acid, and gaseous hydrogen chloride is produced rapidly during the initial reaction. It is necessary to take adequate precautions to protect the skin and eyes from contact with both chemicals. Inhalation of the corrosive vapors should also be avoided, and the reaction must be performed in a well-ventilated fume hood.*

The reaction mixture is stirred and heated to 110° while a flow of N_2 is maintained through the solution, as outlined in Section 68-D. After 5 hr at 110°, the flask is allowed to cool to room temperature, while the nitrogen stream is maintained. It is then cooled further to ~5° in an ice bath, whereupon anhydrous diethyl ether (40 mL) is added cautiously and dropwise to the rapidly stirred solution.

■ **Caution.** *This addition leads to a very exothermic reaction, and due care should be exercised not to add the diethyl ether too quickly. Diethyl ether is toxic and highly flammable. The addition should be performed in a well-ventilated fume hood.*

An oily precipitate forms, and continued scratching of the inside of the reaction flask with a glass rod, while maintaining a stream of nitrogen, results in the formation of the microcrystalline solid.* The complex is filtered on a medium-porosity sintered-glass funnel and is washed with anhydrous diethyl ether (4 × 20 mL). These complexes are more easily hydrolyzed than other triflato complexes to form the aqua complexes. If pure triflato complexes are required, they are generally heated under vacuum at 110° for 24 hr, as described in Section 68-E. Yields are 94% for Ru and 92% for Os.

Anal. Calcd. for $C_{23}F_9H_{16}N_4O_9S_3Ru$: C, 32.10; H, 1.87; N, 6.51. Found: C, 32.07; H, 1.99, N, 6.42. Calcd. for $C_{23}F_9H_{16}N_4O_9S_3Os$: C, 29.08; H, 1.70; N, 5.90. Found: C, 28.97; H, 1.81; N, 5.89.

These complexes may also be prepared by using O_2 as the carrier gas instead of N_2 and using the $[M(bpy)_2Cl_2]$ complexes as starting materials.

F. *cis*-DIAQUABIS(2,2′-BIPYRIDINE-*N,N′*)OSMIUM(III) TRIFLUOROMETHANESULFONATE, *cis*-$[Os(bpy)_2(OH_2)_2](CF_3SO_3)_3$

$$[Os(bpy)_2(OSO_2CF_3)_2](CF_3SO_3) + 2H_2O \rightarrow [Os(bpy)_2(OH_2)_2](CF_3SO_3)_3$$

Procedure

The time required for the preparation of the diaqua complex from the reaction mixture obtained in the previous section is ~2 hr. The complexes may be prepared either from the solid triflato compounds or directly from the reaction solution obtained from the previous section. Aquation via atmospheric moisture occurs in two steps with the monoaqua complex $[Os(bpy)_2(OH_2)(OSO_2CF_3)](CF_3SO_3)_2$ being formed fairly readily.

*The checkers decanted the CF_3SO_3H/ether, added fresh ether, mixed it with the oil, and decanted again. The process was continued until the oil broke. This resulted in yields of 82% and 55%, respectively, for the Ru and Os compounds.

Anal. Calcd. for $C_{23}F_9H_{18}N_4O_{10}S_3Os$: C, 28.54; H, 1.87; N, 5.79. Found: C, 28.64; H, 1.96; N, 5.88.

Further aquation is slow. Water (7 mL) is added dropwise to the reaction mixture containing CF_3SO_3H, diethyl ether, and the oily precipitate. The mixture is cooled in an ice bath and stirred until the crystallization of the red diaqua complexes is complete. The crystals are collected on a medium-porosity sintered-glass filter, air-dried, and washed with anhydrous diethyl ether (5 × 10 mL). Yields are quantitative.

Anal. Calcd. for $C_{23}F_9H_{20}N_4O_{11}S_3Os$: C, 28.05; H, 1.98; N, 5.69; S, 9.76. Found: C, 28.18; H, 1.84; N, 5.52; S, 9.37.

Properties

The triflato complexes are more sensitive to atmospheric moisture than their aliphatic amine analogs. They must be stored over a suitable drying agent and used quickly. The M(II) triflato complexes may be prepared by the reaction of $[M(bpy)_2Cl_2]$ with CF_3SO_3H, but they are extremely sensitive to air oxidation and must be handled in the absence of air. They are also extremely moisture-sensitive in their impure state and therefore are difficult to prepare pure. The

TABLE I Electronic Spectral Properties of *cis*-$[M(bpy)_2L_2]^{n+}$ Complexes

Complex	λ (nm)[a]
$Ru(bpy)_2Cl_2$	293 (sh), 299 (170,000), 378 (9170), 409 (sh), 489 (sh), 555 (9200), 679 (sh)[b,c]
$Os(bpy)_2Cl_2$	299 (~10,000), 383 (11,500), 466 (10,000), 558 (12,000), 749 (sh), 842 (6700)[b]
$[Ru(bpy)_2Cl_2]^+$	254 (23,250), 301 (26,400), 313 (28,100), 370 (4700), 425 (sh), 520 (sh) (340)[c,d]
$[Ru(bpy)_2(OSO_2CF_3)_2]^+$	241 (19,200), 253 (sh), 302 (21,000) 311 (21,200), 345 (sh) (330), 395 (sh), 610 (311), 645 (262)[e]
$[Ru(bpy)_2(OH_2)_2]^{3+}$	247 (13,500), 293 (sh) (11,000), 306 (12,700), 312 (sh) (12,000), 355 (1800), 478 (590), 510 (sh) (310)[f].
$[Os(bpy)_2Cl_2]^+$	244 (32,000), 288 (28,800), 403 (5060), 450 (sh), 525 (sh)[d]
$[Os(bpy)_2(OSO_2CF_3)_2]^+$	245 (31,300), 253 (sh), 294 (sh), 304 (26,100), 313 (26,460), 360 (sh), 418 (sh), 440 (sh), 460 (998)[d]
$[Os(bpy)_2(OH_2)_2]^{3+}$	247 (29,300), 257 (sh), 270 (sh), 292 (22,800), 302 (sh), 312 (sh), 375 (3600), 470 (870), 505 (sh) (510)[f]

[a]Extinction coefficients in parentheses (M^{-1} cm^{-1}).
[b]Reference 30, $CHCl_3$.
[c]There is some dispute as to whether these are monochloro species in solution, reference 31.
[d]Reference 32.
[e]Neat CF_3SO_3H.
[f]4*M* CF_3SO_3H.

diaqua complexes exist in a wide range of oxidation states, with every oxidation state from M(II) (*cis*-$[M(bpy)_2(OH_2)_2]^{2+}$) to M(VI) (*cis*-$[M(bpy)_2(O)_2]^{2+}$) being accessible by control of pH and potential.[18] Spectrophotometric properties of the aqua, triflato, and chloro complexes are summarized in Table I.

The same reactions can be performed with related complexes, such as the bis(1,10-phenanthroline) series of complexes, and all these triflato complexes are useful synthetic intermediates for the syntheses of a large variety of complexes.

References

1. D. A. Buckingham, F. P. Dwyer, and A. M. Sargeson, *Inorg. Chem.*, **5,** 1243 (1966).
2. F. P. Dwyer, N. K. King, and M. E. Winfield, *Aust. J. Chem.*, **12,** 139 (1959).
3. F. P. Dwyer, H. A. Goodwin, and E. C. Gyarfas, *Aust. J. Chem.*, **16,** 544 (1963).
4. D. A. Buckingham, F. P. Dwyer, and A. M. Sargeson, *Aust. J. Chem.*, **17,** 622 (1964).
5. F. P. Dwyer, H. A. Goodwin, and E. C. Gyarfas, *Aust. J. Chem.*, **16,** 42 (1963).
6. D. A. Buckingham, F. P. Dwyer, H. A. Goodwin, and A. M. Sargeson, *Aust. J. Chem.*, **17,** 325 (1964).
7. D. G. Whitten, *Acc. Chem. Res.*, **13,** 83 (1980) and references therein.
8. W. J. Dressick, T. J. Meyer, B. Durham, and D. P. Rillema, *Inorg. Chem.*, **21,** 3451 (1982) and references therein.
9. M. J. Ridd and F. R. Keene, *J. Am. Chem. Soc.*, **103,** 5740 (1981) and references therein.
10. M. S. Thompson and T. J. Meyer, *J. Am. Chem. Soc.*, **103,** 5577 (1981).
11. B. P. Sullivan, R. S. Smythe, E. M. Kober, and T. J. Meyer, *J. Am. Chem. Soc.*, **104,** 4701 (1982).
12. B. P. Sullivan and T. J. Meyer, *Inorg. Chem.*, **19,** 752 (1980).
13. C. D. Ellis, L. D. Margerum, R. W. Murray, and T. J. Meyer, *Inorg. Chem.*, **22,** 1283 (1983).
14. M. S. Thompson and T. J. Meyer, *J. Am. Chem. Soc.*, **104,** 4106 (1982).
15. M. S. Thompson and T. J. Meyer, *J. Am. Chem. Soc.*, **104,** 5070 (1982).
16. B. A. Moyer and T. J. Meyer, *Inorg. Chem.*, **20,** 436 (1981).
17. B. A. Moyer, B. K. Sipe, and T. J. Meyer, *Inorg. Chem.*, **20,** 1475 (1981).
18. K. J. Takeuchi, G. J. Samuels, S. W. Gersten, J. A. Gilbert, and T. J. Meyer, *Inorg. Chem.*, **22,** 1407 (1983) and references therein.
19. S. W. Gersten, G. J. Samuels, and T. J. Meyer, *J. Am. Chem. Soc.*, **104,** 4029 (1982).
20. P. A. Lay, A. M. Sargeson, D. C. Ware, and H. Taube, *Inorg. Synth.*, **24,** 273 (1986).
21. C. F. Liu, N. C. Liu, and J. C. Bailar, Jr., *Inorg. Chem.*, **3,** 1197 (1964).
22. R. A. Krause, *Inorg. Chim. Acta,* **31,** 241 (1978).
23. B. P. Sullivan, D. J. Salmon, and T. J. Meyer, *Inorg. Chem.*, **12,** 3334 (1978).
24. G. Sprintschnick, H. W. Sprintschnick, P. P. Kirsch, and D. G. Whitten, *J. Am. Chem. Soc.*, **99,** 4947 (1977).
25. E. C. Johnson, B. P. Sullivan, D. J. Salmon, S. A. Adeyami, and T. J. Meyer, *Inorg. Chem.*, **17,** 2211 (1978).
26. J. M. Clear, J. M. Kelly, C. M. O'Connell, J. G. Vos, C. J. Cardin, and A. J. Edwards, *J. Chem. Soc., Chem. Commun.*, **1980,** 750.
27. F. P. Dwyer and J. W. Hogarth, *Inorg. Synth.*, **5,** 206 (1957).
28. E. M. Kober, K. A. Goldsby, D. N. S. Narayana, and T. J. Meyer, *J. Am. Chem. Soc.*, **105,** 4303 (1983).

29. N. E. Dixon, W. G. Jackson, G. A. Lawrance, and A. M. Sargeson, *Inorg. Synth.*, **22,** 103 (1983).
30. J. E. Fergusson and G. M. Harris, *J. Chem. Soc. A,* **1966,** 1294.
31. J. A. Arces Sagües, R. D. Gillard, D. H. Smalley, and P. A. Williams, *Inorg. Chim. Acta,* **43,** 24 (1980).
32. G. M. Bryant and J. E. Ferguson, *Aust. J. Chem.*, **24,** 275 (1971).

79. (2,2′-BIPYRIDINE-*N,N′*)(2,2′:6′,2″-TERPYRIDINE-*N,N′,N″*) COMPLEXES OF RUTHENIUM(III)/(II) AND OSMIUM(III)/(II)

Submitted by DAVID C. WARE,* PETER A. LAY,*† and HENRY TAUBE*
Checked by MEI H. CHOU‡ and CAROL CREUTZ‡

The (2,2′-bipyridine-*N,N′*)(2,2′:6′:2″-terpyridine-*N,N′,N″*), (bpy)(trpy) complexes of ruthenium and osmium, like the bis(2,2′-bipyridine-*N,N′*) series, were first studied by Dwyer and coworkers[1–3] and are interesting because more than one oxidation state is substitution inert. More recently, these (bpy)(trpy) species have generated interest because the M(IV) oxo complexes are useful oxidants of organic molecules.[4–6] The (bpy)(trpy) series of complexes have also proved useful in the study of the redox chemistry of coordinated ligands.[7] The (bpy)(trpy) and bis(bpy)[8] complexes containing the coordinated trifluoromethanesulfonato (triflato) ligands are useful precursors to a variety of complexes. The syntheses of these species and the aqua complexes derived from them are reported here.

The complex ion $[Ru(bpy)(trpy)Cl]^+$ was first prepared as the perchlorate salt by the reaction of $Ru(bpy)Cl_4$ with trpy in 25% aqueous ethanol,[2] and more recently[6] in 75% aqueous ethanol. It has also been prepared from $Ru(CO)_2(bpy)Cl_2$, trpy, and trimethylamine oxide in 2-methoxyethanol.[9] A method in which the readily prepared $Ru(trpy)Cl_3$ complex[10] is used as an intermediate is more convenient, and it is this new method that is described here. (2,2′-Bipyridine-*N,N′*)chloro(2,2′:6′,2″-terpyridine-*N,N′,N″*)osmium(II) chloride has been reported[3] and is prepared from $Os(bpy)Cl_4$[11] and trpy in ethylene glycol. (2,2′-Bipyridine-*N,N′*)chloro(2,2′:6′,2″-terpyridine-*N,N′,N″*)osmium(II) chloride may also be prepared by the reaction of $Os(trpy)Cl_3$[3] with bpy in ethylene glycol. However, the insolubility of $Os(trpy)Cl_3$ in this solvent makes the procedure used for the ruthenium analog less desirable, and the literature method has been used without major modification. Reaction of the chloro complexes with hot

*Department of Chemistry, Stanford University, Stanford, CA 94305.
†CSIRO Postdoctoral Fellow.
‡Department of Chemistry, Brookhaven National Laboratories, Upton, NY 11973.

trifluoromethanesulfonic acid (CF_3SO_3H) leads to the evolution of HCl and produces the trifluoromethanesulfonato complexes in nearly quantitative yield. Water readily displaces the triflate ion to produce the aqua complexes.[6] Aqua(2,2′-bipyridine-*N,N′*)(2,2′:6′,2″-terpyridine-*N,N′,N″*)ruthenium(II) was first prepared by the reaction of $AgNO_3$ with [Ru(bpy)(trpy)Cl]Cl.[2] This procedure does not work for the preparation of $[Os(bpy)(trpy)(OH_2)]^{2+}$ since osmium(II) is rapidly oxidized by Ag^+, but recently this complex was prepared by an alternative route similar to that reported here.[6] Aqua(2,2′-bipyridine-*N,N′*)(2,2′:6′,2-terpyridine-*N,N′,N″*)ruthenium(III) has been reported,[4,5] but the synthesis was not described.

A. (2,2′-BIPYRIDINE-*N,N′*)CHLORO-(2,2′:6′,2″-TERPYRIDINE-*N,N′,N″*)RUTHENIUM(II) CHLORIDE HYDRATE, $[Ru(bpy)(trpy)Cl]Cl\cdot 2.5H_2O$

$$2Ru(trpy)Cl_3 + 2bpy + HOCH_2CH_2OH + 2.5H_2O \rightarrow 2[Ru(bpy)(trpy)Cl]Cl\cdot 2.5H_2O + HOCH_2CHO + 2HCl$$

Procedure

The preparation of this complex requires ~1 day. Trichloro(2,2′:6′,2″-terpyridine-*N,N′,N″*)ruthenium(III)[10] (0.200 g, 0.454 mmol), 2,2′-bipyridine (0.0744 g, 0.476 mmol), and dry ethylene glycol (2.5 mL) are placed in a 10-mL round-bottomed flask fitted with a reflux condenser. (■ **Caution.** *2,2′-Bipyridine is a skin irritant. Avoid contact with skin.*) The mixture is stirred and heated at reflux temperature for 2 hr using a silicone oil bath. The resulting deep red solution is cooled to room temperature, and absolute ethanol (4.0 mL) is added. Any unchanged $Ru(trpy)Cl_3$ is removed by filtration at this point. Diethyl ether (8 mL) is added, and the solution is cooled to −10° overnight in a freezer. (■ **Caution.** *Diethyl ether is toxic and highly flammable.*) The dark crystals that form are removed by filtration and are washed with diethyl ether/ethanol (2:1) (2 × 4 mL) and diethyl ether (3 × 8 mL). The washings are combined with the filtrate and, after cooling and treatment as above, yield a second crop. Both crops are dried under vacuum. Yield: 0.2134 g (78%).*

The product may be recrystallized by dissolving it in hot water (15 mL), filtering the solution, and adding 12 *M* HCl (3 mL).

■ **Caution.** *12 M HCl is extremely corrosive. Avoid contact with skin, and avoid inhaling the vapor. The addition is very exothermic and should be performed with caution. This procedure should be performed in a well-ventilated fume hood.*

*Checkers report a yield of 50% using a Dry Ice/acetone bath to crystallize the complex.

The solution is cooled to 0°, and the crystals are removed by filtration, washed with ice cold water (2 × 1 mL) and diethyl ether (3 × 5 mL), and dried under vacuum over P_4O_{10}. The product is hygroscopic. This sample analyzed for 2.5 water molecules of crystallization.

Anal. Calcd. for $C_{25}H_{24}Cl_2N_5O_{2.5}Ru$: C, 49.51; H, 3.99; N, 11.54. Found: C, 49.34; H, 3.50, N, 11.47.

B. (2,2′-BIPYRIDINE-*N,N′*)(2,2′:6′,2″-TERPYRIDINE-*N,N′,N″*)-(TRIFLUOROMETHANESULFONATO-*O*)RUTHENIUM(III) TRIFLUOROMETHANESULFONATE, $[Ru(bpy)(trpy)(OSO_2CF_3)](CF_3SO_3)_2$, AND (2,2′-BIPYRIDINE-*N,N′*)(2,2′:6′,2″-TERPYRIDINE-*N,N′,N″*)(TRIFLUOROMETHANESULFONATO-*O*)OSMIUM(III) TRIFLUOROMETHANESULFONATE, $[Os(bpy)(trpy)(OSO_2CF_3)](CF_3SO_3)_2$

$$2[M(bpy)(trpy)Cl]Cl\cdot xH_2O + 6CF_3SO_3H + O_2 \rightarrow$$
$$2[M(bpy)(trpy)(OSO_2CF_3)](CF_3SO_3)_2 + 4HCl + 2xH_2O + \text{`}H_2O_2\text{'}$$
$$(M = Ru, x = 2.5;\ M = Os, x = 2)$$

Procedure

The time required for the synthesis and isolation of each of these complexes is ~1 day. To $[M(bpy)(trpy)Cl]Cl\cdot xH_2O$ (M = Ru, x = 2.5, 0.50 g, 0.824 mmol; M = Os, x = 2.0, 0.50 g, 0.728 mmol) in a 25-mL two-necked round-bottomed flask is added cautiously, and in a dropwise manner, distilled trifluoromethanesulfonic acid (5 mL, 56 mmol).*

■ **Caution.** *Triflic acid is a very strong protic acid. Gaseous hydrogen chloride is produced rapidly during the initial reaction. It is necessary to take adequate precautions to protect the skin and eyes from contact with both chemicals. Inhalation of the corrosive vapors should be avoided. The reaction must be performed in a well-ventilated fume hood.*

The stirred reaction mixture is heated to 110° while a flow of dry air is maintained through it, as outlined in Section 68-D. The air is dried by passing it through a prebubbler of concentrated H_2SO_4 or neat CF_3SO_3H. After 5 hr at 110°, the gas bubbler is replaced by a vacuum-distillation apparatus and most of the solvent is removed by evaporation at reduced pressure.[12] This step is necessary, since the triflato complex salts are rather soluble in CF_3SO_3H/diethyl ether mixtures.

*The checkers used triflic acid as received.

The flask is allowed to cool to room temperature and is then cooled further in an ice bath, a stream of dry air being maintained over the concentrated solution to prevent contamination by atmospheric moisture. Diethyl ether (10 mL) is added cautiously and in a dropwise manner to the constantly stirred solution.

■ **Caution.** *This addition leads to a very exothermic reaction, and care should be exercised not to add the diethyl ether too quickly. Diethyl ether is toxic and highly flammable. The addition should be performed in a well-ventilated fume hood.*

If an oil forms, it can be induced to solidify by scratching the inner wall of the flask with a glass rod. The complex is filtered on a medium-porosity sintered-glass funnel and is washed with anhydrous diethyl ether. The complex is quite hygroscopic. Water and any occluded $Et_2O \cdot CF_3SO_3H$ are removed by heating under vacuum at 110° for 8 hr. Yields: M = Ru, 0.690 g (89%); M = Os,0.710 g (95%).*

Anal. Calcd. for $C_{28}H_{19}N_5F_9O_9S_3Ru$: C, 35.86; H, 2.04; N, 7.47. Found: C, 35.73; H, 2.10; N, 7.30; Calcd. for $C_{28}H_{19}N_5F_9O_9S_3Os$: C, 32.75; H, 1.87; N, 6.82. Found: C, 32.65; H, 1.97; N, 6.70.

C. (2,2′-BIPYRIDINE-*N,N′*)(2,2′:6′,2″-TERPYRIDINE-*N,N′,N″*)-(TRIFLUOROMETHANESULFONATO-*O*)RUTHENIUM(II) TRIFLUOROMETHANESULFONATE, $[Ru(bpy)(trpy)(OSO_2CF_3)](CF_3SO_3)$

$$[Ru(bpy)(trpy)Cl]Cl \cdot 2.5H_2O + 2CF_3SO_3H \rightarrow [Ru(bpy)(trpy)(OSO_2CF_3)](CF_3SO_3) + 2.5H_2O + 2HCl$$

Procedure

The time required for this reaction is ~1.5 days. To $[Ru(trpy)(bpy)Cl]Cl \cdot 2.5H_2O$ (0.30 g, 0.495 mmol) in a 25-mL two-necked round-bottomed flask is added cautiously, and in a dropwise manner, distilled trifluoromethanesulfonic acid (3 mL, 33.6 mmol).[12]†

■ **Caution.** *Triflic acid is a very strong protic acid. Gaseous hydrogen chloride is produced rapidly during the initial reaction. It is necessary to take adequate precautions to protect the skin and eyes from contact with both chemicals. Inhalation of the corrosive vapors should be avoided. The reaction must be performed in a well-ventilated fume hood.*

*The checkers report that the triflato complexes are obtained by evaporating all of the solvent and then solidifying the oil by repeating trituration and decanting with diethyl ether. Yields: M = Ru, 50%; M = Os, 70%.

†The checkers used a freshly opened commercial sample.

During the addition, a strong flow of dry, deoxygenated nitrogen is passed through the solution. The product is very sensitive to oxidation in neat CF_3SO_3H. The nitrogen is deoxygenated and dried by passing the gas successively through a chromium(II) scrubber and either a concentrated H_2SO_4 or neat CF_3SO_3H bubbler. While the nitrogen stream is maintained, the solution is heated to 100° for 2 hr.†

After the flask has cooled to room temperature, it is placed in an ice bath, and a stream of dry argon is passed through the solution. Diethyl ether (~8 mL) is added cautiously and in a dropwise manner to the constantly stirred solution.

■ **Caution.** *This addition leads to a very exothermic reaction. Care should be exercised not to add the diethyl ether too quickly. Diethyl ether is toxic and highly flammable. The addition should be performed in a well-ventilated fume hood.*

The reaction flask is stoppered and placed in the freezer overnight. The crystals that form are collected by filtration and are washed with diethyl ether (2 × 5 mL). The solid is dried in a vacuum oven at 100° overnight. Yield: 0.373 g (95.6%).

Anal. Calcd. for $C_{27}H_{19}N_5F_6O_6S_2Ru$: C, 41.12; H, 2.43; N, 8.88. Found: C, 40.60; H, 2.50; N, 8.66.

Because of the difficulty of isolating this very air- and moisture-sensitive complex, for synthesis the crude product is best used directly. The solvent is removed using vacuum distillation, and the oily residue is dissolved in either the neat ligand or a sulfolane or acetone solution of the ligand. The products are much less sensitive to oxidation, which facilitates the purification procedure.

D. (2,2′-BIPYRIDINE-*N*,*N*′)(2,2′:6′,2″-TERPYRIDINE-*N*,*N*′,*N*″)-(TRIFLUOROMETHANESULFONATO-*O*)OSMIUM(II)TRIFLUOROMETHANESULFONATE, [Os(bpy)(trpy)(OSO_2CF_3)](CF_3SO_3)

$$[Os(bpy)(trpy)(OH_2)](CF_3SO_3)_2 \cdot H_2O \rightarrow [Os(bpy)(trpy)(OSO_2CF_3)](CF_3SO_3) + 2H_2O$$

Procedure

The time required for this procedure is ~11 hr. Due to the extreme oxygen sensitivity of [Os(bpy)(trpy)(OSO_2CF_3)](CF_3SO_3) in CF_3SO_3H solution, this complex is best prepared in the solid state by the method outlined in Section

†The checkers report that an oily product is obtained by distilling off all the CF_3SO_3H. It is crystallized by repeated trituration and decantation with diethyl ether. Yield: 45%.

68-E. The compound $[Os(bpy)(trpy)(OH_2)](CF_3SO_3)_2$ (0.100 g, 0.1094 mmol) (prepared as in Section F) is ground to a fine powder and dehydrated by heating at 180° under vacuum for 10 hr. The higher temperature is necessary, since at lower temperatures starting material is recovered unchanged. Yield: 0.093 g (97%).

Anal. Calcd. for $C_{27}H_{19}N_5F_6O_6S_2Os$: C, 36.94; H, 2.18; N, 7.98. Found: C, 36.66; H, 2.18; N, 7.92.

E. AQUA(2,2′-BIPYRIDINE-*N,N*′)(2,2′:6′,2″-TERPYRIDINE-*N,N*′,*N*″)-RUTHENIUM(III) TRIFLUOROMETHANESULFONATE TRIHYDRATE, $[Ru(bpy)(trpy)(OH_2)](CF_3SO_3)_3 \cdot 3H_2O$, AND AQUA(2,2′-BIPYRIDINE-*N,N*′)(2,2′:6′,2″-TERPYRIDINE-*N,N*′,*N*″)OSMIUM(III) TRIFLUOROMETHANESULFONATE DIHYDRATE, $[Os(bpy)(trpy)(OH_2)](CF_3SO_3)_3 \cdot 2H_2O$

$$[M(bpy)(trpy)(OSO_2CF_3)](CF_3SO_3)_2 + (x+1)H_2O \rightarrow [M(bpy)(trpy)(OH_2)](CF_3SO_3)_3 \cdot xH_2O$$

$$(M = Ru, x = 3; M = Os, x = 2)$$

Procedure

These complexes are prepared by aquation of the corresponding triflato complexes that are prepared in Section C. This is readily accomplished by allowing the hygroscopic solids obtained from the triflic acid/diethyl ether solutions (before vacuum oven drying) to fully aquate in air for ~1 day. Storage in a vacuum desiccator over P_4O_{10} permits the isolation of the trihydrate and dihydrate complexes for ruthenium and osmium, respectively. The yields are the same as those found for the production of the triflato complexes in Section C.

Anal. Calcd. for $C_{28}H_{27}N_5F_9O_{13}S_3Ru$: C, 33.31; H, 2.70; N, 6.94; S, 9.52. Found: C, 33.28; H, 2.20; N, 6.89; S, 9.59. Calcd. for $C_{28}H_{25}N_5F_9O_{12}S_3Os$: C, 31.11; H, 2.33; N, 6.47. Found: C, 30.84; H, 2.25; N, 6.38.

The aqua complexes may also be prepared from the *isolated* triflato complexes, by following a procedure similar to that given above.

F. AQUA(2,2′-BIPYRIDINE-*N,N*′)(2,2′:6′,2″-TERPYRIDINE-*N,N*′*N*″)OSMIUM(II) TRIFLUOROMETHANESULFONATE HYDRATE, $[Os(bpy)(trpy)(OH_2)](CF_3SO_3)_2 \cdot H_2O$

$$2[Os(bpy)(trpy)(OH_2)](CF_3SO_3)_3 \cdot 2H_2O + Zn \rightarrow 2[Os(bpy)(trpy)(OH_2)](CF_3SO_3)_2 \cdot H_2O + Zn(CF_3SO_3)_2 + 2H_2O$$

Procedure

The aqua osmium(II) complex is prepared by reduction of the aqua osmium(III) complex (prepared as in Section E). Aqua(2,2′-bipyridine-*N,N′*)(2,2′:6′,2″-terpyridine-*N,N′,N″*)osmium(III) trifluoromethanesulfonate dihydrate (0.500 g, 4.063 mmol) is dissolved in 0.33 *M* CF_3SO_3H (30 mL), and the solution is filtered. Two pieces of freshly prepared zinc amalgam are added. The solution is stirred for 5 hr. The zinc is removed, and the solution is cooled in an ice bath. The resulting black crystals are filtered off, washed with ice cold water (2 × 2 mL), and dried under vacuum. Yield: 0.224 g (53%). A second crop may be obtained by addition of fresh zinc amalgam to the filtrate.

Anal. Calcd. for $C_{27}H_{23}N_5F_6O_8S_2Os$: C, 35.48; H, 2.54; N, 7.66. Found: C, 35.46; H, 2.42; N, 7.79.

Properties

The triflato complexes aquate in air in the solid state. To prevent this reaction they must be stored over a suitable drying agent; failing this, they can be regenerated by dehydrating prior to use. The aquation rate for $[Os(bpy)(trpy)(OSO_2CF_3)]^{2+}$ is 8.9×10^{-4} sec^{-1} at 25° in 0.1 *M* CF_3SO_3H. The M(II) triflato species are oxidized by air in neat CF_3SO_3H, but may be handled in air in the solid state. The M(II) triflato species are characterized by the absence of bands due to coordinated water in the IR spectra (~3400 cm^{-1}). Characteristic splitting of IR absorption bands shows the triflate to be both ionic

TABLE I Electronic Spectral Properties of $[M(bpy)(trpy)L]^{3+/2+/+}$ Complexes

Complex	λ (nm) (ε × 10^{-3}, M^{-1} cm^{-1})
$[Ru(bpy)(trpy)Cl]^{+}$[a]	231 (27.0), 269 (29.3), 279 (30.8), 289 (34.2), 312 (34.3), 483 (9.59).
$[Ru(bpy)(trpy)(OH_2)]^{2+}$[b]	231 (26.2), 241 (sh), 253 (sh), 272 (32.5), 279 (34.0), 287 (37.4), 312 (36.7), 475 (9.8), 580 (sh), 605 (sh)
$[Os(bpy)(trpy)(OH_2)]^{2+}$[b]	232 (29.7), 276 (34.5), 290 (36.8), 318 (37.8), 408 (sh), 455 (6.71), 490 (6.93), 500 (9.63), 585 (sh), 752 (2.24)
$[Ru(bpy)(trpy)(OSO_2CF_3)]^{2+}$[c]	244 (32.3), 273 (sh) (24.8), 283 (27.8), 303 (sh), 312 (32.9), 347 (sh), 610 (br) (0.37)
$[Os(bpy)(trpy)(OSO_2CF_3)]^{2+}$[c]	246 (38.2), 256 (38.7), 270 (37.9), 279 (sh), 304 (sh), 313 (30.9), 333 (sh), 376 (sh), 420 (sh), 440 (sh), 475 (0.69), 507 (0.53), 575 (0.59), 605 (sh), 645 (sh)
$[Os(bpy)(trpy)(OH_2)]^{3+}$[d]	243 (32.8), 259 (32.4), 272 (33.5), 280 (sh) (31.7), 304 (22.5), 314 (24.3), 332 (sh), 376 (6.1), 390 (sh) (3.8), 440 (sh) (1.9), 510 (0.94)

[a]0.1 *M* HCl. [b]H_2O. [c]Neat CF_3SO_3H. [d]4 *M* CF_3SO_3H.

and coordinated. The electronic absorption spectra of the chloro, aqua, and M(III) triflato complexes are used for characterization and are reported in Table I.

References

1. D. A. Buckingham, F. P. Dwyer, and A. M. Sargeson, *Inorg. Chem.*, **5,** 1243 (1966).
2. F. P. Dwyer, H. A. Goodwin, and E. C. Gyarfas, *Aust. J. Chem.*, **16,** 42 (1963).
3. D. A. Buckingham, F. P. Dwyer, and A. M. Sargeson, *Aust. J. Chem.*, **17,** 622 (1964).
4. B. A. Moyer, M. S. Thompson, and T. J. Meyer, *J. Am. Chem. Soc.*, **102,** 2310 (1980).
5. M. S. Thompson and T. J. Meyer, *J. Am. Chem. Soc.*, **104,** 4106 (1982).
6. K. J. Takeuchi, M. S. Thompson, D. W. Pipes, and T. J. Meyer, *Inorg. Chem.*, **23,** 1845 (1984).
7. D. W. Pipes and T. J. Meyer, *Inorg. Chem.*, **23,** 2466 (1984).
8. P. A. Lay, A. M. Sargeson, and H. Taube, *Inorg. Synth.*, **24,** 291 (1986).
9. D. St. C. Black, G. B. Deacon, and N. C. Thomas, *Inorg. Chim. Acta,* **65,** L75 (1982).
10. B. P. Sullivan, J. M. Calvert, and T. J. Meyer, *Inorg. Chem.*, **19,** 1404 (1980).
11. D. A. Buckingham, F. P. Dwyer, H. A. Goodwin, and A. M. Sargeson, *Aust. J. Chem.*, **17,** 315 (1964).
12. N. E. Dixon, W. G. Jackson, G. A. Lawrance, and A. M. Sargeson, *Inorg. Synth.*, **22,** 103 (1983).

SOURCES OF CHEMICALS AND EQUIPMENT

Ace Glass, Inc., P.O. Box 688, Vineland, NJ 08360
Ace Scientific Supply Company, 40-A Cotters Lane, East Brunswick, NJ 08816
Air Products and Chemicals, Inc., Specialty Gas Department, Hometown Facility, P.O. Box 351, Tamaqua, PA 18252
Aldrich Chemical Company, 940 West St. Paul Avenue, Milwaukee, WI 53233; P.O. Box 2060, Milwaukee, WI 53201
Alfa Inorganics (*see* Alfa Products)
Alfa Products, Ventron Corp., P.O. Box 299, 152 Andover St., MA 01923
American Scientific Products, 3660 148th Avenue, N.E., Redmond, WA 98052
Applied Science Laboratories, Inc., P.O. Box 140, State College, PA 16801
J. T. Baker Chemical Company, Phillipsburg, NJ 08865
BDH Chemicals, Ltd. (*see* Gallard-Schlesinger)
Brooks Instrument Co., Inc., Hatfield, PA 19440
Crawford Fitting Company, 29500 Solon Rd., Solon, OH 44139
Dresser Industries, 250 East Main Street, Stratford, CT 06497
Eastman Kodak Company, Rochester, NY 14650
Fairfield Chemical Co., P.O. Box 20, Blythewood, SC 29016
Fisher and Porter Co., 51 Warminster Rd., Warminster, PA 18974
Fisher Scientific Co., 2225 Martin Ave., Santa Clara, CA 95050; 7722 Fenton St., Silver Spring, MD 20910; P.O. Box 8740, Rochester, NY 14624; Medford, MA 02155
Fluka Chemical Corporation, Hauppague, NY 11787
The Fredericks Co., Huntington Valley, PA 19006

Gallard-Schlesinger Chemical Manufacturing Company, 584 Mineola Avenue, Carle Place, NY 11514
Halocarbon Products Corp., 82 Burlews Court, Hackensack, NJ 07602
Hamilton Company, P.O. Box 10030, Reno, NV 89510
Hoke, Inc., One Tenakill Park, Cresskill, NJ 07626
Johnson Matthey, Inc., Eagles Landing, P.O. Box 1087, Seabrook, NH 03874; Malvern, PA 19355
Kawecki Berylco Industries, Inc., 220 East 42nd Street, New York, NY 10017
Koch Associates, Inc., 3 Hillside Avenue, Cockeysville, MD 21030
Kontes, Inc., 3045 Teagarden St., San Leandro, CA 94577; P.O. Box 729, Vineland, NJ 08360
Matheson Gas Products, P.O. Box 85, 932 Paterson Plank Road, East Rutherford, NJ 07073
Ozark-Mahoning Co., 1870 So. Boulder, Tulsa, OK 74119
Pallflex Products Corp., Kennedy Drive, Putnam, CT 06260
PCR Research Chemicals, Inc., P.O. Box 1778, Gainesville, FL 32602
Penntube Plastics Co., Madison Avenue and Holley Street, Clifton Heights, PA 19018
Petrach Systems, Levittown, PA 19058
Pfalz and Bauer, Inc., 375 Fairfield Ave., Stamford, CT 06902
Potomac Valve & Fitting, Inc., 1500 East Jefferson St., Rockville, MD 20852
Preiser Scientific, 1500 Algonquin Parkway, Louisville, KY 40201
Pressure Chemical Company, 3419 Smalltown St., Pittsburgh, PA 15201
Reade Manufacturing Co., Ridgeway Blvd., Lakehurst, NJ 08733
Research Organic/Inorganic Chemical Co., 507–519 Main Street, Belleville, NJ 07109
Sargent-Welch, P.O. Box 1026, Skokie, IL 60077
Savillex Corporation, 5325 Highway 101, Minnetonka, MN 55343
Scientific Gas Products, Inc., 2330 Hamilton Blvd, South Plainfield, NJ 07080
Seattle Valve and Fitting, 13417 N.E. 20th Street, Bellevue, WA 98005
Sigma Chemical Company, P.O. Box 14508, St. Louis, MO 63178
Strem Chemicals, Inc., P.O. Box 108, Newburyport, MA 01950; P.O. Box 212, Danvers, MA 01923; 150 Andover St., Danvers, MA 01923
VWR Scientific, Inc., 147 Delta Drive, Pittsburgh, PA 15238; P.O. Box 1004, Norwalk, CA 90650
Warehoused Plastics, Inc., 13 Blackburn St., Toronto, Ont., Canada M4M 2B3
Wateree Chemical Co., P.O. Box 1045, Camden, SC 29020
Whitey Research Tool Co., 5679 Landregan St., Emeryville, CA 94608
Zeus Industrial Products, Inc., Thompson St., Raritan, NJ 08869

INDEX OF CONTRIBUTORS

SUBJECT INDEX

Names used in this Subject Index for Volumes 21–25 are based upon IUPAC *Nomenclature of Inorganic Chemistry,* Second Edition (1970), Butterworths, London; IUPAC *Nomenclature of Organic Chemistry,* Sections A, B, C, D, E, F, and H (1979), Pergamon Press, Oxford, U.K.; and the Chemical Abstracts Service *Chemical Substance Name Selection Manual* (1978), Columbus, Ohio. For compounds whose nomenclature is not adequately treated in the above references, American Chemical Society journal editorial practices are followed as applicable.

Inverted forms of the chemical names (parent index headings) are used for most entries in the alphabetically ordered index. Organic names are listed at the "parent" based on Rule C-10, *Nomenclature of Organic Chemistry,* 1979 Edition. Coordination compounds, salts and ions are listed once at each metal or central atom "parent" index heading. Simple salts and binary compounds are entered in the usual uninverted way, e.g., *Sulfur oxide* (S_8O), *Uranium(IV) chloride* (UCl_4).

All ligands receive a separate subject entry, e.g., *2,4-Pentanedione,* iron complex. The headings *Ammines, Carbonyl complexes, Hydride complexes,* and *Nitrosyl complexes* are used for the NH_3, CO, H, and NO ligands.

FORMULA INDEX

The Formula Index, as well as the Subject Index, is a Cumulative Index for Volumes 21–25. The Index is organized to allow the most efficient location of specific compounds and groups of compounds related by central metal ion or ligand grouping.

The formulas entered in the Formula Index are for the total composition of the entered compound, e.g., F_6NaU for sodium hexafluorouranate(V). The formulas consist solely of atomic symbols (abbreviations for atomic groupings are not used) and arranged in alphabetical order with carbon and hydrogen always given last, e.g., $Br_3CoN_4C_4H_{16}$. To enhance the utility of the Formula Index, all formulas are permuted on the symbols for all metal atoms, e.g., $FeO_{13}Ru_3C_{13}H_{13}$ is also listed at $Ru_3FeO_{13}C_{13}H_{13}$. Ligand groupings are also listed separately in the same order, e.g., $N_2C_2H_8$, 1,2-Ethanediamine, cobalt complexes. Thus individual compounds are found at their total formula in the alphabetical listing; compounds of any metal may be scalled at the alphabetical position of the metal symbol; and compounds of a specific ligand are listed at the formula of the ligand, e.g., NC for Cyano complexes.

Water of hydration, when so identified, is not added into the formulas of the reported compounds, e.g., $Cl_{0.30}N_4PtRb_2C_4 \cdot 3H_2O$.